Springer Proceedings in Complexity

Springer Proceedings in Complexity publishes proceedings from scholarly meetings on all topics relating to the interdisciplinary studies of complex systems science. Springer welcomes book ideas from authors. The series is indexed in Scopus.

Proposals must include the following:

- name, place and date of the scientific meeting
- a link to the committees (local organization, international advisors etc.)
- scientific description of the meeting
- list of invited/plenary speakers
- an estimate of the planned proceedings book parameters (number of pages/articles, requested number of bulk copies, submission deadline)

Submit your proposals to: christoph.baumann@springer.com

More information about this series at http://www.springer.com/series/11637

Christos H. Skiadas · Yiannis Dimotikalis
Editors

12th Chaotic Modeling and Simulation International Conference

Springer

Editors
Christos H. Skiadas
ManLab, Technical University of Crete
Chania, Crete, Greece

Yiannis Dimotikalis
Department of Management
Science and Technology
Hellenic Mediterranean University
Agios Nikolaos, Crete, Lasithi, Greece

ISSN 2213-8684 ISSN 2213-8692 (electronic)
Springer Proceedings in Complexity
ISBN 978-3-030-39514-8 ISBN 978-3-030-39515-5 (eBook)
https://doi.org/10.1007/978-3-030-39515-5

This Springer imprint is published by the registered company Springer Nature Switzerland AG
The registered company address is: Gewerbestrasse 11, 6330 Cham, Switzerland

Contents

Contributors

Andrii V. Abramov Donetsk National University, Donetsk, Ukraine

Olga P. Abramova Donetsk National University, Donetsk, Ukraine

Valeriy S. Abramov Donetsk Institute for Physics and Engineering named after A.A. Galkin, Donetsk, Ukraine

Maricel Agop Department of Physics, Gheorghe Asachi Technical University of Iaşi, Iaşi, Romania

Szymon Bauch Institute of Physics, Polish Academy of Sciences, Warszawa, Poland

S. Behnia Department of Physics, Urmia University of Technology, Urmia, Iran

Šarlota Birnšteinová Faculty of Science, Šafárik University, Ko šice, EU, Slovakia

Gabriel Crumpei Faculty of Psychology and Education Sciences, Alexandru Ioan Cuza University, Iaşi, Romania

D. P. Dowling School of Mechanical and Materials Engineering, University College Dublin, Dublin 4, Ireland

Mauricio Díaz Universidad de Bio Bio, Concepción, Chile

Lucian Eva "Prof. Dr. N. Oblu" Clinical Emergency Hospital Iaşi, Iaşi, Romania

Dalia Ezzeddine Department of Chemistry, American University of Beirut, Beirut, Lebanon

S. Fathizadeh Department of Physics, Urmia University of Technology, Urmia, Iran

Tatiana F. Filippova Ural Federal University and Krasovskii Institute of Mathematics and Mechanics, Ural Branch of Russian Academy of Sciences, Ekaterinburg, Russian Federation

Alina Gavriluţ Faculty of Mathematics, Alexandru Ioan Cuza University, Iaşi, Romania

Michal Hnatič Institute of Experimental Physics SAS, Ko šice, EU, Slovakia; Bogolyubov Laboratory of Theoretical Physics, Joint Institute for Nuclear Research, Dubna, Russian Federation

E. Javan Shour Department of Physics, Urmia University of Technology, Urmia, Iran

Marian Jurčišin Institute of Experimental Physics, Slovak Academy of Sciences, Košice, Slovakia

Eva Jurčišinová Institute of Experimental Physics, Slovak Academy of Sciences, Košice, Slovakia

Shunji Kawamoto Osaka Prefecture University, Sakai, Osaka, Japan

Dmitriy S. Klyuev Samara, Russia

Tatyana P. Konovalyuk Institute of Hydromechanics NASU, Kiev, Ukraine

Tatyana S. Krasnopolskaya Institute of Hydromechanics NASU, Kiev, Ukraine

Abdelbaki Laidoune Health & Safety Institute, Industrial Prevention Research Laboratory (LRPI), University of Batna 2 (Mostafa Ben Boulaid), Fesdis, Batna, Algeria

V. J. Law School of Mechanical and Materials Engineering, University College Dublin, Dublin 4, Ireland

Jiří Lipovský Faculty of Science, Department of Physics, University of Hradec Králové, Hradec Králové, Czechia

Tomáš Lučivjanský Faculty of Science, P. J. Šafárik University, Košice, Slovakia

Oxana G. Matviychuk Krasovskii Institute of Mathematics and Mechanics, Ural Branch of Russian Academy of Sciences and Ural Federal University, Ekaterinburg, Russian Federation

Lukáš Mižišin Bogoliubov laboratory of Theoretical Physics, JINR, Dubna, Moscow Region, Russia;
Institute of Experimental Physics, Slovak Academy of Sciences, Košice, Slovakia

F. Nemati Department of Physics, Urmia University of Technology, Urmia, Iran

Anatoly M. Neshcheret Samara, Russia

Oleg V. Osipov Samara, Russia

Kosmas Papadopoulos Department of Economics, Democritus University of Thrace, Komotini, Greece

Evgeniy D. Pechuk Institute of Hydromechanics NASU, Kiev, Ukraine

Alexander A. Potapov V.A. Kotelnikov Institute of Radio Engineering and Electronics, Russian Academy of Sciences, Moscow, Russia

Richard Remecký Bogoliubov Laboratory of Theoretical Physics, Joint Institute for Nuclear Research, Dubna, Moscow Region, Russian Federation

Volodymyr Rusyn Department of Radio Engineering and Information Security, Yuriy Fedkovych Chernivtsi National University, Chernivtsi, Ukraine

Georges Sarafopoulos Department of Economics, Democritus University of Thrace, Komotini, Greece

Leszek Sirko Institute of Physics, Polish Academy of Sciences, Warszawa, Poland

Christos H. Skiadas ManLab, Technical University of Crete, Chania, Greece

Tatyana V. Sobol Institute of Hydromechanics NASU, Kiev, Ukraine

Siavash H. Sohrab McCormick School of Engineering and Applied Science, Department of Mechanical Engineering, Northwestern University, Evanston, IL, USA

Julia V. Sokolova Samara, Russia

V. M. Somsikov Institute of Ionosphere, Almaty, Kazakhstan;
Al-Farabi, Kazakh National University, Almaty, Kazakhstan

Rabih Sultan Department of Chemistry, American University of Beirut, Beirut, Lebanon

Beatrice Venturi Department of Economics and Business, University of Cagliari, Cagliari, Italy

Michał Ławniczak Institute of Physics, Polish Academy of Sciences, Warszawa, Poland

Active Nanoobjects, Neutrino and Higgs Boson in a Fractal Models of the Universe

Valeriy S. Abramov

Abstract Theoretically the relationships of the main parameters of active nanoobjects with the Higgs boson and the Higgs field in a fractal models of the Universe are investigated. Neutrino, nanoparticles, atomic defects, quantum dots can be as active nanoobjects. The neutrino is characterized by the phenomenon of hysteresis. The estimation of the neutrino rest mass is obtained. Using the example of a silica nanoparticle, trapped in an optical trap and placed in a vacuum, estimates of the limiting frequency of rotation of a particle in a laser field with circular polarization and the size of the nanoparticle are obtained. Using the example of atomic defects in boron nitride nanotubes, we obtained estimates of the wavelengths of quantum emission of separate photons. Super-nonradiative states of physical fields are investigated. The properties of nanoparticles depend on pressure, state of physical vacuum and cosmological parameters.

Keywords Active nanoobjects · Neutrino · Higgs boson · Higgs field · Fractal models of the Universe · Optical traps · Nanoparticles · Frequency of rotation · Physical vacuum · Super-nonradiative states

1 Introduction

The Nobel Prize in Physics in 2018 was awarded to A. Ashkin, J. Moru and D. Strickland for their discoveries in the field of laser technology. A. Ashkin proposed to use the optical traps, developed the technology of "optical tweezers", which allows us to manipulate nanoobjects using a laser [1]. G. Mourou and D. Strickland studied the compression of amplified chirped optical pulses [2]. To compress the pulses, the reflection from diffraction lattices is used. The use of periodic structures and metamaterials [3] leads to various non-linear effects. Femtosecond laser coherent spectroscopy makes it possible to reveal the features of the behavior of active objects in such nonlinear systems [4]. In [5], as an active object a quartz nanoparticle in a

V. S. Abramov (✉)
Donetsk Institute for Physics and Engineering named after A.A. Galkin, Donetsk, Ukraine
e-mail: vsabramov2018@gmail.com

C. H. Skiadas and Y. Dimotikalis (eds.), *12th Chaotic Modeling and Simulation International Conference*, Springer Proceedings in Complexity,
https://doi.org/10.1007/978-3-030-39515-5_1

vacuum trapped in an optical trap was investigated. In this case, the spin of an angular momentum of the laser light is transmitted to the mechanical angular momentum of the particle. The rotation of single 100 nm particles occurs at frequencies of the order of 1 GHz. Atom-like defects in two-dimensional hexagonal boron nitride (hBN) have recently become perspective for quantum informatics. In [6], single-photon quantum emission from atomic defects in boron nitride nanotubes (BNNT) was investigated. This system shows high stability at room temperature, which is attractive for creating various opto-mechanical devices.

On the other hand, such nanoparticles [5, 6] are model nanoobjects [7] for studying the physical properties of vacuum. Taking into account the hypothesis of the hierarchical structure of the Universe [8, 9], we can use these experimental results in cosmology: in the study of anisotropic models [10] of fractal cosmological objects; superradiance [10–12] of gravitational waves, relic photons from binary black holes, neutron stars; the nature of dark matter particles and dark energy [13, 14], the chiral fractal structures of the Universe. Also, there is the task of describing the super-nonradiative states of various fields: gravitational, relic photons, the Higgs field, the neutrino field, the physical vacuum. When describing such various nonlinear physical models, singular points (attractors), lines, surfaces, special volume structures (strange attractors) arise. Many physical properties near these features are stochastic in nature, it becomes necessary to simulate stochastic processes. In [15–17] attractors and the deformation field, the mutual influence of attractors and stochastic processes in coupled fractal multilayer nanosystems were studied.

The aim of this paper is to describe: a model nanoobject and the Higgs field; the neutrino parameter relationships with the Higgs boson and the Higgs field; super-nonradiative states of fractal objects.

2 Description of the Model Nanoobject

To describe the main characteristics of a model nanoparticle, we use the connections of the rest energy of the Higgs boson $E_{H0} = 125.03238$ GeV and the gravitational field $E_G = 12.11753067\,\mu$eV from [10–12]

$$E_{H0}/E_G = \nu^*_{H0}/\nu_{G0} = N_{HG}; \quad E_G/\nu_{G0} = E_{H0}/\nu^*_{H0} = 2\pi\hbar. \tag{1}$$

Here $\hbar$ is Planck's constant. Taking into account (1), we find the parameter $N_{HG} = 1.0318305 \cdot 10^{16}$. The numerical values of characteristic frequencies are equal $\nu_{G0} = 2.9304515$ GHz, $\nu^*_{H0} = 3.0237293 \cdot 10^{25}$ Hz. On the basis of these characteristic frequencies, we find the frequencies of the graviton $\nu_G = \nu_{G0}/N_a$ and the effective Higgs boson $\nu_{H0} = \nu^*_{H0}/N_a$. Here N_a is the Avogadro number. Numerical values are $\nu_G = 4.8634664 \cdot 10^{-15}$ Hz, $\nu_{H0} = 50.182731$ Hz.

For the accelerated expansion of the Universe, in [9], the basic relations of the parameter connections $|\xi_{0H}|^2$ with the rest masses of the effective atom m_H and the Higgs boson M_{H0} are obtained in the form

$$|\xi_{0H}|^2 = m_H/M_{H0} = M_H/m_{H0} = E_H/E_{H0} = R_H/R_{H0};$$
$$M_{Ha} = N_a M_H;\ M'_{H0} = N_a m_{H0};\ R_H = 2GM_{Ha}/c_0^2;\ R_{H0} = 2GM'_{H0}/c_0^2. \quad (2)$$

Here $M_H = N_a m_H$ and $m_{H0} = N_a M_{H0}$ are molar masses of effective atom and Higgs boson; E_H is the rest energy of an effective atom; $R_H,\ R_{H0}$ allow interpretation of Schwarzschild radii of black holes with masses $M_{Ha},\ M'_{H0}$; G is Newton's gravitational constant; c_0 is the light speed in vacuum. Based on (2) in [9], the main spectral parameters of the theory were obtained

$$|S'_{01}| = 0.039541282,\ S'_{02} = 0.03409,\ S'_{03} = 0.460458718,\ S'_{04} = 0.53409. \quad (3)$$

Radii $R_H,\ R_{H0}$ are characteristic parameters that must be considered when modeling active microobjects and using them in laser nanotechnology. There is a need to capture, move, transpose, rotate the nanoparticles by the laser field.

In [5], the rotation of a nanoparticle in the form of a dumbbell was achieved by a laser with circular polarization. Linear polarization laser causes the nanoparticle to vibrate. In our model to estimate the limit frequency of rotation of a nanoparticle, we use the expression $\nu_{rot} = 2Q_{H3}|\chi_{ef}|\nu_{G0}$. Here, the parameter $Q_{H3} = 0.700790572$ determines the positions of the inflection point for the function Q_G, the local minimum for the function V_Q, zero of the function A_Q from [9] for the model of an accelerated expansion of the Universe. The parameter $|\chi_{ef}| = 0.250425279$ describes the effective susceptibility in the anisotropic model of fractal cosmological objects [10, 11]. Based on ν_{rot}, we find the estimate $\nu_{rot} = 1.0285631$ GHz, which is close to the experimentally [5] obtained frequency of 1.029 GHz at pressure $7.2 \cdot 10^{-6}$ mbar. The rotational speed depends on pressure. Experimentally [5] at the pressure $1.1 \cdot 10^{-5}$ mbar, the signal is fixed at 1.31 GHz, which corresponds to the rotation frequency of 655 MHz. In our model [10], the rotation frequency ν'_{c2} and the density of cold dark matter Ω'_{c2} are determined from

$$\nu'_{c2} = \Omega'_{c2}\nu_{G0};\ \Omega'_{c2} = \nu^2_{\gamma b}/4|\lambda|^2 N;\ \upsilon^2_{\gamma b} = \upsilon_0^2/\Omega'_{c2}. \quad (4)$$

Here the frequency $\nu_{\gamma b} = 3.9749732$ ГГц, $\Omega'_{c2} = 0.2240917$, the Fermi effective velocity $\upsilon_{\gamma b} = 14.343536 \cdot 10^6$ cm s^{-1}, associated with neutron stars; the speed is $\upsilon_0 = 6.7835402 \cdot 10^6$ cm s^{-1} in the flat cosmology model. Taking into account ν_{rot}, we find the numerical values $\nu'_{c2} = 656.6899$ MHz, $2\nu'_{c2} = 1.3133798$ GHz, which are close to the above frequencies from the experiment [5]. This indicates the significant role of pressure, when changing the rotation frequency of the nanoparticle, the dependency of its properties on the state of physical vacuum and cosmological parameters.

To describe the other characteristics of the model nanoobject, we write the expressions for the main energies

$$\varepsilon_{G00} = N'_0 E_G|\xi_{0H}|^2;\quad \varepsilon_{G01} = N'_{01} E_G|\xi_{0H}|^2;\quad \varepsilon_{G02} = N'_{02} E_G|\xi_{0H}|^2;$$
$$E_{g00} = z'_\mu \varepsilon_{G00} = z'_\mu n_G \varepsilon_{\mu 00};\quad E_{g01} = z'_\mu \varepsilon_{G01} = z'_\mu n_G \varepsilon_{\mu 01};$$

$$E_{g02} = z'_{\mu}\varepsilon_{G02} = z'_{\mu}n_G\varepsilon_{\mu 02}. \quad (5)$$

Parameters N'_0, N'_{01}, N'_{02} take into account the presence of Bose condensate and the Higgs field [9]. Based on (5) we find the numerical values $\varepsilon_{G00} = 823.57274$ meV, $\varepsilon_{G01} = 836.14777$ meV, $\varepsilon_{G02} = 820.80269$ meV. Taking into account the cosmological redshift $z'_{\mu} = 7.18418108$ [9] and quanta $n_G = 3$, that are transferred by gravitational waves [10–12], on the basis of (6) we find the remaining energies $E_{g00} = 5.9166956$ eV, $E_{g01} = 6.0070369$ eV, $E_{g02} = 5.8249533$ eV; $\varepsilon_{\mu 00} = 274.52425$ meV, $\varepsilon_{\mu 01} = 278.71592$ meV, $\varepsilon_{\mu 02} = 270.26756$ meV. Note that the energy E_{g00} is close to the energy gap of 5.95 eV for atomic defects in boron nitride nanotubes (BNNT) [6], from which the emission of individual photons was experimentally observed.

Based on the energies and parameters from (5), (3), we find the energy spectra

$$\varepsilon_{g0x} = E_{g00}S'_{0x},\ \varepsilon_{g1x} = E_{g01}S'_{0x};\ \varepsilon_{g2x} = E_{g02}S'_{0x}; \quad (6)$$

$$\varepsilon^*_{G0x} = \varepsilon_{G00}S'_{0x};\ \varepsilon^*_{G1x} = \varepsilon_{G01}S'_{0x};\ \varepsilon^*_{G2x} = \varepsilon_{G02}S'_{0x}; \quad (7)$$

$$\varepsilon_{\nu 0x} = \varepsilon_{\mu 00}S'_{0x};\ \varepsilon_{\nu 1x} = \varepsilon_{\mu 01}S'_{0x};\ \varepsilon_{\nu 2x} = \varepsilon_{\mu 02}S'_{0x}. \quad (8)$$

Spectrum (6) allows us to perform estimates for the average diameter of a point defect d_a, the Raman shift $1/\lambda_{RS}$ in the spectrum of BNNT using the formulas

$$2d_a = \lambda_{g24}|\chi_{ef}|;\ \lambda_{RS} = 2\pi n_G(1+\Omega_{0\nu})\lambda_{g14}$$
$$\lambda_{g24} = a_\lambda/\varepsilon_{g24};\ \lambda_{g14} = a_\lambda/\varepsilon_{g14}. \quad (9)$$

Here $a_\lambda = 1239644.01$ meV nm is the constant parameter, $\Omega_{0\nu} = 0.002939801$ is the neutrino density [10–12]. The numerical values $d_a = 49.892844$ nm and $2d_a = 99.785688$ nm are close to the average estimates of the sizes of nanoparticles of 50 nm and 100 nm from [6] and [5], respectively. The Raman shift $\lambda_{RS} = 7304.6019$ nm almost coincides with the value $1/\lambda_{RS} = 1369\,\text{cm}^{-1}$ from experiment [6]. Next, we find the energy difference between the branches of the spectrum (6) $\delta\varepsilon_{g1} = \varepsilon_{g13} - \varepsilon_{g12} - \varepsilon_{g21} = 2330.8865$ meV and the corresponding wavelength $\lambda_{\delta 1} = a_\lambda/\delta\varepsilon_{g1} = 531.8337$ nm. This transition between the branches of the spectrum with a wavelength $\lambda_{\delta 1}$ close to the wavelength of 532 nm of the exciting green laser from [6]. The transition with the energy difference $\delta\varepsilon_{g1}/2 = 1165.4433$ meV and wavelength $2\lambda_{\delta 1} = 1063.6674$ nm is close to the wavelength of the 1064 nm laser used in [6] for optical modulation. In [6], for observing photoluminescence, two different defects in BNNT (two emitters) were obtained. The maximum emission of individual photons from these emitters was obtained at wavelengths of 571 and 569 nm. In our model, the energy differences between the branches of the spectrum (6) and the corresponding wavelengths are equal:

$$\delta\varepsilon_{g2} = \varepsilon_{g13} - 3\varepsilon_{g22} = 2170.2746\,\text{meV},$$

$$\lambda_{\delta 2} = a_\lambda/\delta\varepsilon_{g2} = 571.1923\,\text{nm};$$
$$\delta\varepsilon_{g3} = \varepsilon_{g13} - 3\varepsilon_{g22} + \varepsilon_{g11} - \varepsilon_{g21} = 2177.4744\,\text{meV},$$
$$\lambda_{\delta 3} = a_\lambda/\delta\varepsilon_{g3} = 569.3036\,\text{nm}. \tag{10}$$

These values are close to the experimental data for emitters [6].

The branches of the spectrum (8) are located inside the band gap of the spectrum (6) and depend on temperature. Stable quantum emission of photons is observed in a wide range of temperatures, including room temperature. The upper bound for the temperature is about 800 K [6]. Taking into account the energy spectrum (8), we find an estimate of the upper limit of temperature using the formula

$$T_{\nu s} = a_T(\varepsilon_{\nu 13} + \varepsilon_{\nu 12}) = 799.84687\,\text{K};\quad a_T = 5.802778417\,\text{K}\,(\text{meV})^{-1} \tag{11}$$

In [9–12], in the framework of the anisotropic model, estimates of the temperature $T_r = 2.72548$ K of the background radiation, the dipole anisotropy of the background radiation δT_r, the average value of the temperature fluctuations of the background radiation δT_A were obtained. This background radiation can influence on the basic parameters of active nanoobjects. In the framework of the Dicke theory of superradiance [18–20], we determined the maximum number of active effective particles $N = 17.0073101$ [10–12]. The expressions, which connecting these temperatures and characteristic energies ε_r, ε_{rA}, ε'_{rA} are

$$T_r = T_{rA} + T'_{rA};\ T_{rA} = u^2_{rA}T_r;\ T'_{rA} = \text{v}^2_{rA}T_r;$$
$$T_r = a_T\varepsilon_r;\ u^2_{rA} + \text{v}^2_{rA} = 1;\ 1 - 2u^2_{rA} = (N+1)/z'_{A2};$$
$$T_{rA} = a_T\varepsilon_{rA};\ T'_{rA} = a_T\varepsilon'_{rA};\ N_{rA} = z'_{A2} + N + 1. \tag{12}$$

The numerical values of these parameters are equal: $T_{rA} = 1.3390101$ K, $T'_{rA} = 1.3864699$ K, $\varepsilon_r = 469.685348\,\mu$eV, $\varepsilon_{rA} = 230.7532756\,\mu$eV, $\varepsilon'_{rA} = 238.93207\,\mu$eV, normal redshift $z'_{A2} = 1034.109294$, $N_{rA} = 1052.116604$.

Temperatures of phase transitions T'_A, T_A [10–12] are determined through the number of quanta n_{zA} (for the usual red shift) and $n_{z\mu}$ (for the cosmological red shift z'_μ), the total number of relic radiation quanta N_{ra} by the formulas

$$T'_A = T_A + \delta T_A;\ T_A = n_{zA}T'_A;\ \delta T_A = n_{z\mu}T'_A;\ n_{z\mu} = z'_\mu/N_{ra};$$
$$n_{zA} + n_{z\mu} = 1;\ N_{ra} = z'_{A2} + z'_\mu;\ \delta T_r = Q_{H3}\,T'_{rA}\,\delta T_A/T'_A;\ 2T'_{rA} = N_{rA}T'_A \tag{13}$$

The numerical values of these parameters are: $T'_A = 2.6355822$ mK, $T_A = 2.6173985$ mK, $\delta T_A = 18.183633\,\mu$K, $\delta T_r = 6.7035181$ mK.

Based on the relations from [10–12] for cosmological fractal objects and Dicke superradiance model [18–20], we write the expressions for the ratio of accelerations

$$g_0/g_{SE} = n_G(z'_{A2} - z'_\mu + I_m I_0^{-1})/2. \tag{14}$$

Here $g_0 = 980.665\,\mathrm{cm\,s^{-2}}$ is the acceleration of free fall on the surface of the Earth; g_{SE} is acceleration of the Earth towards the Sun; ratio of maximum I_m to initial I_0 radiation intensities $I_m I_0^{-1} = 81.06580421$. Taking into account the expression (14) we find the numerical value $g_{SE} = 0.590056\,\mathrm{cm\ s^{-2}}$. Taking into account the main parameter $n_{A0} = 58.04663887$ for black holes from [10–12], the parameter N_{HG} from (1), based on the condition

$$g_{SE} N_{HG} = g_{nS} n_{A0} \tag{15}$$

we find the estimate of the gravitational acceleration $g_{nS} = 1.04888 \cdot 10^{12}\mathrm{m\ s^{-2}}$ on the surface of the neutron star, which is coincide with the estimate $10^{12}\ \mathrm{m\ s^{-2}}$ [5].

3 Higgs Field

In the framework of our anisotropic model [10–12], the main parameters are $|\xi_{0H}|^2 = \chi_{11}$ and $(\xi_0')^2 = N_{01}'/N_{02}'$. For the components of the effective susceptibility tensor $\hat{\chi}_{ef}$, we use the notation from [10–12]. The operators $\hat{n}_{\xi 0}$ and $\hat{n}_{\xi 0}^+$ describe the ground and excited states of the Higgs boson field. Averaged values $n_{\xi 0} =< \hat{n}_{\xi 0} >$ and $n_{\xi 0}' =< \hat{n}_{\xi 0}^+ >$ are determined by expressions

$$n_{\xi 0}' - n_{\xi 0} = 1;\ n_{\xi 0} = k_\mu |\chi_{ef}| \cdot |\xi_{0H}|^2 \mathrm{cn}(u_\mu; k_\mu);\ n_{\xi 0}' = (\xi_0')^2. \tag{16}$$

If to use the numerical values of the components of susceptibility χ_{21}, χ_{11} from [10–12], then we obtain the numerical value $n_{\xi 0} = \chi_{21}\,\chi_{11} = 0.031259246$.

In general case, $|\chi_{ef}| = n_{0u}/\chi_{11}$ is a nonlinear function on $n_{\xi 0}$. This allows us to write expressions as follows

$$n_{0u} = \sqrt{2}\{\chi_{11}^4 - n_{\xi 0}(\chi_{11}^2 - n_{\xi 0})\}^{1/2};\ n_{\xi 0} = k_\mu n_{0u} \mathrm{cn}(u_\mu; k_\mu). \tag{17}$$

From (17) it follows: the number of quanta $n_{\xi 0} = n_{0u}\mathrm{sech}(u_\mu)$ at $k_\mu = 1$; $n_{\xi 0} = 0$ at $k_\mu = 0$ or at $\mathrm{cn}(u_\mu; k_\mu) = 0$.

If to introduce the parameters S_{2u}, S_{1u} by the relations

$$S_{2u} = \chi_{11}^2;\ |S_{1u}| = \sqrt{2}\,S_{2u};\ S_{4u} = 0.5 + S_{2u};\ S_{3u} = 0.5 - |S_{1u}|. \tag{18}$$

then we find the basic nonlinear equation for the function n_{0u} on $n_{\xi 0}$, which gives the possibility of choosing the sign of different branches

$$n_{0u}^2 |S_{1u}|^{-2} + n_{\xi 0}(S_{2u} - n_{\xi 0}) S_{2u}^{-2} = 1. \tag{19}$$

Based on the numerical value χ_{11} from [10–12], expressions (18), we obtain the numerical values of the parameters $S_{2u} = 0.033051284$, $|S_{1u}| = 0.046741575$,

$S_{3u} = 0.453258425$, $S_{4u} = 0.533051284$. By analogy with [9], we introduce the Hubble constant $H_{0\nu}$, the main parameter of the theory $Q_{0\nu}$ by the relations

$$Q_{0\nu} = H_{0\nu}/H_{02} = \mathrm{v}_{0\nu}/\mathrm{v}_{02} = 1 + S_{2u};\ \mathrm{v}_{0\nu} = H_{0\nu}L_0 = n_{A0}\,\mathrm{v}_{h2}/2;$$
$$\Omega_{tH0} = Q_{0\nu} + |S_{1u}|;\ \mathrm{v}_q = \mathrm{v}_{01} - \mathrm{v}_{02};\ \mathrm{v}_{2\nu} = \mathrm{v}_{0\nu} - \mathrm{v}_{02};\ \mathrm{v}_{1\nu} = \mathrm{v}_{01} - \mathrm{v}_{0\nu}. \quad (20)$$

Based on the numerical values H_{02}, v_{02}, H_{01}, v_{01}, L_0, n_{A0} from [9], parameters S_{2u}, $|S_{1u}|$ and expressions (20), we find the numerical values: $Q_{0\nu} = 1.0330513$; speeds $\mathrm{v}_{0\nu} = 7.2743002 \cdot 10^6\,\mathrm{cm\ s^{-1}}$, $\mathrm{v}_{h2} = 2.5063640 \cdot 10^5\,\mathrm{cm\ s^{-1}}$; the total density of the Universe (normalized to the critical density) $\Omega_{tH0} = 1.079792859$; speed differences $\mathrm{v}_q = 2.784326 \cdot 10^5\,\mathrm{cm\ s^{-1}}$; $\mathrm{v}_{2\nu} = 2.3273284 \cdot 10^5\,\mathrm{cm\ s^{-1}}$; $\mathrm{v}_{1\nu} = 4.5699756 \cdot 10^4\,\mathrm{cm\ s^{-1}}$. Note that the velocity $\mathrm{v}_{0\nu}$ is related with the characteristic velocity v_{h2} through the main parameter n_{A0} (which determines the spectrum of filling numbers) for black holes.

The numbers of quanta n'_F, n_F for particles of Fermi type from [10–12] are also nonlinear functions on the number of quanta of the Higgs field $n'_{\xi0}$, $n_{\xi0}$

$$n'_F = (S_{2u} - n_{\xi0})/S_{2u};\ n_F = n_{\xi0}/S_{2u};\ n'_F + n_F = 1 \quad (21)$$

From (21) it follows that the neutrino density $\Omega_{0\nu}$, defined in [10–12], is also a nonlinear function on the number of quanta of the Higgs field

$$\Omega_{0\nu} = (n'_F)^2 = (Q_{0\nu} - n'_{\xi0})^2/S_{2u}^2 = (\mathrm{v}_{2\nu} - n_{\xi0}\mathrm{v}_{02})^2/\mathrm{v}_{2\nu}^2. \quad (22)$$

Accounting dependence $\mathrm{v}_{0\nu}$ on n_{A0} from (20) makes it possible to write (22) as

$$\Omega_{0\nu} = (n_{A0} - n'_{\xi0}N_{h2})^2/(n_{A0} - N_{h2})^2 = (n_{0\nu} - n_{\xi0}N_{h2})^2/n_{0\nu}^2;$$
$$N_{h2} = 2\mathrm{v}_{02}/\mathrm{v}_{h2} = 56.189504;\ n_{0\nu} = n_{A0} - N_{h2} = 1.8571352. \quad (23)$$

Expressions (23) determine the neutrino density $\Omega_{0\nu}$ as a function on the number of black hole quanta n_{A0}, the number of Higgs field quanta $n'_{\xi0}$, $n_{\xi0}$, and parameters N_{h2}, $n_{0\nu}$. From (22), (23) it is possible to obtain the limiting values for the neutrino density: $\Omega_{0\nu} = 0.002939801$ at $n'_{\xi0} = 1 + \chi_{21}\,\chi_{11} = 1.031259246$; $\Omega_{0\nu} = 0$ either at $n'_{\xi0} = Q_{0\nu} = 1.033051284$, or at $\mathrm{v}_{2\nu} = n_{\xi0}\mathrm{v}_{02}$, or at $n_{A0} = n'_{\xi0}N_{h2}$, or at $n_{0\nu} = n_{\xi0}N_{h2}$; $\Omega_{0\nu} = 1$ either at $Q_{0\nu} - n'_{\xi0} = S_{2u}$, or at $n_{\xi0} = 0$, or at $n'_{\xi0} = 1$. This indicates to a possible hysteresis behavior of the neutrino density as a function of the given arguments. In [10–12] a connection n_{A0} with the cosmological redshift z'_μ was obtained in the form $n_{A0} + 3/4 = z'_\mu(z'_\mu + 1)$. This allows us to write the dependency (19) in general case

$$n_{0u}^2 n_{0u0}^{-2} - (n_{\xi0} - n_{2u})^2(2n_{2u})^{-2} = z'_\mu(z'_\mu + 1) - n_{A0} = n_Q/n_g;$$
$$n_{2u} = S_{2u}/2;\ n_{0u0} = |S_{1u}|;\ n_g = n_{0u0}^2/n_{2u}^2;\ n_Q = 2n_G = 2 < \hat{c}_G\,\hat{c}_G^+ >. \quad (24)$$

Here $n_g = 8$, $n_Q = 6$, $n_G =< \hat{c}_G \hat{c}_G^+ >= 3$ and $n'_G =< \hat{c}_G^+ \hat{c}_G >= 2$ can be interpreted as the number of quanta of the gluon, quark, excited and ground states gravitational fields, respectively. In this case, expressions (24) exactly go into expressions (19), (22), (23). In the general case, from (24) there follow other possible dependencies of the number of quanta n_{0u}, n_g both on the number of quanta of the Higgs field $n_{\xi 0}$, and on the number of quanta for black holes n_{A0}, cosmological redshift z'_μ. At the same time, functions appear that depend on three arguments, which is typical for bulk fractal structures of the Universe. So, for example, from (24), we can obtain the dependency for the number of quanta of the gluon field $n_g = 2n_G/[z'_\mu(z'_\mu + 1) - n_{A0}]$ at $n_G = 3$ as a function on two arguments z'_μ, n_{A0}.

4 Relationships of Neutrino with Higgs Boson and Higgs Field

To describe the neutrino parameters (the number of solar $N_{0\nu}$ and electronic $N_{e\nu}$ neutrinos; the rest mass m_{HG} and energy ε_{HG} neutrino) and their relationships with the Higgs boson parameters, we will use the anisotropic model [10–12] and expressions (1), (2), (24). Within this model, we find the equation for the characteristic energy $E_{H\nu}$

$$2E_{H0}/E_{e\nu} = 2E_H/\varepsilon_{e\nu} = z'_\mu + 0.5;$$
$$E_{e\nu} = N_{e\nu} E_{H\nu};\ N_{e\nu} = N_{0\nu}/n_G. \tag{25}$$

Here are the numerical values $n_G = 3$, $N_{e\nu} = 1.9997054 \cdot 10^{10}$, $N_{0\nu} = 5.9991163 \cdot 10^{10}$; energies $E_{H\nu} = 1627.3796\,\text{meV}$, $E_{e\nu} = 32.542799\,\text{GeV}$, $\varepsilon_{e\nu} = 5.9162848\,\text{GeV}$. Note, that earlier (in the 60s of the XX century) only electron neutrinos were recorded. At that time, sensors could not detect other types of neutrinos. The creation of the next generation of detectors made it possible to detect contributions from μ and τ-neutrinos to the total number of solar neutrinos $N_{0\nu}$. The action of the tensor $\hat{\chi}_{ef}$ on the energy $E_{H\nu}$ leads to an energy tensor $\hat{\varepsilon}_{H\nu} = \hat{\chi}_{ef} E_{H\nu}$ with the components $\varepsilon_{ij} = \chi_{ij} E_{H\nu}$ $(i, j = 1, 2, 3)$. Based on χ_{ij} [10–12] we find the numerical values

$$\varepsilon_{11} = 295.8578151\,\text{meV}; \varepsilon_{12} = -280.2758586\,\text{meV};\ \varepsilon_{13} = 0;$$
$$\varepsilon_{21} = 279.8164249\,\text{meV};\ \varepsilon_{22} = 295.3728381\,\text{meV};\ \varepsilon_{23} = -23.32509064\,\text{meV}$$
$$\varepsilon_{31} = 16.04139021\,\text{meV};\ \varepsilon_{32} = 16.93321214\,\text{meV}; \varepsilon_{33} = 406.8689518\,\text{meV}. \tag{26}$$

Solving the problem of finding the eigenvalues of the energy ε_R of the energy tensor $\hat{\varepsilon}_{H\nu}$ leads to the nonlinear equation

$$\varepsilon_{33} - \varepsilon_R = \varepsilon_{23}\varepsilon_{32}(\varepsilon_{R3} - \varepsilon_R)/\{[\varepsilon_R - 0.5(\varepsilon_{11} + \varepsilon_{22})]^2 + \varepsilon_{R\gamma}^2\};\ \varepsilon_{HG} = |\varepsilon_{21}\,\varepsilon_{12}|^{1/2};$$

$$\varepsilon_{R3} = |\chi_{ef}|^2 E_{H\nu}/|\xi_{0H}|^2;\ 2\varepsilon_{R\gamma} = [4|\varepsilon_{21}\,\varepsilon_{12}| - (\varepsilon_{11} - \varepsilon_{22})^2]^{1/2} \tag{27}$$

On the basis of (27) we find the numerical values: the neutrino rest energy $\varepsilon_{HG} = 280.0460475$ m eV; energies $\varepsilon_{R\gamma} = 280.0459426$ m eV, $\varepsilon_{R3} = 561.3723777$ meV. The sign of the energies ε_{21}, ε_{12} (or ε_{23}, ε_{32}) is determined by the choice of the sign of the modules k_μ, k'_μ and the value of the angle φ_μ. If $\varepsilon_{21}\varepsilon_{12} < 0$, then $\varepsilon^2_{HG} = -\varepsilon_{21}\varepsilon_{12}$, wherein the energy $2\varepsilon_{R\gamma}$ from (27) is replaced by energy $2\varepsilon^*_{R\gamma} = [(\varepsilon_{11} - \varepsilon_{22})^2 + 4\varepsilon^2_{HG}]^{1/2}$. Numerical value is $\varepsilon^*_{R\gamma} = 280.0461525$ meV. On the basis of the neutrino rest energy ε_{HG} and the cosmological redshift z'_μ, it is possible to determine the characteristic energy $\varepsilon_{z\mu} = z'_\mu(z'_\mu + 2)\varepsilon_{HG}/4 = 4.61941696$ eV, which can be interpreted as the energy of the correlated (coherent) state of field, associated with cosmological redshift z'_μ for relic radiation.

On the basis of the spectra $\varepsilon_{iHG} = 2\varepsilon_{HG}S'_{0i}$, $\varepsilon'_{iHG} = 2|\chi_{ef}|\varepsilon_{HG}S'_{0i}$ $(i = 1, 2, 3, 4)$ we can obtain the energies ε^*_{HG}, ε'_{2HG} from the expressions

$$\varepsilon^*_{HG} = \varepsilon_{HG} + \varepsilon'_{2HG};\ (\varepsilon^*_{HG})^2 - \varepsilon^2_{HG} = z'_{HG}(z'_{HG} + 2)\varepsilon^2_{HG};\ z'_{HG} = \varepsilon'_{2HG}/\varepsilon_{HG} \tag{28}$$

Numeric values are equal: $\varepsilon^*_{HG} = 284.8275525$ meV, $\varepsilon'_{2HG} = 4.781504961$ meV. The parameter $z'_{HG} = 0.017073996$ allows interpretation as a cosmological redshift for the neutrino field ($z'_{HG} < z'_\mu$). The difference of the squares of the energies $(\varepsilon^*_{HG})^2 - \varepsilon^2_{HG} = 2700.94592\ (\text{meV})^2$ practically coincides with the experimental data $2.7 \cdot 10^{-3}\ (\text{eV})^2$ on neutrino oscillations [21, 22]. These experiments proved the existence of a neutrino rest mass and the possibility of changing the type of neutrino. In our model, the neutrino with the rest energy ε_{HG} is replaced by the neutrino with the rest energy ε^*_{HG} due to the presence of a cosmological redshift z'_{HG} for the neutrino field.

5 Super-Nonradiative States of Fractal Objects

The properties of nanoparticles in fractal quantum systems depend on pressure, state of physical vacuum and cosmological parameters. Dynamic models make it possible to describe not only the superradiance of gravitational waves, relic photons from cosmological objects (such as binary neutron stars and black holes) [10–12], but also super-nonradiative states of active objects. In [10–12], for the radiation intensity $J(t)$ (Dicke superradiance model [18–20]), the expression was written

$$J(t) = J_0(a_0 + a_m)[(a_0 - a_m) + 1]. \tag{29}$$

Here J_0 is the initial radiation intensity; in general case $a_0(t)$, $a_m(t)$ depend on time t and other parameters. The super-nonradiative states we call the states with $J(t) = 0$. These states can be realized in the process of development of various transient effects (induction, avalanche, echo, self-induced transparency). From (29) it follows that these states can be realized within the framework of two models.

Model A$_0$. Within this model, we assume $a_0 = -a_m$, where

$$a_m = (z'_{A2})^{1/2};\ a_0^2 = a_m^2 + z'_\mu(z'_\mu + 2)/4;\ a_m^2 = z'_{A2};\ N_{ra} = z'_{A2} + z'_\mu. \quad (30)$$

From (30) it follows that two variants are possible: variant 1 (B1), when $z'_\mu = 0$, then $N_{ra} = z'_{A2}$; variant 2 (B2), when $z'_\mu = -2$, then $N_{ra} = z'_{A2} - 2$.

To describe the super-nonradiative states, we introduce row-vectors $\hat{N}_{d1}$, $\hat{N}_{d2}$ and column-vectors $\hat{N}^+_{d1}$, $\hat{N}^+_{d2}$, respectively, for variants B1 and B2

$$\hat{N}_{d1} = (N_{ra},\ z'_{A2},\ z'_\mu) = (z'_{A2},\ z'_{A2},\ 0);\ \hat{N}_{d2} = (z'_{A2} - 2,\ z'_{A2},\ -2). \quad (31)$$

Next, we find the norms $|N_{d1}|$, $|N_{d2}|$ and the angle θ_{d12}

$$\begin{aligned} &\hat{N}_{d1}\hat{N}^+_{d1} = 2(z'_{A2})^2 = |N_{d1}|^2;\ \hat{N}_{d2}\hat{N}^+_{d2} = 8 + 2z'_{A2}(z'_{A2} - 2) = |N_{d2}|^2; \\ &\hat{N}_{d1}\hat{N}^+_{d2} = \hat{N}_{d2}\hat{N}^+_{d1} = 2z'_{A2}(z'_{A2} - 1); \\ &\cos\theta_{d12} = \hat{N}_{d1}\hat{N}^+_{d2}|N_{d1}|^{-1}|N_{d2}|^{-1} = \sqrt{2}\,(z'_{A2} - 1)[8 + 2z'_{A2}(z'_{A2} - 2)]^{-1/2} \quad (32) \end{aligned}$$

When changing the usual redshift z'_{A2} from (32) follows the possibility of changing the sign $\cos\theta_{d12}$. So with the values $z'_{A2} = 0,\ 1,\ 2$ we get $\cos\theta_{d12} = -0.5,\ 0,\ 0.5$, respectively. Based on (32), we can introduce functions for the distribution density of the Fermi type n_{d12} and n'_{d12}

$$\begin{aligned} &n'_{d12} + n_{d12} = 1;\ n'_{d12} = \cos^2\theta_{d12} = 2(z'_{A2} - 1)^2[6 + 2(z'_{A2} - 1)^2]^{-1}; \\ &n_{d12} = \sin^2\theta_{d12} = 6\,[6 + 2(z'_{A2} - 1)^2]^{-1}; \\ &B_{d12} = n'_{d12} - n_{d12} = [(z'_{A2} - 1)^2 - 3]/[(z'_{A2} - 1)^2 + 3]. \quad (33) \end{aligned}$$

For the parameter B_{d12} from (33), the interpretation is allowed as the difference in population between the states (31), (32). The state with $B_{d12} = 0$ is realized either at $z'_{A2} = 1 + \sqrt{3}$, or at $z'_{A2} = 1 - \sqrt{3}$. At the same time $\cos^2\theta_{d12} = \sin^2\theta_{d12} = 1/2$, that indicates to the presence of the transverse component of the effective vectors from (31), (32). This allows us to perform the interpretation in terms of a super-nonradiative state with a possible chiral (polarization) of the structures from (31), which is formed by the usual red shift z'_{A2}. On the other hand, the values $n_Q = 6$, $n_G = 3$ from (24) can determine the contributions of the quark and gravitational fields to the behavior of the parameters n_{d12}, n'_{d12}, B_{d12}, which follows from (33). Based on (33), you can also introduce functions for the distribution density of Bose type N_{zA}, N'_{zA} and n_{zA}, n'_{zA}

$$N'_{zA} - N_{zA} = 1;\ N_{zA} = (z'_{A2} - 1)^2/3;\ n'_{zA} - n_{zA} = 1;\ n_{zA} = N^{-1}_{zA}. \quad (34)$$

Based on (34) and $z'_{A2} = 1034.109294$ we find the characteristic value of the number of bosons in the equilibrium state $N_{0A} = N_{zA}(z'_{A2}) = 3.557716045 \cdot 10^5$. This allow us to introduce the characteristic energy of the gravitational field $E_{0A} = N_{0A}E_G = 4.311073329\,\text{eV}$ and write down the energy spectrum $E_{0Ai} = 2E_{0A}S_{iu}$ $(i = 1, 2, 3, 4)$. Numerical values of the energies of the spectrum branches are equal: acoustic branches $E_{0A1} = 403.01271\,\text{meV}$, $E_{0A2} = 284.97302\,\text{m eV}$ (near to energy ε^*_{HG}); optical branches $E_{0A3} = 3.9080606\,\text{eV}$, $E_{0A4} = 4.5960463\,\text{eV}$. It should be noted that the energies $\varepsilon_{z\mu}$, E_{0A} are close to the boundary energy 4.3 eV, which is observed during neutrinoless double β-decay in experiments with the ^{136}Xe isotope [23]. This makes it possible to interpret these energies as the rest energy of Majorana neutrino.

On the other hand, on the basis of energies E_{0A}, E_{H0} and conditions $E_{d\nu} = N_{0H\nu}E_{0A}/2 = E_{H0}(S'_{04} + 0.5) = 129.2947339\,\text{GeV}$, can obtain an estimate of the total number of neutrinos $N_{0H\nu} = 5.99826187 \cdot 10^{10}$, which is close to $N_{0\nu}$ from (25). The energy $E_{d\nu}$ is close to the local minimum between local maxima at the dependencies of the number of photons on their energy, which are detected by the detectors in experiments at the Large Hadron Collider (LHC) [24]. Next, we find an estimate of the energy of the acoustic branch of dark relic photons (virtual relic photons in the condensate) $\varepsilon'_{dm2} = N_{0A}\varepsilon_r = 2\varepsilon'_{dm}S_{2u} = 167.0746\,\text{eV}$. From here we find an estimate of the energy of dark matter $\varepsilon'_{dm} = 2.5275057\,\text{keV}$. Between the energies of the spectra branches $\varepsilon'_{dmi} = 2\varepsilon'_{dm}S_{iu}$ and $\varepsilon_{dmi} = 2\varepsilon_{dm}S_{iu}$ for dark matter with ε'_{dm} and $\varepsilon_{dm} = 1.7872164\,\text{keV}$ conditions take place

$$\begin{aligned}\varepsilon'_{dm} &= \varepsilon'_{dm4} - \varepsilon'_{dm2} = \varepsilon'_{dm3} + \varepsilon'_{dm1};\\ \varepsilon_{dm} &= \varepsilon_{dm4} - \varepsilon_{dm2} = \varepsilon_{dm3} + \varepsilon_{dm1}.\end{aligned} \tag{35}$$

The numerical values of the energies of the spectra branches are equal: acoustic branches $\varepsilon'_{dm1} = 236.27919\,\text{eV}$, $\varepsilon_{dm1} = 167.07461\,\text{eV}$, $\varepsilon_{dm2} = 118.13959\,\text{eV}$; optical branches $\varepsilon'_{dm3} = 2.2912265\,\text{keV}$, $\varepsilon'_{dm4} = 2.6945803\,\text{keV}$, $\varepsilon_{dm3} = 1.6201417\,\text{keV}$, $\varepsilon_{dm4} = 1.9053559\,\text{keV}$. Direct experiments (DAMA/LIBRA, CoGeNT, CRESST-II collaborations) [13] by observing the spectrum and the angular distribution γ-radiation, the modulation spectrum from the galactic center show the following features: at the background of stochastic behavior the main local maximum near 2.4 keV and two local main minimums near 1.9, 2.7 keV are observed. In our model, the calculated energy values ε_{dm4}, ε'_{dm4}, which correspond to the local minima (potential wells) of the optical energy branches for dark matter with rest energy ε_{dm}, ε'_{dm}, practically coincide with the position of the indicated local minima in the experimental modulation spectrum. The energy value $\varepsilon'_{dm3} + \varepsilon_{dm2} = 2.4093661\,\text{keV}$ almost coincides with the position of the main local maximum. In this case, the energy ε_{dm2}, that lies on the acoustic branch of the energy spectrum carries information about the transmitted pulse of γ-radiation (about the half-width of the radiation line). Thus, information on the presence of super-nonradiative states, dark matter, dark relic photons can be extracted from the experimental spectra of the γ-radiation by the presence of local minima at the background of the stochastic behavior of signals.

In a similar way, we can extract information on super-nonradiative states of a gravitational field by the presence of local minima with energies on optical branches E_{0A3}, E_{0A4} (deformation field), transmitted pulses of gravitational field (stress field) with energies on acoustic branches E_{0A1}, E_{0A2} and rest energy E_{0A}.

Model A$_1$. In this model, we assume $a_0 = a_m - 1$. This condition makes it possible to introduce distribution density functions a_m, a'_m for particles of Fermi type and distribution density functions N_{zg}, N'_{zg}, n_{zg}, n'_{zg}, for particles of Bose type

$$a_m + a'_m = 1;\ 2a_m = 1 - b_m;\ 2a'_m = 1 + b_m;$$
$$4b_m = z'_\mu(z'_\mu + 2);\ z'_{A2} = N_{ra} - z'_\mu; \tag{36}$$

$$N'_{zg} - N_{zg} = 1;\ N_{zg} = (1 + b_m)/(1 - b_m);$$
$$N'_{zg} = 8/[4 - z'_\mu(z'_\mu + 2)]; \tag{37}$$

$$n'_{zg} - n_{zg} = 1;\ n_{zg} = (1 - b_m)/(1 + b_m);$$
$$n'_{zg} = 8/[4 + z'_\mu(z'_\mu + 2)]. \tag{38}$$

The parameter $b_m = a'_m - a_m$ (which can be interpreted as the population difference for particles of Fermi type) from (36) confirms the presence of a superstate, which is associated with the cosmological redshift z'_μ. On the other hand, the occupation numbers N'_{zg}, n'_{zg} from (37), (38) confirm the possibility of describing non-radiative states (dark matter) using the gluon field ($n_g = 8$), renormalized by contributions due to the cosmological redshift z'_μ. Expressions (36)–(38) are written when the condition $-1 \le b_m \le 1$ is obtained. At increase z'_μ the parameter b_m becomes greater than 1. In this case, the distribution density functions for particles of Fermi type (36) are transferred to the distribution density functions for particles of Bose type $a'_m - |a_m| = 1$; density distribution functions for particles of Bose type (37) are transferred to other density distribution functions for particles of Bose type $|N_{zg}| - |N'_{zg}| = 1$; the distribution density functions for particles of Bose type (38) are transferred to the distribution density functions for particles of Fermi type $n'_{zg} + |n_{zg}| = 1$. When $z'_\mu = 7.18418108$ we will obtain the numerical values: $|a_m| = 7.747602495$, $|N'_{zg}| = 0.129072187$, $|n_{zg}| = 0.885682963$, $n'_{zg} = 0.114317037$. This makes it possible to determine a frequency $\nu'_{zg} = n'_{zg}\nu_{G0} = 335.00053$ MHz close to the frequency 330 MHz at which dark matter dominates from observations of radiofilaments [13]. Thus, within the framework of this **model A$_1$**, it is possible to describe the processes of mutual transformations of particles of Bose and Fermi types. This can be used in neutrino physics, neuromedicine (to describe neuromediators). Detailed analysis of the **model A$_1$** requires a separate study.

6 Conclusions

On the basis of the hypothesis of the hierarchical structure of the Universe, relations between the parameters of fractal cosmological objects such as black holes, neutron stars, and the model nanoobject were obtained. As an active nanoobject, the quartz nanoparticle, trapped in an optical trap and placed in a vacuum is considered. The size estimate of this nanoparticle and the limiting frequency rotation in a laser field with circular polarization are obtained. It is shown that the rotation frequency depends on the properties of the physical vacuum. As other active nanoobjects, the atomic defects in boron nitride nanotubes are considered. The estimate of wavelengths of the quantum emission of separate photons from such emitters are obtained.

In the framework of the anisotropic model, the dependencies of the components of the effective susceptibility for active objects as nonlinear functions on the number of quanta of the Higgs field, black holes are obtained. It is shown that the hysteresis behavior of the neutrino density as a function of the two specified arguments is possible. The additional dependency on the cosmological redshift leads to nonlinear functions on three arguments, that is characteristic for the bulk fractal structures of the Universe.

The relationship between the neutrino parameters and the Higgs boson and the Higgs field allow us to make estimates of the number, mass, and rest energy of the neutrino.

The super-nonradiative states of active fractal objects are studied within the **model** $\mathbf{A_0}$ (taking into account the usual redshift z'_{2A}). It is shown, that information on the presence of super-nonradiative states, dark matter, dark relic photons can be extracted from experimental spectra of γ-radiation by the presence of local minima on the background of the stochastic behavior of signals. In the framework of the **model** $\mathbf{A_1}$ (taking into account the cosmological redshift z'_{μ}) the distribution density functions for particles of Fermi and Bose types are introduced.

References

1. A. Ashkin, *Optical Trapping and Manipulation of Neutral Particles Using Lasers*. World Scientific Publishing Company (2006)
2. D. Strickland, G. Mourou, Compression of amplified chirped optical pulses. Opt. Commun. **56**(3), 219–221 (1985)
3. Y.S. Kivshar, N.N. Rozanov (Eds.), *Nonlinearities in Periodic Structures and Metamaterials* (Fizmatlit, Moscow, 2014)
4. V.V. Samartsev, V.G. Nikiforov, *Femtosecond Laser Spectroscopy*, ed. by M. Trovant (2017)
5. R. Reimann, M. Doderer, E. Hebestreit et al., GHz rotation of an optically trapped nanoparticle in vacuum. arXiv:1803.11160v2 [physics.optics] 21 Jul 2018, 5 p
6. J. Ahn, Z. Xu, J. Bang et al., Stable emission and fast optical modulation of quantum emitters in boron nitride nanotubes. arXiv:1806.06146v1 [quant-ph] 15 Jun 2018, 4 p
7. V.S. Abramov, Active nanoelements with variable parameters in fractal quantum systems. Bull. Rus. Acad. Sci. Phys. **82**(8), 1062–1067 (2018)

8. V. Abramov, Higgs field and cosmological parameters in the fractal quantum system, in *XI international symposium on photon echo and coherent spectroscopy (PECS-2017)*. EPJ Web Conf., vol. 161, no. 02001, 2 p (2017)
9. V.S. Abramov, Cosmological parameters and Higgs boson in a fractal quantum system. CMSIM J. **4**, 441–455 (2017)
10. V.S. Abramov, Anisotropic model and transient signals from binary cosmological objects: black holes, neutron stars. Bull. Donetsk Nat. Univ. A, **1**, 55–68 (2018)
11. V.S. Abramov, Superradiance of gravitational waves, relic photons from binary black holes, neutron stars. Bull. Rus. Acad. Sci. Phys. **83**(3), 364–369 (2019)
12. V.S. Abramov, Gravitational waves, relic photons and Higgs boson in a fractal models of the Universe, in *"11th Chaotic Modeling and Simulation International Conference". Springer Proceedings in Complexity*, ed. by C.H. Skiadas, I. Lubashevsky (Springer Nature Switzerland AG, 2019), pp. 1–14
13. D. Hooper, The empirical case for 10-GeV dark matter. Dark Universe **1**, 1–23 (2012)
14. P.K. Suh, Dark matter and energy in the universe of symmetric physics. IJARPS **5**, 19–34 (2018)
15. O.P. Abramova, A.V. Abramov, Attractors and deformation field in the coupled fractal multilayer nanosystem. CMSIM J. **2**, 169–179 (2017)
16. O.P. Abramova, Mutual influence of attractors and separate stochastic processes in a coupled fractal structures. Bull. Donetsk Nat. Univ. A, 1, 50–60 (2017)
17. O.P. Abramova, A.V. Abramov, Effect of ordering of displacement fields operators of separate quantum dots, elliptical cylinders on the deformation field of coupled fractal structures, in *"11th Chaotic Modeling and Simulation International Conference". Springer Proceedings in Complexity*, ed. by C.H. Skiadas, I. Lubashevsky (Springer Nature Switzerland AG, 2019), pp. 15–27
18. R.H. Dicke, Coherent in spontaneous radiation processes. Phys. Rev. **93**(1), 99–110 (1954)
19. R. Bonifacio, P. Schwendimann, F. Haake, Quantum statistical theory of superradiance I. Phys. Rev. A **4**(1), 302–313 (1971)
20. R. Bonifacio, P. Schwendimann, F. Haake, Quantum statistical theory of superradiance II. Phys. Rev. A **4**(3), 854–864 (1971)
21. Y. Fukuda et al., Evidence for oscillation of atmospheric neutrinos. Phys. Rev. Lett. **81**(8), 1562–1567 (1998)
22. Q.R. Ahmad et al., Direct evidence for neutrino flavor transformation from neutral-current interactions in the sudbury neutrino observatory. Phys. Rev. Lett. **89**(1(011301)), 1–6 (2002)
23. R.M. Barnett et al., Review of particle physics. Phys. Rev. D **54**, 1 (1996)
24. S. Carroll, *The Particle at the End of the Universe*. Publ. by Dutton, New York (2012)

Coupled Fractal Structures with Elements of Cylindrical Type

Olga P. Abramova and Andrii V. Abramov

Abstract By the numerical modelling method the behavior of the deformation field of the coupled fractal structure with elements of cylindrical type was investigated. It is shown that the presence of variable parameters (semiaxes, modules) lead to the stochastic behavior of the complex deformation field. Complex zero displacement field operators for both separate and coupled elements of the structure are introduced. It is shown that the transposition of separate operators in the zero operator for a coupled structure leads to the appearance of a nonzero complex deformation field. At the same time, noise tracks appear on the background of stochastic peaks. The noise track is a stochastic ring, inside of which there is a regular area.

Keywords Coupled fractal structures · Complex deformation field · Complex zero operators · Ordering of operators · Numerical modelling

1 Introduction

At the present time, to create a quantum computer, the actual tasks are quantum processing of information and quantum computing [1–4]. To solve these problems it is necessary to use quantum mechanics.

Classical computers encode information in bits that can be in one of two states, 0 or 1. The realization of state 0 is achieved by adding two signals with the same amplitude and shape, but with opposite phases.

Quantum computers encode information in qubits [5, 6]. The physical systems that realise qubits can be any objects having two quantum states. Different nanostructures and metamaterials [7] can be chosen as active objects. These active objects can exhibit stochastic properties, which complicates the practical realization of qubits for quantum computers.

Recently, further development was given to nanophysics, condensed matter physics [8, 9], neural networks [10], which use nanoobjects as active objects. The

O. P. Abramova (✉) · A. V. Abramov
Donetsk National University, Donetsk, Ukraine
e-mail: oabramova@ua.fm

C. H. Skiadas and Y. Dimotikalis (eds.), *12th Chaotic Modeling and Simulation International Conference*, Springer Proceedings in Complexity,
https://doi.org/10.1007/978-3-030-39515-5_2

most important task is to study the mutual influence, collective interactions, dynamics of such nanoobjects. In this case, there are possible effects of various topological phase transitions associated with the shape and dimension of nanoobjects, the formation of new coupled structures (clusters). One example of this kind can be model multilayer nanosystems and bulk fractal structures appearing there [11–14]. In this case, it is possible to change the form, the fractal dimension of separate elements in the coupled fractal structure.

The relevance of such studies is confirmed by the award of the Nobel Prize in Physics for 2016 to M. Kosterlitz, D. Thouless, D. Haldane for theoretical discoveries of topological phase transitions and topological phases of matter.

On the other hand, the brain of living organisms is a specific neural network [10]. The active cells respond to an intersection of nodes of imaginary hexagonal lattice, that is the passage of certain distance in a certain direction. The Nobel Prize 2014 on physiology and medicine was awarded to J. O'Keefe, M.-B. Moser and E. Moser for the discovery of cells system in the brain that determine the position in space [15–17]. J. O'Keefe has opened active neurons or "place cells". M.-B. Moser and E. Moser have opened coordinate neurons or "grid-cell". These medical discoveries make it possible to use model concepts of nanoobjects in coupled fractal structures to also describe neural networks.

In [11–14, 18, 19], the features of the stochastic state of the deformation field of coupled fractal multilayer nanosystems, nanotraps and quantum dots taking into account the variation of the variable parameters were studied. The study of the mutual influence and ordering of separate structures, attractors, stochastic processes in the coupled fractal structure is described in [13, 14, 19]. In this case, the effects associated with the shape of objects are investigated using attractors of both separate and coupled structures. Attractors can be separate points, lines, surfaces and their superposition [13, 14]. The presence of stochastic attractors leads to the appearance of stochastic properties of active objects. The behavior features of such active objects (coupled fractal structures) is the appearance of complex deformation fields, which is associated with the presence of variable parameters and the ordering effect of separate displacement field operators in the total deformation field. The task of realization of state 0 for such active objects with stochastic properties is appears. This task can be realized by introducing complex zero operators [19].

The aim of this work is to study the complex deformation field, the influence of the ordering of separate operators of displacement fields, the introduction of complex zero operators in coupled fractal structures with cylindrical elements.

2 Fractal Structure with Two Elements of Cylindrical Type

We consider a coupled fractal structure (**I**, **II**), consisting of two fractal elliptic cylinders ($i = 1, 2$), located in a bulk discrete lattice $N_1 \times N_2 \times N_3$, whose nodes are given by integers n, m, j. By analogy with [18, 19], nonlinear equations for the dimensionless displacement function $u = u_2$ (the second branch of the eight possible

[18]) of the lattice node are

$$u = u_2 = \sum_{i=1}^{2} u_{Ri}; \quad u_{Ri} = R_i k_{ui}^2 (1 - 2\mathrm{sn}^2(u_2 - u_{0i}, k'_{ui})); \tag{1}$$

$$k_{ui}^2 = (1 - \alpha_i)/Q_i; \quad k'_{ui} = (1 - k_{ui}^2)^{1/2}; \quad p_{0i} = p'_{0i} + p'_{1i}n + p'_{2i}m + p'_{3i}j; \tag{2}$$

$$Q_i = p_{0i} - b_{1i}(n - n_{0i})^2/n_{ci}^2 - b_{2i}(m - m_{0i})^2/m_{ci}^2 - b_{3i}(j - j_{0i})^2/j_{ci}^2. \tag{3}$$

Here u_0 is the constant (critical) displacement; α_i is the fractal dimension of the deformation field u along the axis Oz ($\alpha_i \in [0, 1]$); variable modules k_{ui} are functions of indices n, m, j nodes of the bulk discrete lattice. The choice of the positive sign of the module k'_{ui} is associated with the choice of the second branch of the displacement function $u = u_2$ [18]. Functions Q_i determine the form of the fractal structure, the type of attractors and take into account the interaction of the nodes of both in the main plane of the discrete rectangular lattice $N_1 \times N_2$ as well as interplane interactions. The parameters $p_{0i}, p'_{3i}, b_{1i}, b_{2i}, b_{3i}, n_{0i}, n_{ci}, m_{0i}, m_{ci}, j_{0i}, j_{ci}$ characterize different fractal structures. The parameters R_i determine the orientation of the deformation fields of separate structures in the coupled system. In general case, these parameters may depend on the layer index j and the dimensionless time t. Taking into account for interplanar interactions (dependencies on the index j) leads to a change in the constant semi-axes of the elliptic cylinders (n_{ci}, m_{ci}) in (3) to the variables (n'_{ci}, m'_{ci}) semi-axes

$$(n'_{ci})^2 = a_{1i} n_{ci}^2; \quad (m'_{ci})^2 = a_{2i} m_{ci}^2. \tag{4}$$

By modelling the variable parameters a_{1i}, a_{2i} from (4) are given by the expressions

$$\begin{aligned} a_{1i} &= 1 + (p'_{1i}n + p'_{2i}m + p'_{3i}j)/p'_{0i} - b'_{3i}(j - j_{0i})^2/(j_{ci}^2 p'_{0i}), \\ a_{2i} &= 1 + (p'_{1i}n + p'_{2i}m + p''_{3i}j)/p'_{0i} - b''_{3i}(j - j_{0i})^2/(j_{ci}^2 p'_{0i}). \end{aligned} \tag{5}$$

Then, taking into account (5), the functions Q_i from (3) take the form

$$Q_i = a_{0i}(p'_{0i} - b_{1i}(n - n_{0i})^2/(n'_{ci})^2 - b_{2i}(m - m_{0i})^2/(m'_{ci})^2), \quad a_{0i} = 1. \tag{6}$$

Note, that when solving real problems, dependencies (5) should be determined experimentally.

By numerical modelling, it was assumed, that

$$\begin{aligned} &N_1 = 240, N_2 = 240, \alpha_i = 0.5, u_{0i} = 29.537, p_{0i} = 1.0423, b_{1i} = b_{2i} = 1, \\ &n_{0i} = 121.1471, m_{0i} = 120.3267, n_{c1} = m_{c1} = 57.4327, n_{c2} = m_{c2} = 28.7164, \\ &j_{0i} = 31.5279, j_{ci} = 11.8247, p'_{1i} = p'_{2i} = 0, p'_{3i} = 0.1368, p''_{3i} = 0.1547, \\ &b'_{31} = b''_{31} = 0, b'_{32} = 1, b''_{32} = -1. \end{aligned} \tag{7}$$

In the considered coupled fractal structure (**I**, **II**), the elliptical cylinder (**I**) is external, and the elliptical cylinder (**II**) is internal. For the above parameters, they are both circular, and the radius of the cross section of the first (**I**) is twice the radius of the cross section of the second (**II**).

The presence of variable semi-axes leads to significant changes in the behavior of the deformation field u of the coupled system from (1)–(3): in addition to the real part of the displacements Reu, the imaginary part Imu appears. The resulting stochastic surface essentially depends on the order of the separate terms in the sum (1), which can be interpreted as separate operators, describing the total stochastic process [19]. The appearance of a stochastic state is also associated with the presence of variable modules k_{ui}, k'_{ui} from (2).

For the considered elliptic cylinders with parameters $b'_{31} = b''_{31} = 0$ from (7), the displacement function u does not depend on the index j. If these parameters are not zero, then the elliptic cylinders turn into either ellipsoids or hyperboloids [11].

Therefore, in this work the iterative procedure on index n simulates a stochastic process on a rectangular lattice with sizes $N_1 \times N_2$.

As examples Figs. 1, 2 show the behavior Reu and Imu the dimensionless displacement function u of coupled structures (**I**(+), **II**(+)) (Fig. 1), (**I**(+), **II**(−)) (Fig. 2), consisting of two fractal circular cylinders with constant semi-axes: structure (**I**) is

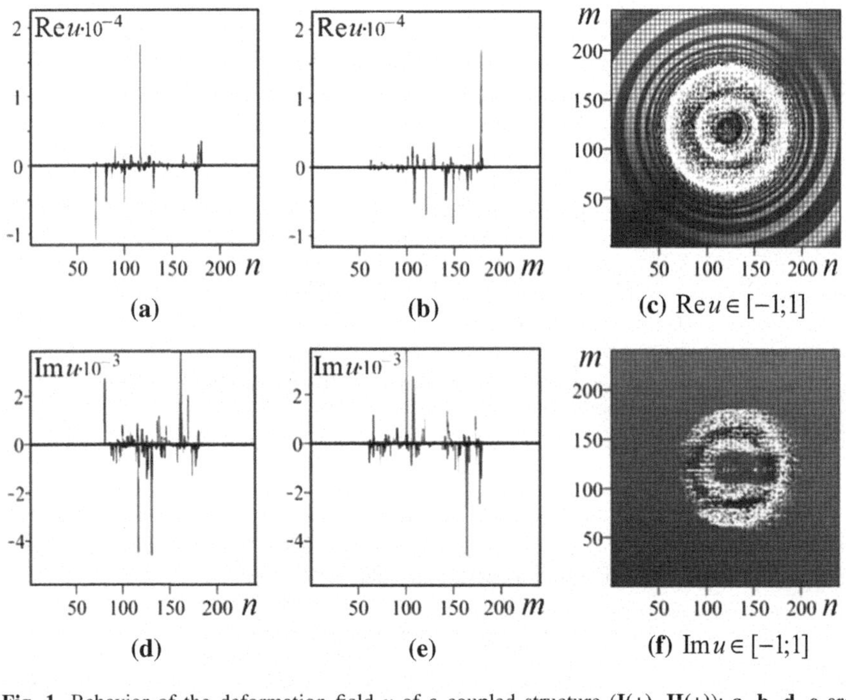

Fig. 1 Behavior of the deformation field u of a coupled structure (**I**(+), **II**(+)): **a**, **b**, **d**, **e** are projections on the planes nOu, mOu; **c**, **f** are cross-sections (top view)

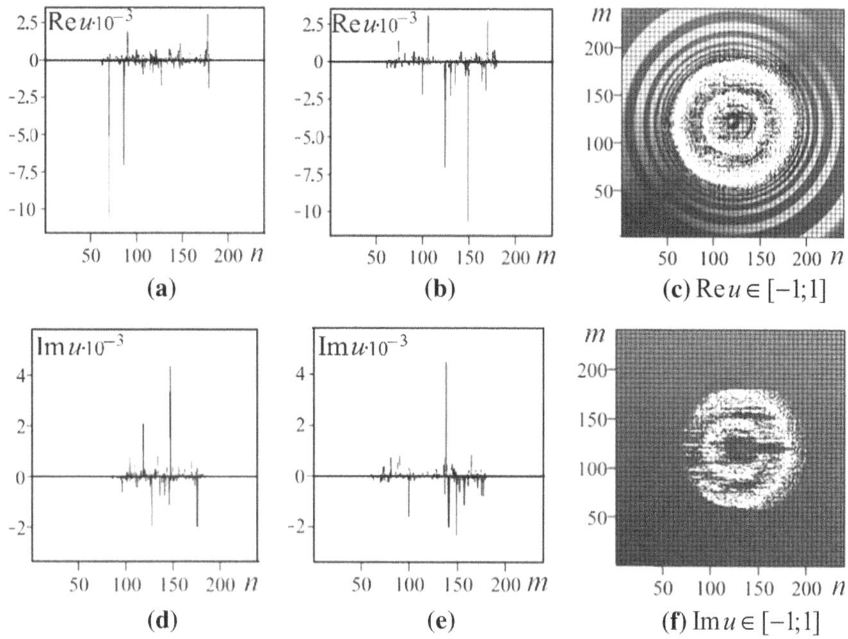

Fig. 2 Behavior of the deformation field u of a coupled structure (**I(+), II(−)**): **a**, **b**, **d**, **e** are projections on the planes nOu, mOu; **c**, **f** are cross-sections (top view)

outer cylinder with semi-axes $n_{c1} = m_{c1}$; structure (**II**) is inner cylinder with semi-axes $n_{c2} = m_{c2}$. In a coupled system (**I(+), II(+)**), the deformation fields of the separate structures (**I**), (**II**) are oriented in the same way ($R_1 = 1, R_2 = 1$). In a coupled system (**I(+), II(−)**), the deformation fields of the separate structures (**I**), (**II**) have opposite directions ($R_1 = 1, R_2 = -1$). The coupled fractal structure (**I(+), II(+)**) in expressions (1) corresponds to the ordered operator

$$u = u_{R1} + u_{R2}. \tag{8}$$

The coupled fractal structure (**I(+), II(−)**) in expressions (1) corresponds to the ordered operator

$$u = u_{R1} - u_{R2}. \tag{9}$$

The deformation fields of such structures are essentially stochastic along the boundaries of the circular cylinders and between them. At this, the amplitudes of the peaks Reu for the fractal structure (**I(+), II(+)**) (Fig. 1a) are an order of magnitude larger (higher) than the amplitudes of the peaks Reu for the structure (**I(+), II(−)**) (Fig. 2a). The wave-like behavior Reu (Figs. 1c, 2c) outside the boundary of the outer cylinder is explained by the presence of variable modules k_{ui}, k'_{ui} in expressions (1).

The imaginary part of the displacement function Imu (Figs. 1d–f, 2d–f) indicates to the presence of an effective damping. In this case, the character of damping for the structures (**I(+), II(+)**) (Fig. 1f) and (**I(+), II(−)**) (Fig. 2f) is different.

Transposition separate operators of displacement fields in a coupled cylindrical structure (**I(+), II(+)**) leads to the structure (**II(+), I(+)**), which corresponds to the ordered operator

$$u = u_{R2} + u_{R1}. \tag{10}$$

In this case, the stochastic process changes, which leads to a nonzero deviation

$$\delta u = (u_{R1} + u_{R2}) - (u_{R2} + u_{R1}) \neq 0 \tag{11}$$

of the deformation field, which is realized in the iterative process.

Complex deviations Re$\delta u \neq 0$, Im$\delta u \neq 0$ are observed (Fig. 3). On the background of stochastic peaks Imδu, the presence of a noise track is typical (Fig. 3d, e). At this, the noise level (noise amplitude Imδu) is approximately equal to $5 \cdot 10^{-24}$, what is near the theoretical value $0.497889116 \cdot 10^{-23} = 3/N_a$, where $N_a = 6.025438 \cdot 10^{23}$ is the Avagadro number.

This noise track is a stochastic ring, the inner region of which is a regular region (Fig. 3f). For Reδu on the background of stochastic behavior, the presence of a main

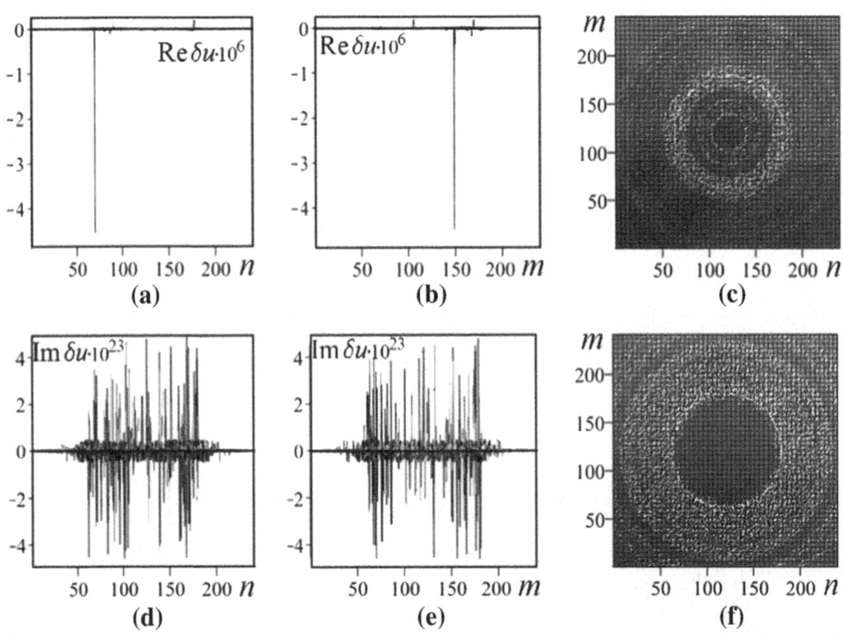

Fig. 3 The behavior of the deviation δu from (11): **a**, **b**, **d**, **e** are projections on the planes nOu, mOu; cross-sections (top view) **c** Re$\delta u \in [-10^{-8};\ 10^{-8}]$, **f** Im$\delta u \in [-10^{-23};\ 10^{-23}]$

peak down with a large amplitude of the order $4.5 \cdot 10^{-6}$ is characteristic (Fig. 3a, b). In the cross-section for Reδu the features of the type of stochastic rings with a regular inner region are observed (Fig. 3c).

Similar results take place for other orders of the sequence of separate displacement field operators in the coupled cylindrical structure (**I(+), II(+)**)

$$\delta u = (u_{R1} + u_{R2}) - (u_{R1} + u_{R2}) \neq 0; \quad \delta u = (u_{R1} + u_{R2}) + (-u_{R1} - u_{R2}) \neq 0;$$
$$\delta u = u_{R1} + (u_{R2} - u_{R2}) - u_{R1} \neq 0; \quad \delta u = u_{R1} + u_{R2} - u_{R1} - u_{R2} \neq 0. \quad (12)$$

Zero operators for coupled cylindrical structures (**I, II**) and (**II, I**), respectively, have the form

$$\delta u_{\mathrm{I,II}} = (u_{R1} - u_{R1}) + (u_{R2} - u_{R2}) = 0; \quad \delta u_{\mathrm{II,I}} = (u_{R2} - u_{R2}) + (u_{R1} - u_{R1}) = 0;$$
$$\Delta u_{\mathrm{I,II}} = u_{R1} - u_{R1} + u_{R2} - u_{R2} = 0; \quad \Delta u_{\mathrm{II,I}} = u_{R2} - u_{R2} + u_{R1} - u_{R1} = 0. \quad (13)$$

Here zero operators $\delta u_{\mathrm{I,II}}$, $\delta u_{\mathrm{II,I}}$ are different from zero operators $\Delta u_{\mathrm{I,II}}$, $\Delta u_{\mathrm{II,I}}$ by the presence of round brackets, and, therefore, by the organization of the stochastic process.

Coupled structures (**I(+), I(−)**), (**II(+), II(−)**), consisting of two separate fractal circular cylinders (with the same semi-axes $n_{ci} = m_{ci}$), but with the opposite direction of the deformation fields, correspond to ordered zero operators

$$\Delta_1 u = u_{R1} - u_{R1} = 0; \quad \Delta_2 u = u_{R2} - u_{R2} = 0. \quad (14)$$

According to expressions (14), the ordered operator

$$\Delta u = \Delta_1 u + \Delta_2 u = u_{R1} - u_{R1} + u_{R2} - u_{R2} = 0 \quad (15)$$

is identically zero. However, changes in the order (transposition) of separate operators in the zero operators (15) for a coupled structure leads to operators that are not zero (Fig. 4).

$$\delta u = u_{R1} + u_{R2} - u_{R1} - u_{R2} \neq 0; \quad (16)$$

$$\delta u = u_{R1} + u_{R2} - u_{R2} - u_{R1} \neq 0; \quad (17)$$

$$\delta u = u_{R1} - u_{R2} + u_{R2} - u_{R1} \neq 0; \quad (18)$$

$$\delta u = u_{R1} - u_{R2} - u_{R1} + u_{R2} \neq 0. \quad (19)$$

These operators (16)–(19) lead to the appearance of a complex deformation field. On Fig. 4, as an example, the complex functions of deviations Re$\delta u \neq 0$, Im$\delta u \neq 0$

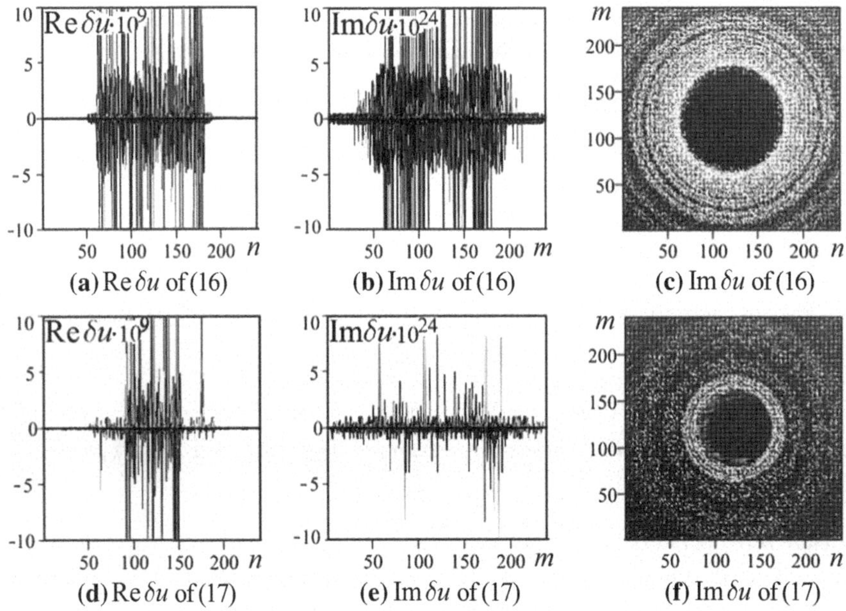

Fig. 4 The behavior of the deviations δu, the projection on the plane nOu: Re$\delta u \in [-10^{-8}; 10^{-8}]$ (**a**, **d**)—mOu; Im$\delta u \in [-10^{-23}; 10^{-23}]$ (**b**, **e**); cross-section Im$\delta u \in [-10^{-25}; 10^{-25}]$ (**c**, **f**) (top view)

are given, for the operators from (16) and (17). The character of behavior of the deviations of the complex deformation field, which correspond the operators from (16) and (17), is significantly different, which is confirmed both for Reδu (Fig. 4a, d) and for Imδu (Fig. 4b, c, e, f). On the background of stochastic peaks Imδu, the appearance of two noise tracks is typical (Fig. 4b, e). The noise level of the first track is near the noise level of the track from (Fig. 3d, e). The noise level of the second track (approximately equal $5 \cdot 10^{-25}$) is an order of magnitude less than the noise level of the first track. These noise tracks are stochastic rings, the insides of which are regular regions (Fig. 4c, f).

3 Fractal Structure with Three Elements of Cylindrical Type

We consider a coupled fractal structure (**I, III, II**), consisting of three fractal elliptic cylinders ($i = 1, 3, 2$), located in a bulk discrete lattice $N_1 \times N_2 \times N_3$. By analogy with expressions (1)–(3), nonlinear equations for the dimensionless displacement function u of the lattice node are

$$u = \sum_{i=1}^{3} u_{Ri}; \quad u_{Ri} = R_i k_{ui}^2 (1 - 2\text{sn}^2(u_2 - u_{0i}, k'_{ui})). \tag{20}$$

All other parameters are described by expressions (2), (3).

The singular points (attractors) of the deformation field of such a coupled fractal structure are located on the surface, the core of which is determined from the condition

$$Q_1 \cdot Q_3 \cdot Q_2 = 0. \tag{21}$$

Expressions for functions Qi are defined in (3). If the surface (21) is crossed by a plane $j = j_k$, then we obtain the equation of the isolines.

Fractal structures with parameters from (7) were considered as model examples.

The cylindrical structures (**I**) and (**II**) were chosen the same as in the previous Sect. 2. To describe the effects of stretching and compression of the cylindrical structure (**III**), we consider three variants for the choice of parameters for semi-axes:

$$\text{variant 1} : n_{31} = m_{c3} = 43.0746; \tag{22}$$

$$\text{variant 2} : n_{31} = m_{c3} = 50.2537; \tag{23}$$

$$\text{variant 3} : n_{31} = m_{c3} = 35.8955. \tag{24}$$

Figure 5 shows the attractors (Fig. 5a, d, g) for each of the three variants and the cross-section Reu (Fig. 5b, e, h), Imu (Fig. 5c, f, i). In the first variant (Fig. 5a–c) the cylindrical structure (**III**) is in the middle between the cylindrical structures (**I**) and (**II**), which corresponds to the initial state of the structure (**III**). In the second variant (Fig. 5d–f) the cylindrical structure (**III**) is located closer to the cylindrical structure (**I**), which corresponds to the stretched state of the structure (**III**). In the third variant (Fig. 5g–i) the cylindrical structure (**III**) is located closer to the cylindrical structure (**II**), which corresponds to the compressed state of structure (**III**).

The attractors of the coupled fractal structure (**I, III, II**) are continuous curves close to concentric circles (variant 1). Pronounced angular features (kinks) are observed near the main diagonal from the point (0, 240) to the point (240, 0), which is due to the mutual influence of the separate structures on each other (Fig. 5a).

When the attractor of structure (**III**) approaches to the attractor of structure (**I**) near the main diagonal on the isolines of structure (**III**) and (**I**) there are discontinuities, intersections, and formation of "islands" type features (variant 2). In this case, the isoline (attractor) of the structure (**II**) remains continuous and becomes smoother (Fig. 5d).

When the attractor of structure (**III**) approaches the structure (**II**) attractor, on the isolines of structures (**III**) and (**II**) there are gaps, intersections, formation of features like "islands" not only near the main diagonal, but along the all length (variant 3).

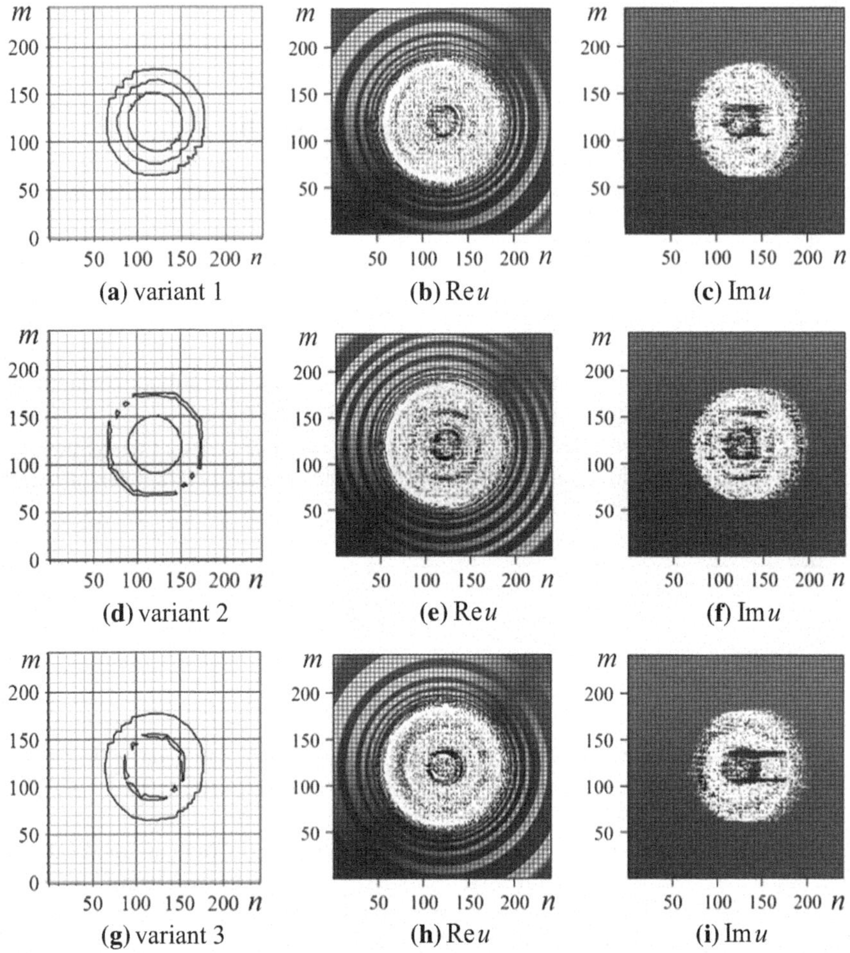

Fig. 5 Behavior of attractors (**a**, **d**, **g**) and displacement functions u: cross sections Re$u \in [-1; 1]$ (**b**, **e**, **h**); Im$u \in [-1; 1]$ (**c**, **f**, **i**) (top view)

At the same time, the isoline (attractor) of structure (**I**) remains continuous with pronounced angular features (kinks) near the main diagonal (Fig. 5g).

The wave-like behavior Reu (Fig. 5b, e, h) outside the boundary of the outer cylinder of structure (**I**) is explained by the presence of variable modules k_{ui}, k'_{ui} in the expressions of the displacement functions u of the coupled structure (**I, III, II**) and depends on the choice of variants (22)–(24).

As structure (**III**) approaches to the structure (**I**) (variant 2), the number of waves increases (Fig. 5e) as compared with variants 1 and 3 (Fig. 5b, h). The inner region Reu is characterized by a pronounced stochastic behavior, which localized in regions of a ring-like type.

The imaginary part of the displacement function Imu of the coupled structure (**I, III, II**) indicates the presence of effective damping (Fig. 5c, f, i). At the same time, the nature of the damping is significantly different and depends on the location of the structure (**III**) relative to the structures (**I**) and (**II**).

4 Conclusions

The behavior of attractors and the deformation field for a fractal coupled structure consisting of cylindrical elements is investigated. Various forms of attractors indicate the influence of separate structures on each other.

Using the example of coupled fractal structures with two and three circular cylinders with different semi-axes it is shown that the deformation field is complex, essentially stochastic along the boundaries of the circular cylinders and between them. The wave-like behavior of the real part of the displacement function Reu outside the boundary of the outer cylinder is explained by the presence of variable modules k_u, k'_u. The imaginary part of the displacement function Imu indicates the presence of effective damping.

The explicit form of the resulting stochastic surface essentially depends on the order of the sequence of the separate terms in expressions (1) or (20) for the displacement function, which can be interpreted as separate operators describing the total stochastic process. We introduce ordered zero operators. It is shown that changes in the order of the sequence of separate operators in the zero operator lead to complex deviations Re$\delta u \neq 0$, Im$\delta u \neq 0$. At the same time, on the background of stochastic peaks Imδu the presence of noise tracks is characteristic. These noise tracks are various stochastic rings, the inside regions of which are regular regions.

Using the example of a coupled fractal structure (**I, III, II**) with three cylindrical elements, the effects of compression (variant 3) and stretching (variant 2) of the initial state (variant 1) of the structure (**III**) are investigated. It is shown that the effects of compression and stretching are accompanied by changes in wave-like behavior Reu and effective damping Imu.

The results can be used by the description of neural networks with variable parameters, in medicine by the simulation of blood vessels, for quantum information processing.

References

1. M. Nielsen, I. Chuang, *Quantum Computation and Quantum Information* (Cambridge University Press, New York, 2010)
2. D. Boumeister, A. Eckert, A. Zeilinger, *Physics of Quantum Information* (Springer, New York, 2001)
3. Y. Ozhigov, Quantum computers speed up classical with probability zero. Chaos, Solitons Fractals **10**, 1707–1714 (1999)

4. D. Castelvecchi, Quantum computers ready to leap out of the lab. Nature **541**, 9–10, 5 Jan 2017
5. A.N. Omelyanchuk, E.V. Ilyichev, S.N. Shevchenko, *Quantum Coherent Phenomena in Josephson Qubits* (Naukova Dumka, Kiev, 2013)
6. M.N. Fedorov, I.A. Volkov, Yu.M. Mikhailova. Coutrites and kukwarts in spontaneous parametric scattering of light, correlation and entanglement of states. JETP **142** (1(7)), 20–43 (2012)
7. Y.S. Kivshar, N.N. Rozanov (Eds.), *Nonlinearities in Periodic Structures and Metamaterials* (Fizmatlit, Moscow, 2014)
8. G. Tosi, G. Christmann, N.G. Berloff et al., Sculpting oscillators with light within a nonlinear quantum fluid. Nat. Phys. **8**(3), 190–194 (2012)
9. U. Schneider, L. Hackermüller, J. Ronzheimer et al., Fermionic transport and out-of-equilibrium dynamics in a homogeneous Hubbard model with ultracold atoms. Nat. Phys. **8**(3), 213–218 (2012)
10. K. Heyman, The map in the brain: grid cells may help us navigate. Science **312**, 680–681 (2006)
11. V.S. Abramov, Quantum dots in a fractal multilayer system. Bull. Rus. Acad. Sci. Phys. **81**(5), 625–632 (2017)
12. V.S. Abramov, Active Nanoelements with variable parameters in fractal quantum systems. Bull. Rus. Acad. Sci. Phys. **82**(8), 1062–1067 (2018)
13. O.P. Abramova, A.V. Abramov, Attractors and deformation field in the coupled fractal multilayer nanosystem. CMSIM J. **2**, 169–179 (2017)
14. O.P. Abramova, Mutual influence of attractors and separate stochastic processes in a coupled fractal structures. Bull. Donetsk Nat. Univ. A, **1**, 50–60 (2017)
15. M. Fyhn, S. Molden, M.P. Witter, E.I. Moser, Spatial representation in the entorhinal cortex. Science **305**, 1258–1264 (2004)
16. T. Hafting, M. Fyhn, S. Molden, M.-B. Moser, Microstructure of spatial map in the entorhinal cortex. Nature **436**, 801–806 (2005)
17. F. Sargolini, M. Fyhn, T. Hafting et al., Conjunctive representation of position, direction, and velocity in the entorhinal cortex. Science **312**, 758–762 (2006)
18. V.S. Abramov, Alteration of the stochastic state of the deformation field in the model multilayer nanosystem. Bull. Donetsk Nat. Univ. A, **2**, 81–89 (2014)
19. O.P. Abramova, A.V. Abramov, Effect of ordering of displacement fields operators of separate quantum dots, elliptical cylinders on the deformation field of coupled fractal structures, in *"11th Chaotic Modeling and Simulation International Conference". Springer Proceedings in Complexity*, ed. by C.H. Skiadas, I. Lubashevsky (Springer Nature Switzerland AG, 2019), pp. 15–27

Brain Dynamics Explained by Means of Spectral-Structural Neuronal Networks

Maricel Agop, Alina Gavriluţ, Gabriel Crumpei and Lucian Eva

Abstract In this chapter, we propose a mathematical-physical model, starting from the morphological-functional assumption of the fractal brain, by activating brain non-differentiable dynamics through the determinism-nondeterminism inference of the responsible mechanisms.

Keywords Fractal brain · Brain dynamics · Neuronal networks · Neuropsychological mechanisms · Brain geodesics

1 Introduction

In the most general representation, the biological systems can be divided into three different categories such as open, dissipative and nonlinear systems. In our opinion, the "specialization" process of any biological structure (for instance, differentiation process) is based on the legitimate alternation between chaos and order of mutual states. This behavior is defined by the living matter multivalent logic and its com-

12th CHAOS Conference Proceedings, 18–22 June 2019, Chania, Crete, Greece.

M. Agop
Department of Physics, Gheorghe Asachi Technical University of Iaşi, Iaşi, Romania
e-mail: m.agop@yahoo.com

A. Gavriluţ (✉)
Faculty of Mathematics, Alexandru Ioan Cuza University, Iaşi, Romania
e-mail: gavrilut@uaic.ro

G. Crumpei
Faculty of Psychology and Education Sciences, Alexandru Ioan Cuza University, Iaşi, Romania
e-mail: crumpei.gabriel@yahoo.com

L. Eva
"Prof. Dr. N. Oblu" Clinical Emergency Hospital Iaşi, Iaşi, Romania
e-mail: elucian73@yahoo.com

C. H. Skiadas and Y. Dimotikalis (eds.), *12th Chaotic Modeling and Simulation International Conference*, Springer Proceedings in Complexity,
https://doi.org/10.1007/978-3-030-39515-5_3

munication codes. Further to this representation, we can say now that this visible interdisciplinary work aims to explain how mathematical knowledge can be used to describe, predict and control the phenomena observed in some biological systems. From a functional point of view and not only, we understand here the biological systems in the particular sense of the discipline known under a variety of names such as: "complexity theory", "self-organization theory", "chaos theory" or "nonlinear dynamics". Our choice was not easy but in what follows we will use the latter name, nonlinear dynamics, respectively.

Assuming that the biological systems are fractal systems, a few aspects of natural dynamics in biological structures, especially in human brain, are studied.

Many phenomena exhibiting complex patterns and structures can be easily observed in brain. These phenomena can be understood by means of a multidisciplinary paradigm called complexity.

There are many diverse methods of analysis of the dynamics of complex systems, particularly of those addressed to brain. However, due to the fact that they use only differentiability in the study of systems, all of them involve sophisticated and sometimes ambiguous models (Kändel brain's standard models [1]). In our opinion, a way of analysis necessary in the dynamics of complex systems, especially those related to brain, must respect the recent results related to the harmony between morphology and functioning of a system. Thus, if we only concentrate to brain, from structural point of view, the nature is abundant in examples.

The standard approach, by extrapolating neuron performing way to the description of the performance of the whole brain (Kändel [1]), did not produce expected results, because the neuron is part of a network, and from the complex systems' viewpoint, the properties of the constitutive elements could be recaptured in the properties of the whole system only as emergent.

On the other hand, the tackling from a quantum perspective, as proposed by different authors (Atmanspacher [2, 3]), did not allow for specifying of some functional models of the brain, because at every scale the emergent phenomenon generates new properties which can be recaptured as emergent in the scales that follow.

Consequently, the performance of brain as an assembly, from the psychic life point of view, seems to be best approached from the perspective of complex systems, with their potential components, non-differential and noncausal from the structural point of view, between which there is a chaotic part structured via attractors, and highlighted within a phase space.

From the very beginning we point out that, both from brain structural (morphological) and from functional (processing) points of view, examples are brought in order to substantiate the idea set forth as the starting one, namely that the brain dynamics at every level are dictated by the brain functional-structural coherence.

For example, Caserta et al. [4, 5] identify fractal shapes with a fractal dimension of aproximatively 1.7, and this could be explained by a diffusion limited aggregation model (Witten and Sander [6]). A fractal structure was observed by Kniffki et al. [7] for the branching dendrite patterns of thalamic neurons in Golgi impregnated specimens. Moreover, as one can easily observe, the entire neuronal network has a fractal structure.

The fractality is also manifest when the functionality of the brain is considered. In fact, a great body of laws governing such a functionality at any scale of resolution, prove to be reducible to a power-type law. As a matter of fact, power laws are to be found both in the functionality and in the structure of brain, which substantiates the idea to suggested in the present chapter, namely that only the structural-functional unity of the brain can lead to a sound explanation of the complex phenomena met at this level (Werner [8, 9]).

The recording of the brain functions highlighted an electric and magnetic activity correlated with certain brain functions. This electromagnetic activity is spectral in nature, and as such assumes a spectral functionality and processing. One can thereby conclude that the morphology and the fractal structure of the brain should be duplicated by a fractal functioning and processing (de Valois and de Valois [10, 11]—processing of the visual image, von Békési [12]—processing of proprioceptive sensoriality).

Also, Lowen and Teich [13, 14] consider that the fractal action potential patterning in auditory nerve could be explained by fractal activity in the ion channels of the sensory organs feeding into the auditory nerve.

Also, the muscular fibers contract nonlinearly due to fractal type mechanisms, which allows for the nonlinear adaptation of muscular contraction to the environmental necessities (sharp transients from rest to maximum function, functional reserve for continuous effort via nonlinear training in time of the muscle fibers).

Our chapter sets out to tackling the old problem of brain-mind duality, whereby the description of brain can be approached according to the laws of physics and chemistry, while the mind cannot. We are thus allowed conclude that the two aspects of the duality mind-brain actually embody a structural and functional unity that can be modeled physically and mathematically, and can be analyzed according the the modern scientific paradigms. The psychic life is thus represented by a complex dynamics of exchange between the neuronal network and the spectral network.

In the present chapter, the brain's dynamics through spectral-structural neuronal networks are analyzed. Thus, by the determinism-nondeterminism inference in brain dynamics, we quit either the classical determinism [15, 16], or the quantum nondeterminism [17, 18]. Moreover, the fractal type brain [4, 13, 14, 19–22] is both morphologically and functionally specified by activating the non-differentiable type brain dynamics [23–26].

With our approach we finally aim at the solution of the old problem of the duality brain-mind. Our physical-mathematical model offers the vision that both the mind and the brain do form a functional, describable through the dynamics existing between the spectral and neural networks lying at the foundation of the psychic life.

2 The Mathematical Model

The brain is, morphologically, a fractal (see the examples from Introduction). Moreover, its own space (the one generated by the brain) is structurally, a fractal (Mandelbrot [23]). In such space, the only possible functionalities (which are compatible with the brain structure) are achieved on continuous but non-differentiable curves [23–30].

Brain's structural-functional compatibility (structural-functional duality) as a source of the cerebral dynamics at any scale is thus imposed.

Accepting the structural-functional duality of the brain, the trajectory of motion realized on the structural component must be identified with an element from the functional part. If, according to Caserta et al. [4, 5], we admit that the anharmonic oscillations of the neurofibrils would be the source of the functional part of the brain, then the curve describing the motion of a neurofibrile is a continuous but non-differentiable curve (Werner [8, 9]). So this motion takes place in a fractal space, the one generated by the fractal structure of the brain and thus, one can identify it with the geodesic of the associated fractal space. At yet another scale, the neuron can be identified with its corresponding geodesic. More generally, the wave is identified with the corpuscle, the corpuscle motion in the field of its associated wave being obviously a continuous but non-differentiable curve (fractal curve), whence the idea of geodesic. We shall detail these considerations in what follows:

By "brain dynamics" we understand the application between structural component ("space" variables) and functional component ("time" variables). Due to the fact that the dynamics reflects different levels of application, the time gets in through scale resolution and is denoted by δt. We shall use the notation dt where we refer to the usual time as in the hamiltonian dynamics; δt will be defined through a special substitution principle.

All these lead to the following consequences, which will be explained in detail throughout the present chapter:

(i) Any continuous and non-differentiable curve of the brain dynamics (brain non-differentiable curve) is explicitly scale resolution dependent δt. This means the curve length tends to infinity when δt tends to zero;

We recall that a curve is non-differentiable if it satisfies the Lebesgue theorem: its length becomes infinite when the scale resolution goes to zero (Mandelbrot [2]). In consequence, it exhibits the self-similarity property in every one of its points and this can be understood as a property of holography [2, 3, 31].

(ii) The physics of the brain phenomena can be related to the behavior of a collection of functions during the zoom operation of the scale resolution δt. Therefore, by means of the substitution principle, δt is identified with dt, i.e., $\delta t \equiv dt$. In consequence, it is considered as an independent variable;

The fractal variables are dynamical variables, which depend on space coordinates and time, but also on the scale resolution. Then, a difference should be made for instance between the variables describing the dynamics at the nanoscale (induced by neurofibrils), the dynamics at the synapse level, nerve impulse transmission at the dendrite level (for details see Kniffki et al. [7]). It is the global structural-functional coherence of our brain, which, in our opinion, determines a permanent interdependence between these variables;

(iii) One can describe the brain dynamics by means of fractal variables. The functions depend both on the space-time coordinates and the scale resolution. In consequence, two derivatives of the variable field $Q(t, dt)$ can be defined in any point of the non-differentiable curve:

$$\frac{d_+Q(t,dt)}{dt} = \lim_{\Delta t \to 0_+} \frac{Q(t+\Delta t, \Delta t) - Q(t, \Delta t)}{\Delta t}$$

$$\frac{d_-Q(t,dt)}{dt} = \lim_{\Delta t \to 0_-} \frac{Q(t, \Delta t) - Q(t-\Delta t, \Delta t)}{\Delta t}. \quad (1)$$

(the sign + corresponds to the forward process and the sign— corresponds to the backwards one);

(iv) The differential of the spatial coordinate field $dX^i(t, dt)$ can be expressed as the sum of the two differentials. One of them is not scale resolution independent (the differential part $d_\pm x^i(t)$). The other one is scale resolution dependent (the fractal part $d_\pm \xi^i(t)$), i.e.,

$$d_\pm X^i(t, dt) = d_\pm x^i(t) + d_\pm \xi^i(t, dt); \quad (2)$$

(v) The non-differentiable part of the brain spatial coordinate field verifies the fractal equation:

$$d_\pm \xi^i(t, dt) = \lambda^i_\pm (dt)^{1/D_F}. \quad (3)$$

Here, $\lambda^i_\pm$ are constant coefficients through which the fractalization type is specified. D_F means the fractal dimension of the brain non-differentiable curve. Let us note that any definition (Kolmogorov or Hausdorff-Besicovici fractal dimensions [23–26]) is acceptable for D_F.

The functionality of the cerebral processes implies dynamics on geodesics which have various fractal dimensions: For $D_F = 2$, quantum type functionalities are generated (percolation in living neural networks, tunneling in neurofibrils etc.). For $D_F < 2$, correlative type functionalities are generated. For $D_F > 2$, non-correlative type functionalities can be observed (limited or unlimited diffusions-branching pattern of dendritic trees of retina neurons by a diffusion limited aggregation [4–7]);

(vi) One can recover the differential time reflection invariance of any dynamical variable by combining the derivates d_+/dt and d_-/dt in the non-differentiable operator

$$\frac{\hat{d}}{dt} = \frac{1}{2}\left(\frac{d_+ + d_-}{dt}\right) - \frac{i}{2}\left(\frac{d_+ - d_-}{dt}\right). \quad (4)$$

This can be obtained using the procedure [32]. If we apply the non-differentiable operator to the brain spatial coordinate field one gets the brain complex velocity field

$$\hat{V}^i = \frac{\hat{d}X^i}{dt} = V_D^i - V_F^i \tag{5}$$

with

$$\begin{aligned} V_D^i &= \frac{1}{2}(v_+^i + v_-^i),\ V_F^i = \frac{1}{2}(v_+^i - v_-^i) \\ v_+^i &= \frac{d_+x^i + d_+\xi^i}{dt},\ v_-^i = \frac{d_-x^i + d_-\xi^i}{dt}. \end{aligned} \tag{6}$$

The real part V_D^i is differentiable and scale resolution independent (differentiable velocity field), and the imaginary part V_F^i is non-differentiable and scale resolution dependent (fractal velocity field);

(vii) The average values of the brain fluid variables must be considered in the sense we previously mentioned. Therefore, the average of $d_\pm X^i$ writes

$$\langle d_\pm X^i >\equiv d_\pm x^i \tag{7}$$

with

$$\langle d_\pm \xi^i >= 0; \tag{8}$$

(viii) One can describe the brain dynamics by means of a covariant derivative, whose explicit form is obtained in the following:

Now, let us consider that the non-differentiable curves are immersed in a three-dimensional space (the brain dynamics consciousness is three-dimensional) and that X^i are the spatial coordinate field of a point on the non-differentiable curve. Let us consider a variable field $Q(X^i, t)$. Let also be the following Taylor expansion up to the second order

$$d_\pm Q(X^i, t) = \partial_t Q dt + \partial_i Q d_\pm X^i + \frac{1}{2}\partial_l \partial_k Q d_\pm X^l d_\pm X^k. \tag{9}$$

As a consequence, forward and backward value of (9) have the form:

$$\begin{aligned} \langle d_\pm Q\rangle &= \langle \partial_t Q dt\rangle + \langle \partial_i Q d_\pm X^i\rangle \\ &+ \frac{1}{2}\langle \partial_l \partial_k Q d_\pm X^l d_\pm X^k\rangle. \end{aligned} \tag{10}$$

If one supposes that the average value of variable field Q and its derivatives coincide with themselves and the differentials $d_\pm X^i$ and dt are independent, then the average of their products coincides with the product of averages. In consequence, (10) has the form:

$$d_{\pm}Q = \partial_t Q dt + \partial_i Q \langle d_{\pm}X^i \rangle + \frac{1}{2}\partial_l \partial_k Q \langle d_{\pm}X^l d_{\pm}X^k \rangle. \tag{11}$$

By (3) one writes

$$\langle d_{\pm}\xi^i d_{\pm}\xi^l \rangle = \pm \lambda_{\pm}^i \lambda_{\pm}^l (dt)^{(2/D_F)-1} dt, \tag{12}$$

(the sign + corresponds to $dt > 0$, while the sign—corresponds to $dt < 0$).

Consequently, (11) has the form

$$\begin{aligned} d_{\pm}Q &= \partial_t Q dt + \partial_i Q d_{\pm}x^i + \frac{1}{2}\partial_i \partial_l Q d_{\pm}x^i d_{\pm}x^l \\ &\pm \frac{1}{2}\partial_i \partial_l Q [\lambda_{\pm}^i \lambda_{\pm}^l (dt)^{(2/D_F)-1} dt]. \end{aligned} \tag{13}$$

If we divide by dt and neglect the terms which contain differential factors ([23–30]) we obtain:

$$\frac{d_{\pm}Q}{dt} = \partial_t Q + v_{\pm}^i \partial_i Q \pm \frac{1}{2}\lambda_{\pm}^i \lambda_{\pm}^l (dt)^{(2/D_F)-1} \partial_i \partial_l Q. \tag{14}$$

These relations enable us to define:

$$\frac{d_{\pm}}{dt} = \partial_t + v_{\pm}^i \partial_i \pm \frac{1}{2}\lambda_{\pm}^i \lambda_{\pm}^l (dt)^{(2/D_F)-1} \partial_i \partial_l. \tag{15}$$

In these conditions, using (4),(5) and (15), we calculate $\hat{d}/dt$. It gets:

$$\frac{\hat{d}Q}{dt} = \partial_t Q + \hat{V}^i \partial_i Q + \frac{1}{4}(dt)^{(2/D_F)-1} D^{lk} \partial_l \partial_k Q, \tag{16}$$

with

$$\begin{aligned} D^{lk} &= d^{lk} - i\overline{d}^{lk} \\ d^{lk} &= \lambda_+^l \lambda_+^k - \lambda_-^l \lambda_-^k, \overline{d}^{lk} = \lambda_+^l \lambda_+^k + \lambda_-^l \lambda_-^k. \end{aligned} \tag{17}$$

The relation (16) enables us to introduce the covariant derivative

$$\frac{\hat{d}}{dt} = \partial_t + \hat{V}^i \partial_i + \frac{1}{4}(dt)^{(2/D_F)-1} D^{lk} \partial_l \partial_k. \tag{18}$$

3 Brain Geodesics

Let us consider now the scale covariance principle (the physics laws are invariant we respect to scale transformations) and postulate that the passage from the differentiable mathematical model to the non-differentiable mathematical model can be achieved by replacing the standard time derivative d/dt by the non-differentiable operator $\hat{d}/dt$. In this way, this operator has the role of the covariant derivative. Precisely, it is used to write the fundamental equations of brain dynamics under the same form as in the classical (differentiable) case. In these circumstances, in we apply the operator (18) to the complex velocity field (5), the brain geodesics have the following form in the presence of an external constraint given by the scalar potential U:

$$\frac{\hat{d}\hat{V}^i}{dt} = \partial_t \hat{V}^i + \hat{V}^l \partial_l \hat{V}^i + \frac{1}{4}(dt)^{(2/D_F)-1} D^{lk} \partial_l \partial_k \hat{V}^i = -\partial^i U. \tag{19}$$

One can interpretate (19) as follows: the local acceleration $\partial_t \hat{V}^i$, the convection $\hat{V}^l \partial_l \hat{V}^i$, the dissipation $D^{lk} \partial_l \partial_k \hat{V}^i$ and the forces induced by the external constraints $\partial^i U$ make their balance in any point of the brain non-differentiable curve.

If the fractalization is obtained by stochastic processes of Markov type [23–26], then

$$\lambda_+^i \lambda_+^l = \lambda_-^i \lambda_-^l = 2\lambda \delta^{il}, \tag{20}$$

where δ^{ij} is Kronecker's symbol:

$$\delta^{ij} = \begin{cases} 1, & i = j \\ 0, & i \neq j. \end{cases}$$

In this case, the equation of brain geodesics has the form

$$\frac{\hat{d}\hat{V}^i}{dt} = \partial_t \hat{V}^i + \hat{V}^l \partial_l \hat{V}^i - i\lambda (dt)^{(2/D_F)-1} \partial^l \partial_l \hat{V}^i = -\partial^i U. \tag{21}$$

Even more, if we separate the motions on differential and fractal scale resolutions, we get:

$$\begin{aligned} \frac{\hat{d}V_D^i}{dt} &= \partial_t V_D^i + V_D^l \partial_l V_D^i - [V_F^l - \lambda (dt)^{(2/D_F)-1} \partial^l] \partial_l V_F^i = -\partial^i U \\ \frac{\hat{d}V_F^i}{dt} &= \partial_t V_F^i + V_D^l \partial_l V_F^i + [V_F^l - \lambda (dt)^{(2/D_F)-1} \partial^l] \partial_l V_D^i = 0. \end{aligned} \tag{22}$$

4 Spectral Functionality Through Schrödinger Type Brain Geodesics

For irrotational motions

$$\varepsilon_{ikl}\partial^k \hat{V}^l = 0, \tag{23}$$

with ε_{ikl}, the Lévy-Civita pseudo-tensor, let us choose $\hat{V}^i$ in the form which makes this definition an identity

$$\hat{V}^i = -2i\lambda(dt)^{(2/D_F)-1}\partial^i \ln \Psi, \tag{24}$$

(here $\ln \Psi$ means the scalar potential of the complex velocity field).

If we substitute (24) in (21), we get:

$$\begin{aligned}\frac{\hat{d}\hat{V}^i}{dt} &= -2i\lambda(dt)^{(2/D_F)-1}\{\partial_t\partial^i \ln \Psi - i[2\lambda(dt)^{(2/D_F)-1}(\partial^l \ln \Psi \partial_l) \cdot \partial^i \ln \Psi + \\ &+ \lambda(dt)^{(2/D_F)-1}\partial^l\partial_l\partial^i \ln \Psi]\} = -\partial^i U.\end{aligned} \tag{25}$$

By

$$\begin{aligned}\partial^l\partial_l \ln \Psi + \partial_i \ln \Psi \partial^i \ln \Psi &= \frac{\partial_l\partial^l \Psi}{\Psi} \\ \partial^i\left(\frac{\partial^l\partial_l \Psi}{\Psi}\right) &= 2(\partial^l \ln \Psi \partial_l)\partial^i \ln \Psi + \partial^l\partial_l\partial^i \ln \Psi\end{aligned} \tag{26}$$

(25) has the form:

$$\begin{aligned}\frac{\hat{d}\hat{V}^i}{dt} &= -2i\lambda(dt)^{(2/D_F)-1}\partial^i\left[\partial_t \ln \Psi - 2i\lambda(dt)^{(2/D_F)-1}\frac{\partial^l\partial_l \Psi}{\Psi}\right] \\ &= -\partial^i U.\end{aligned} \tag{27}$$

If we integrate this equation up to an arbitrary phase factor, which may be set to zero by a suitable choice of phase of Ψ, one gets:

$$\lambda^2(dt)^{(4/D_F)-2}\partial^l\partial_l \Psi + i\lambda(dt)^{(2/D_F)-1}\partial_t \Psi - \frac{U}{2}\Psi = 0. \tag{28}$$

The relation (28) is a Schrödinger type equation (Schrödinger brain geodesics) and it implies the following:

(i) According to [26], Ψ is a wavefunction. It has a direct physical signification only by means of $|\Psi|^2$ as probability density;

(ii) The unpredictable character of the brain dynamics is specified through the wave properties of the neuronal medium (or, neuronal network). In this way, brain's spectral type functionality is provided;

(iii) The mechanisms that are responsible of brain's spectral functionality are of quantum type only when they are extrapolated for various scale resolutions (such as tunneling, percolation, entanglement states—see Werner [8] etc.).

5 Structural Functionality Through Brain Geodesics of Hydrodynamic Type

If $\Psi = \sqrt{\rho}\exp(iS)$, where $\sqrt{\rho}$ is the amplitude and S is the phase of Ψ, then the complex velocity field (5) has the form

$$\begin{aligned}\hat{V}^i &= 2i\lambda(dt)^{(2/D_F)-1}\partial^i \ln \Psi\\ V_D^i &= 2i\lambda(dt)^{(2/D_F)-1}\partial^i S\\ V_F^i &= 2i\lambda(dt)^{(2/D_F)-1}\partial^i \ln \rho.\end{aligned} \tag{29}$$

If we substitute (29) into (19) and we separate the real and imaginary parts, up to an arbitrary phase factor which may be set to zero by a suitable choice of the phase of Ψ, one gets:

$$\partial_t V_D^i + (V_D^l \partial_l) V_D^i = -\partial^i (Q + U) \tag{30}$$

$$\partial_t \rho + \partial^i (\rho V_D^i) = 0 \tag{31}$$

where Q is the specific non-differentiable potential

$$Q = -2\lambda(dt)^{(4/D_F)-2}\frac{\partial^l \partial_l \sqrt{\rho}}{\sqrt{\rho}} = -\frac{V_F^l V_{Fl}}{2} - \lambda(dt)^{(2/D_F)-1}\partial_i V_F^i. \tag{32}$$

In consequence,

(i) Any neuronal cell permanently interacts with a fractal medium, identified with the neuronal network by means of the specific non-differentiable potential (32). The physics fractal medium is prone to computational properties [33];
(ii) One can identity the neuronal network with a fractal fluid (non-differentiable fluid), whose dynamics is described by the fractal hydrodynamical model;
(iii) The fractal velocity field V_F^i does not represent actual motion. However, it contributes to the specific momentum transfer and to the brain energy focus. All these confer spectral representability to brain functioning through what we call *neuronal network.* One can clearly see this from the absence of V_F^i from the states density conservation law and also from its role in the variational principle [24–26];

(iv) While the brain energy is stored as mass motion and potential energy, some is available elsewhere and only the total is conserved. The conservation of the energy and the specific momentum ensure reversibility and the existence of eigenstates, but deny a Lévy type motion (Lévy [34]) of brain interaction with an external medium;

(v) The predictable character of the brain activity is specified through the corpuscular properties of the neuronal network. Thus, the brain's corpuscular type functionality is provided;

(vi) The mechanisms that are responsible of brain's structural functionality are of a hydrodynamical type, but when they are extrapolated for various scale resolutions (shock waves, solitons–Jackson [35] etc.);

(vii) The specific non-differentiable potential coordinates the transitory functionality (the spectral-structural functionality).

6 Communication Language Generation

Both functionalities (either the one which is responsible of the brain activity unpredictable character, or the other one which is responsible of the brain activity predictable character) act simultaneously. By their interconditioning there result either brain coherence or brain incoherence. Indeed, let us admit that both in the brain representation of Schrödinger type and in the brain representation of hydrodynamical fractal type, the external constraint is proportional with the states density, i.e., $U = 2a|\Psi|^2 = 2a\rho$, with $a = \text{const}$. Then for the stationay case ($\partial_t \Psi = 0$ and $\partial_t \rho = 0$, $V_D^i = \partial^i S = 0$) the equation (28) becomes:

$$\lambda^2 (dt)^{(4/D_F)-2} \partial^l \partial_l \Psi + E_\Psi \Psi - a|\Psi|^2 \Psi = 0, \tag{33}$$

while the equations (30)–(32) get by integration the form:

$$Q + U = -2\lambda^2 (dt)^{(4/D_F)-2} \frac{\partial^l \partial_l \sqrt{\rho}}{\sqrt{\rho}} + 2a\rho = 2E_\rho \tag{34}$$

with E_Ψ and E_ρ constants having specific energy significances.

By the substitutions:

$$\frac{(E_\Psi)^{1/2}}{\lambda k_\Psi} (dt)^{1-(2/D_F)} (k_{\Psi x} x + k_{\Psi y} y + k_{\Psi z} z) = \xi_\Psi$$

$$\frac{(E_\rho)^{1/2}}{\lambda k_\rho} (dt)^{1-(2/D_F)} (k_{\rho x} x + k_{\rho y} y + k_{\rho z} z) = \xi_\rho$$

$$\Psi = \left(\frac{E_\Psi}{a}\right)^{1/2} f, \sqrt{\rho} = \left(\frac{E_\rho}{a}\right)^{1/2} h$$

$$\mathbf{k}_{\Psi}^2 = k_{\Psi x}^2 + k_{\Psi y}^2 + k_{\Psi z}^2, \mathbf{k}_{\rho}^2 = k_{\rho x}^2 + k_{\rho y}^2 + k_{\rho z}^2, \quad (35)$$

(33) and (34) reduce to the equations of Ginzburg-Landau type (Jackson [35]):

$$\frac{d^2 f}{d\xi_{\Psi}^2} = f^3 - f \quad (36)$$

for spectral functionality, respectively

$$\frac{d^2 h}{d\xi_{\rho}^2} = h^3 - h, \quad (37)$$

for structural functionality, where $\mathbf{k}_{\Psi}$ and $\mathbf{k}_{\rho}$ are the brain wave vectors.

Using the methodology from Timofte et al. [30], these previous equations admit *either* the infinite energy solutions

$$|\Psi|^2 = \frac{2s_{\Psi}^2}{1+s_{\Psi}^2} sn^2 \left(\frac{\xi_{\Psi} - \xi_{\Psi_0}}{\sqrt{2}}; s_{\Psi} \right) \quad (38)$$

$$\rho = \frac{2s_{\rho}^2}{1+s_{\rho}^2} sn^2 \left(\frac{\xi_{\rho} - \xi_{\rho 0}}{\sqrt{2}}; s_{\rho} \right) \quad (39)$$

(sn are Jacobi's elliptic functions of modules s_{Ψ} and s_{ρ} (Armitage [36]))

$$s_{\Psi} = \frac{1-(1-2c_{\Psi})^{1/2}}{1+(1-2c_{\Psi})^{1/2}}, s_{\rho} = \frac{1-(1-2c_{\rho})^{1/2}}{1+(1-2c_{\rho})^{1/2}} \quad (40)$$

with $\xi_{\Psi_0}, \xi_{\rho 0}, c_{\Psi}, c_{\rho}$ integration constants, *or* the finite energy solutions (kink solutions [37])

$$|\Psi|^2 = \tanh\left(\frac{\xi_{\Psi} - \xi_{\Psi_0}}{\sqrt{2}} \right)$$

$$\rho = \tanh\left(\frac{\xi_{\rho} - \xi_{\rho 0}}{\sqrt{2}} \right)$$

obtained through the degeneration of the elliptic functions sn in the modules s_{Ψ} and s_{ρ}, i.e.,

$$s_{\Psi} \to 1 \text{ for } c_{\Psi} \to 1/2$$

$$s_{\rho} \to 1 \text{ for } c_{\rho} \to 1/2. \quad (41)$$

7 Implications of the Toda Network Model

Now, some conclusions are obvious:

(i) Both the probability density fluctuations

$$\delta|\Psi|^2 = 1 - \frac{1+s_\Psi^2}{2s_\Psi^2}|\Psi|^2 = cn^2\left(\frac{\xi_\Psi - \xi_{\Psi_0}}{\sqrt{2}}; s_\Psi\right) \tag{42}$$

and the states density ones

$$\delta\rho = 1 - \frac{1+s_\rho^2}{2s_\rho^2}\rho = cn^2\left(\frac{\xi_\rho - \xi_{\rho_0}}{\sqrt{2}}; s_\rho\right) \tag{43}$$

obtained based on the infinite energy solutions express oneselves through the cnoidal oscillations modes of the spatial coordinates fields ξ_Ψ and ξ_ρ, where cn are Jacobi's elliptic functions of modules s_Ψ and s_ρ (Armitage [36]);

(ii) Generally speaking, as it also results from Jackson [35], the cnoidal oscillation modes are equivalent to one-dimensional Toda type lattices of nonlinear oscillators (Toda [38, 39]. Moreover, according to [15, 16, 18], their mapping implies Toda type neuronal networks. That is why, based on the above results, we shall be able to define two Toda type neuronal networks, one of them being specific to the spectral functionality (and which will be called the *spectral neuronal network*) and the other one being specific to the structural functionality (and which will be called the *structural neuronal network*);

(iii) Since both functionalities, the spectral one and the structural one, define the same physical object, these imply the identities:

$$\xi_\Psi \equiv \xi_\rho, E_\Psi \equiv E_\rho, \mathbf{k}_\Psi \equiv \mathbf{k}_\rho, |\Psi|^2 \equiv \rho. \tag{44}$$

Consequently, the probability density fluctuations are identified with the states density fluctuations, which specifies the brain coherence (or brain compatibility) of the two neuronal networks (the spectral one and the structural one). Such a situation implies mathematically the functionality of the elliptic functions equivalence theorem (Armitage [36]) and, in consequence, it implies the existence of certain algebraic relations among the variables which define brain dynamics on the two neuronal networks, particularly,

$$\delta|\Psi|^2 = F(\delta\rho). \tag{45}$$

By the algebraic relations, self-structuring, communication codes (languages) of the physical object result;

(iv) According to [28, 29], the finite energy solutions (see (41) and (42)) can be also obtained by field theories with spontaneous symmetry breaking, which also implies strange topologies [29, 40]. In any of these topologies one can always define through the associated topological charges, two distinct stable physical states. Now, according to our results, we shall have on one hand a topology specific to the spectral functionality (called spectral topology) which will define the spectral states through the spectral topological charge and, on the other hand, the topology specific to the structural functionality (called structural topology) which will define the structural states through the structural topological charge. Moreover, due to their physical object status, the two topologies, the spectral one and the structural one act simultaneously, influencing one each other. Practically speaking, we have a unique topology which encompasses both of them in the form of their direct product (Willard [40]). This has as a consequence the existence of four distinct stable states. In our opinion, these states could be associated with the nucleotide base from the human DNA structure (Brown [41]).

8 Hypothesis in Brain Functioning

The implementation of the functional structure of complex systems to psychic life, can explain a series of classical concepts circulated during te last century. Thus the unconscious from psychoanalysis can be associated with the unpredictable, non-causal and potential part from the structure of the complex system, while the conscience, as well as the unconscious behavioural patterns (superego of psychoanalysis) can be associated with the structured, causal and deterministic part of the complex system. Between the two parts, there exist a pemanent dynamics through attractors, descriptible in the phase space. The chaoticity existing between the two components is absolutely necessary for the functioning of the brain. When it is affected by repetitive cycles (epilepsy crisis) the conscience is blurred.

It is well known that a one-dimensional Toda type network of nonlinear oscillators can be attributed to cnoidal oscillation modes. Furthermore, by mapping it, a neural network can be induced [35, 42]. Since the "identity" of any biological structure is dictated by the morphological-functional "compatibility", in this status the "coherence" duplication of two neural networks is involved, namely the structural morphological specific neural network, and the spectral functional specific neural network. In such a framework the communication code among the complex system structural units assimilated to a biological structure is also generated, a code of an algebraic nature, taking into account the Elliptic Functions Equivalency Theorem (Armitage [36]);

Toda network and the electromagnetic brain theory hypothesis lead to new conclusions concerning different mental processes that have not been fully understood yet (see Crumpei et al. [43]).

In what concerns the electromagnetic field theories (or "EM field theories") of consciousness, they propose that consciousness results when a brain produces an electromagnetic field with specific characteristics. We should mention Pockett [44] and McFadden [45–47] which have proposed EM field theories, while Uttal [48] has criticized McFadden's and other field theories.

If we locate consciousness in the brain's EM field, rather than the neurons, then we have the advantage of neatly accounting for how information located in millions of neurons scattered through the brain can be unified into a single conscious experience (sometimes called the binding or combination problem): the information is unified in the EM field. Thus, EM field consciousness could be considered to be "joined-up information".

9 Concluding Remarks

With our approach we finally aim at the solution of the old problem of the duality brain-mind. Our physical-mathematical model offers the vision that both the mind and the brain do form a functional, describable through the dynamics existing between the spectral and neural networks lying at the foundation of the psychic life. This could explain (see Crumpei et al. [43] for details) some problems Chomski [17], Ramachandran [49] etc. were taking about.

We made obviously that the brain dynamics can be represented by a Bohmian mechanics (Bohm [31]) exactly the way the regular Schrödinger wave mechanics is represented. Only here there are now hidden parameters. Their place is explicitly filled by neuronal medium.

The following conclusions are immediate:

(i) The morphological-functional assumption of the fractal brain (in the most general sense of this concept described by Mandelbrot [23]) induces a mathematical model by activating brain's non-differentiable dynamics. Thus, one gives up either on the classical determinism, or on the quantum nondeterminism;
(ii) Through the mathematical model, the scale covariance principle induces brain geodesics in the velocity representation, situation in which any neuron from the network is substituted with the brain geodesics themselves;
(iii) Through the Schrödinger type representation of the brain geodesics, the spectral (wave) functionality of the brain dynamics is explicited, which implies specific mechanisms of tunneling type, entanglement states type etc., while through the hydrodynamical representation, the structural (corpuscular) functionality of the brain dynamics is explicited, which induces specific mechanisms of shock waves type, solitons type etc.;
(iv) The inference of these two representations sets forth the transitory (spectral-structural) functionality, which is controlled by the specific non-differentiable potential;
(v) If the external constraint is proportional with the states density, then the stationary dynamics through their fluctuations activate both the spectral neuronal network and the structural neuronal network. These classes of neuronal networks result by the mapping of certain classes of one-dimensional Toda type networks. Usually, the one-dimensional Toda type networks are associated with

certain cnoidal oscillation modes that, in our situation, can be specific either to the spectral character, or to the corpuscular one;

(vi) The structural-functional compatibility (coherence) of these two classes generates classes of algebraic type communication codes;

(vii) The spectral-structural compatibility of the neuronal networks only on the solitonic component simultaneously activates the functionality of a strange topology (which is the direct product of the spectral topology and the structural topology). It induces four distinct physical states through the associated topological charges. Thus, the quadruple logic elements are generated.

References

1. E. Kändel, *Principles of Neural Science*, 5th edn. (McGraw-Hill Companies, 2013)
2. H. Atmanspacher, *Quantum Approaches to Consciousness* (The Stanford Encyclopaedia of Philosophy, 2011)
3. H. Atmanspacher, W. Fach, A structural-phenomenological typology of mind-matter correlation. J. Anal. Psychol. **58**, 219–244 (2013)
4. F. Caserta, W.D. Eldred, E. Fernandez, R.E. Hausman, L.R. Stanford, S.V. Bulderev, S. Schwarzer, H.E. Stanley, Determination of fractal dimension of physiologically characterized neurons in two and three dimensions. J. Neurosci. Methods **56**, 133–144 (1995)
5. F. Caserta, H.E. Stanley, W.D. Eldred, G. Daccord, R.E. Hausman, J. Nittman, Physical mechanisms underlying neurite outgrowth: a quantitative analysis of neuronal shape. Phys. Rev. Lett. **64**, 95–98 (1990)
6. T.A. Witten Jr., L.M. Sander, Diffusion-limited aggregation, a kinetic critical phenomenon. Phys. Rev. Lett. **47**, 1400–1403 (1981)
7. K.D. Kniffki, M. Pawlak, C. Vahle-Hinz, Scaling behavior of the dendritic branches of thalamic neurons. Fractals **1**, 171–178 (1993)
8. G. Werner, Perspectives on the neuroscience of cognitions and consciousness. BioSystems **87**, 82–95 (2007)
9. G. Werner, Consciousness related neural events viewed as brain state space transition. Cogn. Neurodyn. **3**, 83–95 (2009)
10. R.L. de Valois, K.K. de Valois, *Spatial Vision, Oxford Psychology series No. 14*, (Oxford University Press, New York, 1988)
11. R.L. de Valois, K.K. de Valois, A multi-stage color model. Vision Res. **33**, 1053–1065 (1993)
12. G. von Békési, Problems relating psychological and electrophysiological observations in sensory perception. Perspect. Biol. Med. **11**, 179–194 (1970)
13. S.B. Lowen, M.C. Teich, Fractal Auditory Nerve Firing Patterns May Derive From Fractal Switching in Sensory Hair Cell Ion Channels, in *Proceedings of the AIP Conference on American Institute of Physics*, vol. 285 eds. P.H. Handel, A.L. Chung (1993), pp. 745–748
14. S.B. Lowen, M.C. Teich, *Fractal based Point Processes* (Wiley, New York, 2005)
15. N.B. Karayiannis, A.N. Venetsanopoulos, Artificial neural networks, learning algorithms, performance evaluation, and applications. Springer Int. Eng. Comput. Sci. **209** (1993)
16. A. Slavova, *Cellular Neural Networks: Dynamics and Modeling, Mathematical Modeling: Theory and Applications*, vol. 16, (Springer, Berlin, 2003)
17. N. Chomski, *Language and the Study of Mind* (Sansyusya Publishing, Tokyo, 1982)
18. T.W.S. Chow, *Neural Networks and Computing. Learning Algorithms and Applications, Series in Electrical and Computer Engineering* (2007)

19. M.H. Díaz, F.M. Córdova, L. Cañete, F. Palominos, F. Cifuentesa, C. Sánchez, M. Herrera, Order and chaos in the brain: fractal time series analysis of the EEG activity during a cognitive problem solving task, Proc. Comput. Sci. Inf. Technol. Quant. Manage. **55**, 1410–1419 (2015)
20. H. Diaz, F. Córdova, Harmonic fractals in the brain: transient tuning and synchronic coordination in the quasi-chaotic background of ongoing neural EEG activity. Proc. Comput. Sci. **17**, 403–411 (2013)
21. A.J. Ibáñez-Molina, S. Iglesias-Parro, Fractal characterization of internally and externally generated conscious experiences. Brain Cogn. **87**, 69–75 (2014)
22. A. Khodabakhsh, A. Mehran, R. Majid, A.M. Esfandiar, S. Firoozeh, Brain activity of women is more fractal than men. Neurosci. Lett. **535**, 7–11 (2013)
23. B. Mandelbrot, *The Fractal Geometry of Nature* (W. H. Freeman and Company, New York, 1983)
24. L. Nottale, *Fractal Space-Time and Microphysics: Towards a Theory of Scale Relativity* (World Scientific, Singapore, 1993)
25. L. Nottale, *Scale Relativity and Fractal Space-Time: A New Approach to Unifying Relativity and Quantum Mechanics* (Imperial College Press, London, UK, 2011)
26. L. Nottale, Scale relativity: a fractal matrix for organization in nature. Electron. J. Theor. Phys. **4**, 187–274 (2007)
27. M. Agop, N. Forna, I. Casian-Botez, New theoretical approach of the physical processes in nanostructures. J. Comput. Theor. Nanosci. **5**, 483–489 (2008)
28. M. Agop, A. Gavriluţ, G. Crumpei, B. Doroftei, Informational non-differentiable entropy and uncertainty relations in complex systems. Entropy **16**, 6042–6058 (2015)
29. M. Agop, A. Gavriluţ, G. Ştefan, B. Doroftei, Implications of non-differentiable entropy on a space-time manifold. Entropy **17**, 2184–2197 (2015)
30. A. Timofte, I. Casian-Botez, D. Scurtu, M. Agop, System dynamics control through the fractal potential. Acta Phys. Pol. A **119**, 304–311 (2011)
31. D. Bohm, A suggested interpretation of the quantum theory in terms of "hidden" variables. Phys. Rev. **85**, 166–179 (1952)
32. J. Cresson, Scale relativity theory for one dimensional non differentiable manifolds. Chaos Solitons Fractals **14**, 553–562 (2002)
33. V.S. Bîrlescu, M. Agop, M. Craus, Computational properties of a fractal medium. Int. J. Quantum Inf. **12**, 22 (2014). https://doi.org/10.1142/S0219749914500221
34. P. Lévy, *Theories de l'Addition Aléatoires* (Gauthier-Villars, Paris, 1937)
35. E.A. Jackson, *Perspectives on Nonlinear Dynamics* (Cambridge University Press, Cambridge, 1992)
36. J.V. Armitage, W.F. Eberlein, *Elliptic Functions* (Cambridge University Press, Cambridge, 2006)
37. M. Chaichian, N.F. Nelipa, *Introduction to Gauge Field Theories* (Springer, Berlin, Heidelberg, 1984)
38. M. Toda, *Theory of Nonlinear Lattices* (Springer, New York, Berlin, 1981)
39. M. Toda, Nonlinear lattice and soliton theory. IEEE Trans. CAS **30**, 542–554 (1983)
40. S. Willard, *General Topology*, (Addison-Wesley Pub. Co., Reading, MA, 1970) ISBN 0486434796, Retrieved 2013
41. T.A. Brown, *Genomes*, 2nd edn. (Wiley-Liss, Oxford, 2002). ISBN -10: 0-471-25046-5
42. J. Gardiner, R. Overall, J. Marc, The fractal nature of the brain. NeuroQuantology **8**(2), 137–141 (2010)
43. G. Crumpei, A. Gavriluţ, I. Crumpei Tanasă, M. Agop, *New Paradigms on Information, Mind and Reality from a Transdisciplinary Perspective* (Junimea Publishing House, Iaşi, 2016)
44. S. Pockett, *The Nature of Consciousness: A Hypothesis*, Writers Club Press (2000)
45. J. McFadden, The conscious electromagnetic information (Cemi) field theory: the hard problem made easy? J. Conscious. Stud. **9**(8), 45–60 (2002)
46. J. McFadden, Synchronous firing and its influence on the brain's electromagnetic field: evidence for an electromagnetic field theory of consciousness. J. Conscious. Stud. **9**(4), 23–50 (2002)

47. J. McFadden, *The CEMI Field Theory: Seven Clues to the Nature of Consciousness*, ed. by J.A. Tuszynski (Springer, The Emerging Physics of Consciousness, Berlin, 2006), pp. 385–404
48. W.R. Uttal, *Neural Theories of Mind: Why the Mind-Brain Problem May Never Be Solved* (Erlbaum, Mahwah, NJ, 2005)
49. V.S. Ramachandran, *Mirror Neurons and Imitation Learning as the Driving Force Behind the Great Leap Forward in Human Evolution*. Edge Foundation. Retrieved 19 Oct 2011

Effect of Long-Range Spreading on Two-Species Reaction-Diffusion System

Šarlota Birnšteinová, Michal Hnatič and Tomáš Lučivjanský

Abstract Two-species reaction diffusion system $A + B \to A$ and $A + A \to (\emptyset, A)$ is studied in presence of long-range spreading. Long-range hops are described by Lévy flights, i.e. by a probability distribution that decays in d dimensions with a distance r according to a power-law function $r^{-d-\sigma}$. Critical dimension $d_c = \sigma$ depends on the control parameter for the Lévy flights $\sigma < 2$, and ε expansion is now performed in the form $\varepsilon = \sigma - d$. The renormalization group is applied in order to determine the time dependence of the density of reacting particles.

Keywords Reaction-diffusion systems · Long-range spreading · Perturbative renormalization group · Lévy flights

1 Introduction

Generic reaction-diffusion models describe a multitude of phenomena in various disciplines, from population dynamics in ecology, competition of bacterial colonies in microbiology, dynamics of magnetic monopoles in the early universe in cosmology, to the stock market in economy, opinion exchange in sociology, etc [1].

12th CHAOS Conference Proceedings, 18–22 June 2019, Chania, Crete, Greece.

Š. Birnšteinová (✉)
Faculty of Science, Šafárik University, Moyzesova 16,
040 01 Ko šice, EU, Slovakia
e-mail: sarlota.birnsteinova@student.upjs.sk

M. Hnatič
Institute of Experimental Physics SAS, Watsonova 47,
040 01 Ko šice, EU, Slovakia

Bogolyubov Laboratory of Theoretical Physics, Joint Institute
for Nuclear Research, 141980 Dubna, Russian Federation

T. Lučivjanský
Faculty of Science, P. J. Šafárik University, Šrobarova 2,
041 54 Košice, Slovakia

C. H. Skiadas and Y. Dimotikalis (eds.), *12th Chaotic Modeling and Simulation International Conference*, Springer Proceedings in Complexity,
https://doi.org/10.1007/978-3-030-39515-5_4

We consider a system with two particle species A and B, with corresponding diffusion constants D_{nA} and D_{nB}, undergo the reactions

$$A + A \to \begin{cases} \emptyset, & \text{with probability } p, \\ A, & \text{with probability } 1 - p, \end{cases} \tag{1}$$
$$A + B \to A.$$

Survival probability of B particles in the stationary $d = 1$ case is equivalent to the calculating the fraction of spins that have not flipped up to time t in the q-state Potts model evolving via zero temperature Glauber dynamics, where $q = p/(1-p)$ [2, 3]. The more general problem, where both particle species are mobile has been studied in [4–6]. Rate equations for reaction process (1) take the forms

$$\frac{\partial a}{\partial t} = D_{nA}\nabla^2 a - \lambda a^2, \quad \frac{\partial b}{\partial t} = D_{nB}\nabla^2 b - \lambda' ab. \tag{2}$$

Result for particle density A valid for $d > 2$ decreases with time as $a \sim 1/(\lambda t)$ [4, 5, 7]. The density of B particles for $d > 2$ decays with time as $b \sim t^{-\theta}$, where exponent θ depends on $\delta = D_{nB}/D_{nA}$ and p [3, 4, 6]. For dimensions $d \leq 2$ fluctuations become important and θ is no longer given by the rate Eq. (2). The field theoretic approach using the renormalization group formalism is used here to calculate the decay exponents below the critical dimension.

In this study of the two-species reaction diffusion process we exchange the ordinary diffusing Brownian particles with super-diffusing Lévy flying particles [8]. Superdiffusive behavior is observed in nature in various forms such as turbulence [9, 10], epidemic spreading [11] or transport phenomena [12, 13]. Considering Lévy flights enables us to continuously varies an upper critical dimension with the Lévy index [14], which we introduce later.

Distribution of the spreading distance r of the Lévy particle is given by

$$P(r) \sim r^{-d-\sigma}, \tag{3}$$

where d denotes the spatial dimension of the system. The exponent σ is a free parameter that controls the characteristic shape of the distribution. It should be emphasized that σ does not introduce any new length scale, rather it changes the scaling properties of the underlying (anomalous) diffusion process [15]. Fourier transform of the probability distribution has following form

$$P(k) = \mathrm{e}^{-D_A k^\sigma}, \quad P(k) = \mathrm{e}^{-D_B k^\sigma}, \ 0 < \sigma \leq 2, \tag{4}$$

where D_A, D_B scale the distributions. Considering anomalous diffusion modeled by Lévy flights results to the following substitutions

$$D_{nA}\nabla^2 \to D_{nA}\nabla^2 + D_A\nabla^\sigma, \quad D_{nB}\nabla^2 \to D_{nB}\nabla^2 + D_B\nabla^\sigma, \tag{5}$$

in Eq. (2). The analysis of the canonical dimensions shows, that for the case $\sigma < 2$ ordinary diffusion terms $\propto \nabla^2$ are superficially irrelevant compared to anomalous diffusion terms $\propto \nabla^\sigma$. Both ordinary diffusion terms and anomalous diffusion terms must be included for $\sigma \to 2$ [16, 17]. Here, we are not interested in such case, and we assume that $\sigma \ll 2$ and discard the ordinary diffusion terms.

2 The Field Theory Approach

Two species reaction-diffusion model (2) with Lévy flights can be reformulated in terms of the master equation using Doi [18, 19] and Peliti [20] formalism to map the system into Schrödinger-like equation. Then the continuous limit of the model leads to the field-theoretic action [7, 21]

$$\begin{aligned} \mathcal{S} = \int_0^\infty \mathrm{d}t \int_{-\infty}^\infty \mathrm{d}^d x \Big[&\psi_A^\dagger(\partial_t - \nabla^\sigma)\psi_A + \psi_B^\dagger(\partial_t - \delta\nabla^\sigma)\psi_B \\ &+ \lambda_0 \psi_A^\dagger \psi_A^2 + \lambda_0 \psi_A^{\dagger 2}\psi_A^2 + \lambda_0' Q \psi_B^\dagger \psi_A \psi_B + \lambda_0' \psi_A^\dagger \psi_B^\dagger \psi_A \psi_B \Big] \\ &- \int \mathrm{d}^d k (\psi_A^\dagger a_0 + \psi_B^\dagger b_0). \end{aligned} \tag{6}$$

In order to get the simpler form of the action, we rescale time t to absorb the anomalous diffusion constant D_A and a new parameter $\delta = D_B/D_A$ was defined. First line of the action (6) represents an anomalous diffusion process of particle species A and B. Second line in (6) represents reaction processes, while last line corresponds to Poison initial conditions with intensities a_0 and b_0. A new parameter $Q = 1/(2-p)$ was introduced. The following rescaling of the fields have been done: $\psi_A \to Q\psi_A$, $\psi_A^\dagger \to \psi_A^\dagger/Q$, since a coefficient $1/Q$ appeared in term $\psi_A^\dagger \psi_A^2$.

3 Perturbation Theory

The action (6) of the studied model has a convenient form for using the standard Feynman diagrammatic technique. The inverse matrix of the quadratic part of the action (6) determines the form of the bare propagators. These propagators correspond to lines connecting interaction vertices. In time-momentum and frequency-momentum representation the propagators have the forms

$$\begin{aligned} \langle \psi_A \psi_A^\dagger \rangle(t,k) &= \mathrm{e}^{-k^\sigma t}, \quad \langle \psi_A \psi_A^\dagger \rangle(\omega,k) = \frac{1}{-i\omega + k^\sigma}, \\ \langle \psi_B \psi_B^\dagger \rangle(t,k) &= \mathrm{e}^{-\delta k^\sigma t}, \quad \langle \psi_B \psi_B^\dagger \rangle(\omega,k) = \frac{1}{-i\omega + \delta k^\sigma}. \end{aligned} \tag{7}$$

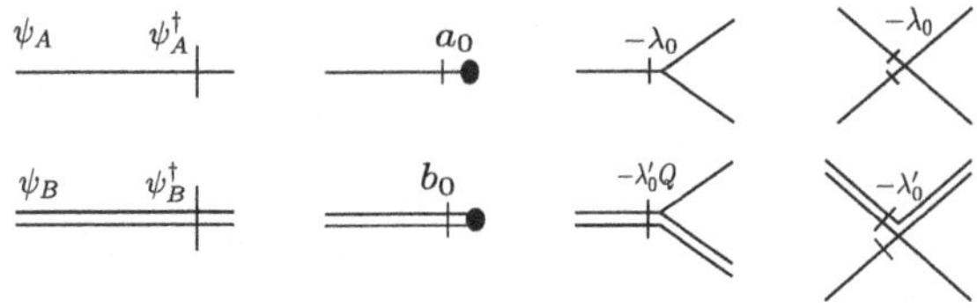

Fig. 1 Propagators, initial conditions and vertices of the theory described by action (6)

The vertex factors can be extracted from the non-quadratic part of the action (6), the reaction precesses and initial conditions. The vertex functions have the following form [22]

$$V_m(t_1, \boldsymbol{x}_1 \ldots t_m, \boldsymbol{x}_m; \phi) = \frac{\delta^m \mathcal{S}}{\delta\phi(t_1, \boldsymbol{x}_1) \ldots \delta\phi(t_m, \boldsymbol{x}_m)}, \tag{8}$$

where ϕ can be an arbitrary field of the model (6). Graphical elements of the perturbation theory for the field-theoretic action (6) are depicted in Fig. 1.

The perturbative expansion of density of B particles can be constructed using the Feynman diagrams shown in Eq. (9).

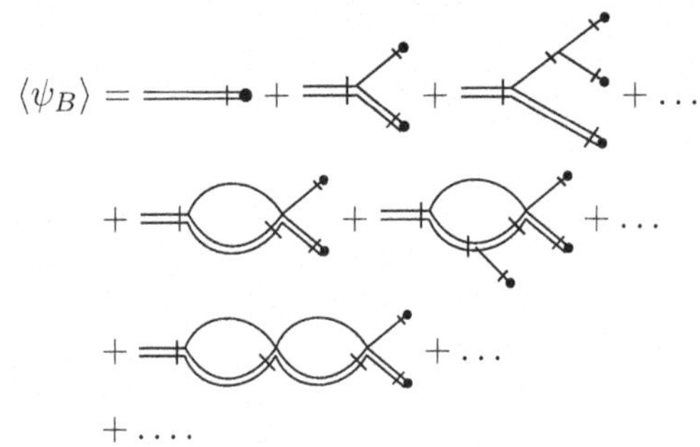

(9)

In the language of Feynman diagrammatic technique we need diagrams with outgoing $\langle \psi_B \psi_B^\dagger \rangle$ propagator. There are infinite number of diagrams, which contribute to densities of B particles. It is convenient to group these diagrams according to number of loops, which results after renormalization in an $\varepsilon = \sigma - d$ expansion [7, 23].

The perturbative expansion of the A species density is constructed in the same manner as for the B particle. The evolution of the density $\langle \psi_A \rangle$ is independent of the statistics of B particles and result corresponds to the time dependence in [23].

It is necessary to take into account all possible contribution from the initial conditions to the expansion in every number of loops (9). First we need to calculate mean-field (zero-loop) contributions to the densities of the A and B particle species (10). These correspond the first line in Eq. (9) and they are denoted with a dashed line and a double dashed line for A and B particles (10), respectively.

$\langle\psi_A\rangle_{mf} \equiv$ 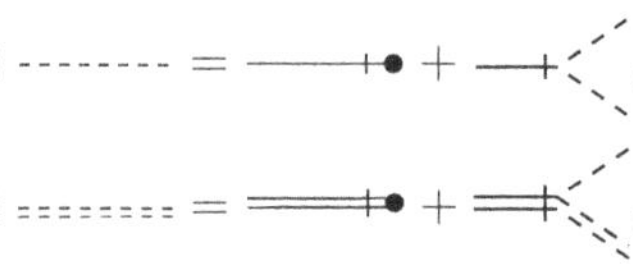

$\langle\psi_B\rangle_{mf} \equiv$

(10)

where mf is denoting a mean-field approximation. Previous expressions lead to known time dependences of A and B densities [6]

$$\langle\psi_A(t)\rangle_{mf} = \frac{a_0}{1+a_0\lambda_0 t}, \quad \langle\psi_B(t)\rangle_{mf} = \frac{b_0}{(1+a_0\lambda_0 t)^{\lambda_0' Q/\lambda_0}}. \tag{11}$$

Next step is to derive the mean-field propagators G^{mf}_{AA} and G^{mf}_{BB} in the following form

$G^{mf}_{AA}(p,t_2,t_1) \equiv \langle\psi_A(-p,t_2)\psi^\dagger_A(p,t_1)\rangle_{mf} =$ 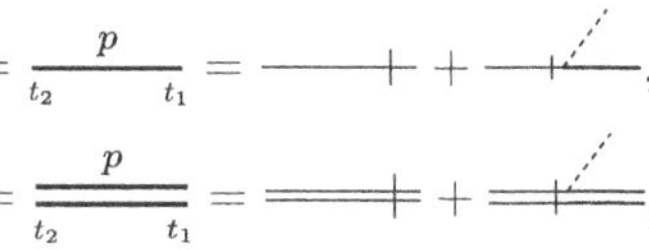

$G^{mf}_{BB}(p,t_2,t_1) \equiv \langle\psi_B(-p,t_2)\psi^\dagger_B(p,t_1)\rangle_{mf} =$

(12)

where the dashed lines correspond to all possible contribution of initial terms (10). Mean field propagators (12) have the following form

$$\begin{aligned} G^{mf}_{AA}(p,t_2,t_1) &= \mathrm{e}^{-p^\sigma(t_2-t_1)}\left(\frac{1+\lambda_0 a_0 t_1}{1+\lambda_0 a_0 t_2}\right)^2, \\ G^{mf}_{BB}(p,t_2,t_1) &= \mathrm{e}^{-\delta p^\sigma(t_2-t_1)}\left(\frac{1+\lambda_0 a_0 t_1}{1+\lambda_0 a_0 t_2}\right)^{\lambda_0' Q/\lambda_0}. \end{aligned} \tag{13}$$

One loop contributions to density B (9) can be expressed by three diagrams Fig. 2 using mean-field sums (10) and propagators (12).

They lead to the following expressions

$$\begin{aligned} \text{Fig. 2a} &= \frac{b_0\lambda_0'^2 Q}{\lambda_0(a_0\lambda_0 t)^{\lambda_0' Q/\lambda_0}}\frac{2(1+\delta)^{-d/\sigma}\Gamma[d/\sigma]}{\sigma(4\pi)^{d/2}\Gamma[d/2]}t^{\varepsilon/\sigma}\frac{\sigma^3}{(\varepsilon+\sigma)\varepsilon^2}, \\ \text{Fig. 2b} &= -\frac{b_0\lambda_0' Q}{(a_0\lambda_0 t)^{\lambda_0' Q/\lambda_0}}\frac{2^{1-d/\sigma}\Gamma[d/\sigma]}{\sigma(4\pi)^{d/2}\Gamma[d/2]}t^{\varepsilon/\sigma}\frac{2\sigma^5}{\varepsilon^2(\varepsilon+\sigma)^2(\varepsilon+2\sigma)}, \end{aligned}$$

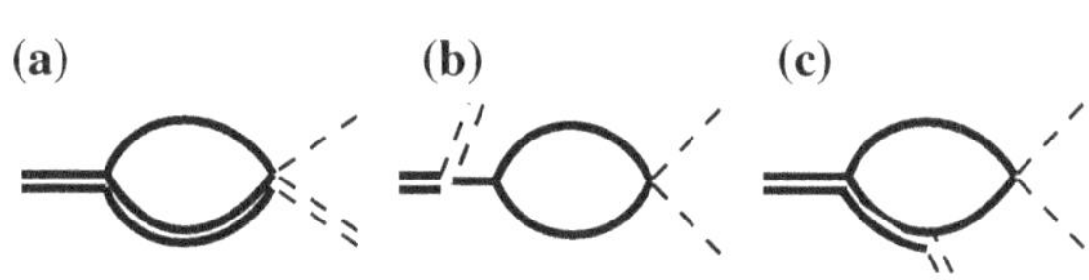

Fig. 2 One-loop diagrams contributing to the perturbative expansion of the B particle density (9)

$$\text{Fig. 2c} = -\frac{b_0\lambda_0'^2 Q^2}{\lambda_0(a_0\lambda_0 t)^2}\frac{2\Gamma(d/\sigma)}{\sigma(4\pi)^{d/2}\Gamma(d/2)}\int_0^t \mathrm{d}t_2\int_0^{t_2}\mathrm{d}t_1\left(\frac{1+a_0\lambda_0 t_1}{1+a_0\lambda_0 t_2}\right)^2 \times$$
$$\frac{1-\delta}{(2t_1-t_2(1+\delta))^{1+d/\sigma}}\left[\mathrm{B}\left(\frac{2t_1-t_2(1+\delta)}{t_1(1-\delta)}, 1+d/\sigma, 1-d/\sigma\right)\right.$$
$$\left. - \mathrm{B}\left(\frac{2t_1-t_2(1+\delta)}{t_2(1-\delta)}, 1+d/\sigma, 1-d/\sigma\right)\right], \tag{14}$$

where

$$\mathrm{B}(z;x,y) = \int_0^z \mathrm{d}t\, t^{x-1}(1-t)^{y-1},\ \ \Gamma(z) = \int_0^\infty x^{z-1}\mathrm{e}^{-x}\mathrm{d}x. \tag{15}$$

4 Renormalization

Renormalization group method is used to analyze the infra-red scaling ($t \to \infty$, $x \to \infty$) of the model (6), while effectively eliminating divergences arising from loop expansion of (9) for large t in $d \le d_c$. In order to renormalize the model, we need first to identify all possible superficial divergences. This can be done employing standard power counting scheme [24] for dynamical models, whose each quantity F has two canonical dimensions, in respect to the wave number d_F^k and to the frequency d_F^ω. Normalization of these dimensions can be written in the following way

$$\mathrm{d}_\omega^\omega = 1,\ \mathrm{d}_t^\omega = 0,\quad \mathrm{d}_k^k = -\mathrm{d}_x^k = 1,\quad \mathrm{d}_k^\omega = 0,\ \mathrm{d}_\omega^k = -\sigma. \tag{16}$$

Total canonical dimension for quantity F is: $\mathrm{d}_F = \mathrm{d}_F^k + \sigma\mathrm{d}_F^\omega$. The canonical dimensions of fields and parameters are listed in Table 1. The model is logarithmic when canonical dimensions of all the coupling constants $\{\lambda_0, \lambda_0'\}$ vanish simultaneously. This situation occurs for $\varepsilon = \sigma - d$, we can see that the upper critical dimension of the model is $d_c = \sigma$, rather then $d_c = 2$ of the model without anomalous diffusion term [6]. Renormalization of the model leads to the following renormalization of parameters [21, 22], where vertices takes the simple form of a geometric sum (21) as have been observed in similar models [3, 4, 7, 21].

$$\begin{aligned}\lambda_R &= \frac{\lambda}{1+\lambda/\lambda^*},\\ \lambda_R' &= \frac{\lambda'}{1+\lambda'/\lambda'^*},\end{aligned} \tag{17}$$

with dimensionless parameter $\lambda = \lambda_0\mu^{-\varepsilon}$, with some arbitrary momentum scale μ and the values of the fixed points

$$\lambda^* = (C\Gamma(\epsilon/\sigma))^{-1} \approx 2^{\sigma-1}\pi^{\sigma/2}\epsilon\Gamma\left(\frac{\sigma}{2}\right) + O\left(\epsilon^2\right), \tag{18}$$

Table 1 Canonical dimensions of the fields and parameters of the model (6) in d dimensional space

F	ψ_A, ψ_B	$\psi_A^\dagger, \psi_B^\dagger$	λ_0, λ_0'	Q	a_0, b_0	δ
d_F^k	d	0	$\sigma - d$	0	d	0
d_F^ω	0	0	0	0	0	0
d_F	d	0	$\sigma - d$	0	d	0

$$\lambda'^* = (C'\Gamma(\epsilon/\sigma))^{-1} \approx (\delta + 1)2^{\sigma-1}\pi^{\sigma/2}\epsilon\Gamma\left(\frac{\sigma}{2}\right) + O\left(\epsilon^2\right), \tag{19}$$

where

$$C = \frac{\sigma^{-1}}{(4\pi)^{d/2}}\frac{\Gamma(d/\sigma)}{\Gamma(d/2)}2^{2-d/\sigma}, \quad C' = \frac{\sigma^{-1}}{(4\pi)^{d/2}}\frac{\Gamma(d/\sigma)}{\Gamma(d/2)}\frac{2}{(1+\delta)^{d/\sigma}}. \tag{20}$$

The diagrams contributing to the renormalized annihilation rate λ_0' can be written in this way

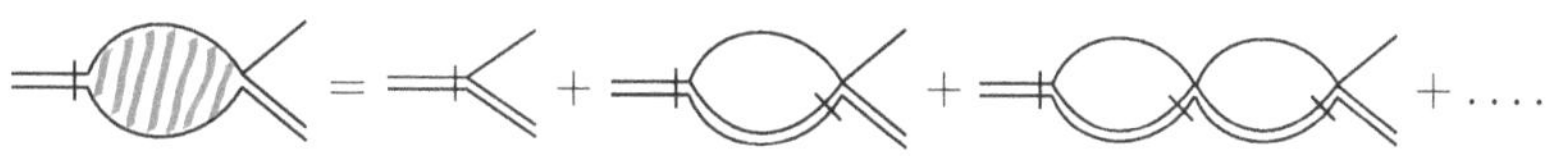

(21)

5 Summary

We study fluctuation effects in the two-species reaction-diffusion system $A + B \to A$ and $A + A \to (A, \emptyset)$. Instead of usually assumed short-range diffusion spreading of the reactants we consider long-range hops, which cause diffusion to become anomalous. For description of such transport we use approach based on Lévy distribution. When the diffusion is anomalous, the critical dimension depends continuously on the control parameter for the Lévy distribution. The model is studied in the limit of irrelevant ordinary diffusion terms employing field-theoretic approach, which allows us to analyze large-scale behavior of reaction process. The multiplicative renormalizability of the model has been proven. We have derived the necessary expressions for calculating the decay exponent of the B particle density to one-loop approximation, and have prepared a good starting point for further analysis of a different correlation functions.

References

1. U.C. Täuber, *Critical Dynamisc: a Field Theory Approach to Equilibrium and Non-Equilibrium Scaling Behavior* (Cambridge University Press, Cambridge, 2014)
2. B. Derrida, V. Hakim, V. Pasquier, Exact first-passage exponents of 1D domain growth: relation to a reaction-diffusion model. Phys. Rev. Lett. **75**, 751 (1995)
3. R. Rajesh, O. Zaboronski, Survival probability of a diffusing test particle in a system of coagulating and annihilating random walkers. Phys. Rev. E **70**, 036111 (2004)
4. M. Howard, Fluctuation kinetics in a multispecies reaction-diffusion systems. J. Phys. A: Math. Gen. **29**, 3437–3460 (1996)
5. S. Krishnamurthy, R. Rajesh, O. Zaboronski, Persistence properties of a system of coagulating and annihilating random walkers. Phys. Rev. E **68**, 046103 (2003)
6. B. Vollmayr-Lee et al., Anomalous dimension in a two-species reaction-diffusion system. J. Phys. A: Math. Theor. **51**, 034002 (2018)
7. B.P. Lee, Renormalization group calculation for the reaction $kA \to \emptyset$. J. Phys. A: Math. Gen. **27**, 2633–2652 (1994)
8. J.-P. Bouchaud, A. Georges, Anomalous diffusion in disordered media: statistical mechanisms, models and physical applications. Phys. Rep. **195**, 127 (1990)
9. M.F. Shlesinger, B.J. West, J. Klafter, Lévy dynamics of enhanced diffusion: application to turbulence. Phys. Rev. Lett. **58**, 1100 (1987)
10. T.H. Solomon, E.R. Weeks, H.L. Swinney, Observation of anomalous diffusion and Lévy flights in a two-dimensional rotating flow. Phys. Rev. Lett. **71**, 3975 (1993)
11. H. Janssen, K. Oerding, F. van Wijland, H. Hilhorst, Lévy-flight spreading of epidemic processes leading to percolating clusters. Eur. Phys. J. B. **7**, 137 (1999)
12. A. Ott, J.P. Bouchaud, D. Langevin, W. Urbach, Anomalous diffusion in "living polymers": a genuine levy flight? Phys. Rev. Lett. **65**, 2201 (1990)
13. R. Metzler, J. Klafter, The restaurant at the end of the random walk: recent developments in the description of anomalous transport by fractional dynamics. J. Phys. A-Math. Gen. **37**, R161 (2004)
14. D. Vernon, M. Howard, Branching and annihilating Lévy flights. Phys. Rev. E **63**, 041116 (2001)
15. H. Hinrichsen, M. Howard, A model for anomalous directed percolation. Eur. Phys. J. B **7**, 635–643 (1999)
16. J. Honkonent, MYu. Nalimov, Crossover between field theories with short-range and long-range exchange or correlations. J. Phys. A: Math. Gen. **22**, 751–763 (1989)
17. H.-K. Janssen, O. Stenull, Field theory of directed percolation with long-range spreading. Phys. Rev. E **78**, 061117 (2008)
18. M. Doi, Second quantization representation for classical many-particle system. J. Phys. A **9**, 1465 (1976)
19. M. Doi, Stochastic theory of diffusion-controlled reaction. J. Phys. A **9**, 1479 (1976)
20. L. Peliti, Path integral approach to birth–death processes on a lattice. J. Phys. (Paris) **46**, 1469 (1985)
21. U.C. Täuber, M. Howard, B.P. Vollmayr-Lee, Applications of field-theoretic renormalization group methods to reaction-diffusion problems. J. Phys. A: Math. Gen. **38**, R79–R131 (2005)
22. A.N. Vasilev, *The Field Theoretic Renormalization Group in Critical Behaviour Theory and Stochastic Dynamics* (Chapman Hall/CRC Press, Florida, 2004)
23. D.C. Vernon, Long range hops and the pair annihilation reaction $A + A \to \emptyset$: renormalization group and simulation. Phys. Rev. E **68**, 041103 (2003)
24. L.T. Adzhemyan, A.N. Vasil'ev, Y.M. Pis'mak, Renormalization group approach in the turbulence theory. Teor. Mat. Fiz. **57**, 268 (1983)

Special Sensitive System via Furstenberg Family and Its Applications

Mauricio Díaz

Abstract In this article we going to study the Sensitive system that can be described by a usual geometric form using a sub collection of elements of a Furstenberg family. We prove that a MDS without equicontinuous points has the form $(F_{ip} \cap F_{ps} \cap F_{pd1})^*$. At the end, we going to see some examples thrown basic simulations.

1 Introduction y Some Preliminaries

A dynamical system (DS) denoted as a pair (X,T) is a continuous map T on a compact metric space X and the dynamic is obtained by iterating the map. Let $\mathbb{N}$ the semi group of non negative integer and $\mathbb{Z}$ an additive group of integers, the DS induces an action of $\mathbb{Z}^+$ on X for all x $\in X$ and it's denoted as n, with n $\in \mathbb{Z}$, such that $T^0(x) = x$, $T^{n+1}(x) = T(T^n(x))$. For A $\subset$ X and k $\in \mathbb{N}$ we denote by $T^k(A)$ the image of A under T^k when k $\in \mathbb{N}$ and preimage under T^k when k is negative. We say that for subsets A,B $\subset X$, we define the hitting time set

$$N(A, B) = \left\{n \in \mathbb{N} : T^{-n}(A) \cap B \neq \phi\right\}$$

We can identify A as a singleton $\{x\}$, writing $N(x, U)$ for $N(\{x\}, U)$. We say that A $\subset X$ is $+$invariant when $T(A) \subset A$ or equivalently A $\subset T^{-1}(A)$ and $-$invariant when $T^{-1}(A) \subset A$. Being A a$+$ invariant, the complement is $-$invariant. A is invariant when $T(A) = A$ and weakly $-$invariant when A $\subset T(A)$ or equivalently if for any x $\in A$ there's some x' $\in A$, such that $T(x') = x$. If A is $+$invariant, the restricted DS (A,T_A) is called a subsystem of (X,T). We call the system topologically transitive (TT) when for every open and nonempty subset U, $\bigcup_{i=0}^{\infty} T^i(U)$ is dense in X, or equivalently if for every pair U,V $\subset X$ then N(U,V) is nonempty and infinite (i.e X is perfect) [1]. Let the DS and a point x $\in X$, the T-orbit of x is $O(x) = \{T^n(x) : n \in \mathbb{Z}\}$. If the T-orbit of x is iterated by a semi group of non negative integer(i.e n $\in \mathbb{N}$), the

M. Díaz (✉)
Universidad de Bio Bio, Concepción, Chile
e-mail: mauricio.diazraby@gmail.com

C. H. Skiadas and Y. Dimotikalis (eds.), *12th Chaotic Modeling and Simulation International Conference*, Springer Proceedings in Complexity,
https://doi.org/10.1007/978-3-030-39515-5_5

subset V $\in O(x)$ is called forward ward orbit; otherwise V is called backward orbit. For every x $\in X$ the set of limit points of orbits ω(x) is denote as $\bigcap_{i=0}^{\infty}\overline{\{T^n(x) : n \geq i\}}$ and it's invariant ($\overline{A}$ is the closure of A). The point x is a transitive point if for every nonempty open subset U $\in X$, $N(x, U)$ is nonempty. The set of transitive points is denoted as $Trans(X, T)$.The point x is called a recurrent point (RT) if x $\in \omega(x)$ or equivalently if for every open and non empty U $\subset X$, there's some n > 0, such that $T^n(x) \in U$; x is a strongly recurrent (SR) point if for every non empty open subset U, there's some N > 0, such that if $T^n(x) \in U$ then $T^{n+m}(x) \in U$ for some $0 < m \leq N$; x is a regularly recurrent (RR) point if for every open nonempty U $\subset X$, there exists some N > 0 such that for all $k > 0$, $T^{kN}(x) \in U$. A DS is DO if there exists a point x $\in X$, such that the T-orbit of x is dense in X (i.e TT is equivalent to DO if (X,T) is TDS i.e TT is equivalent to DO if TDS is a TDS); also the set of transitive points is nonempty; $Trans(X, T)$ with $\omega(x) = X$ is a dense G_δ subset of X (A subset G is G_δ if G is represented as a countable intersection of open sets under a sequence $\{U_i : i \in \mathbb{N}\}$). A DS is minimal if for every open and nonempty set U $\subset X$ and every point x $\in X$, the hitting time set $N(x, U)$ is syndetic (L $\subset \mathbb{N}$ is syndetic if for some N $\in \mathbb{N}$, for every sub interval with length N, it meets with L or equivalently for every point x $\in X$, the T-orbit is dense in X or is in a dense G_δ subset. A mixing topological(MT) system is equivalent to say that for every pair U,V $\subset X$ the set $N(U, V)$ is co finite. A co finite subset A is when its relative complement is finite. A weak mixing system means that for every N $\in \mathbb{N}$, the product system (X^N, T^N) is (TT).If (X,T) is topologically mixing then it's weak mixing (WM). A point x $\in X$ is equicontinuous point if for each $\epsilon > 0$, there's $\delta > 0$ such that $d(x, y) < \delta \Rightarrow d(T^n x, T^n y) < \epsilon$ for any $n \in \mathbb{Z}^+$ and each y $\in B_\delta(x)$. A TT DS with an equicontinuous and transitive point is called almost equicontinuous DS (AE).

To respect Devaney, let X a compact metric space, and a continuous self map T, A chaotic system (X,T) is define satisfying three properties [2], (Transitivite DS (TT)) $\forall U, V \subset X, \exists n \in \mathbb{Z}, T^n U \cap V \neq \phi$, (Dense Periodic points) the set of periodic points is dense in X $\overline{(Per(T))} = X$ (A periodic point is a point x $\in X$, there's some prime period n $\in \mathbb{Z}^+ : T^n x = x$) and (T is Sensitive dependence of initial condition (SDIC)) $\exists\epsilon > 0, \forall x \in X, \forall x \in U, \exists y \in U, \exists n \in \mathbb{Z}^+, d(f^n(x), f^n y) > \epsilon$. Two famous theorem had been proven, first Let (X,d) a metric space, (TT) and Dense Periodic points $\Rightarrow$ (SDIC) and [3] proved that for a (TT), T is almost continuous iff T is not (SDIC). Before of that, Li and York did their own definition, in which let (X,T) be a Topological Dynamical System (TDS). The system (X, T) is called topologically transitive, weakly transitive if $T \times T$ is transitive on X $\times$ X, strongly mixing if being $\{i \in S : i \geq N\}$, for N $\in S$, there's two nonempty open subsets u and v, such that $T^i u \cap v \neq \phi$ and minimal if there's not trivial subsystems.

Let (X,T) be a TDS, with a metric d and T a surjective self-continuous map in X. Let (x,y) $\in X \times X$ is said to be Li-York pair if

$$\liminf_{n\to\infty} d(T^n x, T^n y) = 0$$

$$\limsup_{n\to\infty} d(T^n x, T^n y) > 0$$

A set $S_c \subset X$ is a scrambled set. A set is scrambled set if any pair of distinct points $\{x, y\} \subseteq S_c$ has (x,y) as a Li-York pair. A (X,T) is chaotic in sense of Li-York if X contain an uncountable scrambled subset.

Now let (X, B, μ, T) a MDS, where B is a Borel σ-algebra, a measurable cover of X is a finite family of B-measurable sets whose union is X. Let C a collection of finite open covers of X and P any finite cover of a compact space X, for some k $\in$ $[0, N-1]$ with N $\in$ $\mathbb{N}$, $T^{-k}(P) = \{T^{-k}(A) : A \in P\}$ and $P^{T,N} = \bigvee_{i=0}^{N-1} T^{-k}(P)$,being

$$U \vee V = \{u' \cap v' : u' \in U, v' \in V\}$$

given two open covers U and V. Let N(P) as the minimum cardinality in any finite sub covers of P. The topological complexity function of the finite cover P of the DS is denoted as $N(P^{T,N})$. A DS is called scattering if any finite cover P by non dense open sets has unbounded complexity. An unbounded complexity is when $N(P^{T,N}) \rightarrow \infty$. (X,T) is 2-scattering system when it the same property under a product system (X^2, T^2) and only for 2-set covers. A If DS is weakly mixing imply DS is scattering and scattering implies 2-scattering. 2-scattering is total transitive. (X,T) is totally transitive if (X,T) is (TT) for every n $\in$ $\mathbb{N}$. In [4] Huang and Ye proved that let T: X $\rightarrow X$ be 2-scattering. Then the proximal set is dense in $X \times X$. Then T has a dense scrambled set. Later Akin and Kolyada proved that if (X,T) is WM then its Li-York sensitive, which it open to problems associated with the study of almost HY-systems or uniformly spaces. By now, our focus is the relation between Sensitive systems (global and local) with the set of actions, whose describes the changes of states of the DS.

2 SDIC and Its Relation with Furstenberg Families

As we seen previously, in this chapter we going to see relations between T with global and local SDIC and their relation with a set of actions. Assuming that the global DS is TT, then we can use a set of action where the elements are infinite and proper, then we can use the hitting time set as an element of the set of action. Many authors had obtained results comparing local Chaos with a set of action as a Furstenberg family (i.e see results in [5, 6]), but almost all the results are associated to local analysis, because neither Li-York chaotic system nor any local chaotic system as DC or F-chaos implies Devaney chaos [7]. In this section we going to explore some results obtained of Furstenberg family in DS with two different sensitive DS.

2.1 SDIC Global and Other Points

Previously we see that a DS is SDIC if $\exists \epsilon > 0$, $\forall x \in X$, $\forall x \in U$, $\exists y \in U$, $\exists n \in \mathbb{Z}^+$, $d(f^n(x), f^n y) > \epsilon$ or in each point x there's a point y as nearby as being possible, such that for some infinite and proper set of action, we obtain results so different to the expectations, which is clearly the opposite of (AE) DS, where for any set of action, the iterations are inside an open ball (i.e case of metric space with TT DS). In [1], Glaisner and Weiss studied SDIC in E-system and M-system with positive and null topological entropy and they proved that a SDIC TDS must being in DS with null topological entropy if it has an infinite P-system. After in [8] Li and Ye proved that for a non-trivial DS. If there's a subsystem (Y,T) of (X,T) such that $(X \times X, T \times T)$ is transitive, then (X,T) being scattering, (Y, T_Y) a minimal DS and $(X \times Y, T \times T_Y)$ being TT, its is densely uniformly chaotic (DUC) and also they saw that a DS is DUC if its TT and has a periodic point of period p, then its closed T^p invariant subset X_0 of X, such that (X_0, T^p) is DUC and $X = \bigcup_{j=0}^{p-1} T^j X_0$. A minimal system can be (AE) or (SDIC). In comparison, first let (X,T), X is a non-trivial Hausdorff uniform space, T is an uniform continuous map, $T : X \to X$. A non empty collection U of subsets of $X \times X$ is an uniform structure, $r \in U$ is called entourages. (X,T) is SDIC if there exists an entourage $E \in U$ (called sensitive entourage), such that for any $x \in X$ and any $D \in U$, there exists $y \in D(x)$ and $n \in \mathbb{Z}^+$, such that $(T^n x, T^n y) \notin E$ (i.e see [9]). It's so easy to see that the different results in DS with SDIC are because choosing one subset of X, it loose properties or their dynamic hasn't being described using geometric analysis nor topological properties unless it would being TT or has transitive points, but in the section of simulation, we going to talk about this point. Studying it by points of X, we can see that no-where iteration of action in $\mathbb{Z}^+$ might made projection of images in similar open neighborhood, but there's subsets of $\mathbb{Z}^+$ that can stay in an unique neighborhood, but in this case the hard part is to describe the set of action linked to that effect. Some propositions about SDIC are the next:

Proposition 2.1 *Let (X,T) is a TDS with SDIC, the following properties are equivalent*

1. *Let E_{ij} a subset of X and there's $n > 1$ with $i, j \in \mathbb{N} \cap [1, n]$, there's a function $\phi_{ij,i',j'} : E_{ij} \to E_{i',j'}$ for some i', j', such that it's not bijective unless $i = i'$ and $j = j'$*
2. *(X,T) is not (AE)*
3. *Let F_{ij} an infinite and proper subset of $\mathbb{Z}^+$ and there's $n > 1$ with $i, j \in \mathbb{N} \cap [1, n]$, there's a function $\lambda_{ij,i',j'} : F_{ij} \to F_{i',j'}$ for some i', j', such that it's not bijective unless $i = i'$ and $j = j'$.*

Remark 1 This case is more easy to see in expansive DS, but clearly this proposition clarifies the previous about the impossibility of take a full action, unless it has equicontinuous points, because $\mathbb{Z}^+$ is a subset of itself.

We'll return with this proposition in the next section.

Proposition 2.2 *Let (X,T) being a (DUC), then the following properties are implies:*

1. *(X,T) is densely Li-York chaotic*
2. *(X,T) has MDS-type 1 properties in a TT with a periodic point*
3. *There's a non distal T-invariant measure.*

Is easy checking the readings [8, 10, 11]. MDS-type 1 is a TDS with T-invariant under a positive measure, it has perfect subsets in X, being X a Hausdorff space and X contain compact and metric subsystems.

2.2 SDIC Local

When we talk about local Sensitive, there's many definitions. Huang and Ye said that T is sensitive iff there's $\epsilon > 0$, such that for any neighborhood of any $x \in X$ contains a point y, such that the trayectories of x and y are separated by ϵ for infinitely many times, that is

$$\limsup_{n\to\infty} d(T^n x, T^n y) > 0.$$

Later, Akin and Kolyada added the proximal property for infinitely many times (i.e $\liminf_{n\to\infty} d(T^n x, T^n y) = 0$), but not all Li-York sensitive are Li-York chaotic and in this case is important include the scrambled set. In [8], Jana Hantakova worked with a case with Li-York sensitive but not chaotic. About proximality, the Theorem 3.11 in [11] said that a DS is proximal iff it has a fixed point which is the unique minimal point in X. A set $S_c \subset X$ is a scrambled set. A set is scrambled set if any pair of distinct points $\{x, y\} \subseteq S_c$ has (x,y) as a Li-York pair. A (X,T) is chaotic in sense of Li-York if X contain an uncountable scrambled subset. A scrambled set is independent for a Li-York chaotic system. About [11], let (X,T) as a minimal system with X infinite, then if (X,T) is WM then there's a (X,T), such that its Li-York chaotic. Let a (X,T) with Li-York pair and it admits a dense, uncountable scrambled set, then (X,T) is a densely Li-York chaotic. A DS is WM if its densely Li-York chaotic. Recall that (x,y) is asymptotic if $\lim_{n\to\infty} d(T^n x, T^n y) = 0$, and its proximal if $\liminf_{n\to\infty} d(T^n x, T^n y) = 0$ and its distal if $\liminf_{n\to\infty} d(T^n x, T^n y) > 0$. A DS has (x,y) as Li-York pair if its proximal but not asymptotic. Let X a compact metric space, Δ_X is the diagonal of Cartesian product $X \times X$. Also Δ_X can be denoted as $\{(x, x) : x \in X\}$. Let N > 1, we define (AS) the space defined by the collection of distinct points in X^N where it has asymptotic pairs, (Pr) the space defined by the collection of distinct points in X^N that it has proximal pairs and (Dis) the space defined by the collection of distinct points with distal pairs, then we obtain the next results.

Lemma 2.3 *Let (X,T) a WM and* $\Delta_k = \left\{(x, y) : d(x, y) < \frac{1}{k}\right\}$*, then for some* $R \subset X \times X$

1. $(AS) = \bigcap_{n\geq 1} \bigcup_{k\geq 1} \bigcap_{i\geq k} (T \times T)^{-i} \overline{\Delta_n}$
2. $(Pr) = \bigcap_{n\geq 1} \bigcup_{i\geq 1} (T \times T)^{-i} \Delta_n$
3. $(Pr)/(AS) \subset X \times X.$

Associated with local Chaos, we have the next theorems:

Theorem 2.4 *Let (X,T) be a non-periodic transitive DS. If there exists a periodic point, then it's Li-York chaos*

and associated with the Lemma 2.3.

Theorem 2.5 *Let (X,T), the following conditions are equivalent*

1. *(X,T) is SDIC*
2. *There exists* $\epsilon > 0$ *such that (AS) is a first category subset of* $X \times X$
3. *There exists* $\epsilon > 0$ *for every* $x \in X$, *the set (AS)(x) of* $y \in X$, *such that* $(x,y) \in (AS)$ *is a first category of X*
4. *There's* $\epsilon > 0$ *such that for every* $x \in X$, $x \in \overline{X/(AS)(x)}$
5. *There exists* $\epsilon > 0$ *such that the set of* $(x,y) \in X \times X$, *such that* $\limsup_{n\to\infty} d(T^n x, T^n y) > 0$ *is dense of* $X \times X$.

Those theorems and some application you can find in [4, 10, 12].
One of them is associated with the next theorem.

Theorem 2.6 *Let (X,T) a non-trivial DS, if it's scattering, then there's a dense Mycielski scrambled set*

In this case we can study Li-York if there's a 2-scattering system.
In uniformly spaces, we can define Li-York of the next form:

- About (X,T), X is a non-trivial Hausdorff uniform space, T is an uniform continuous map, $T : X \to X$. A non empty collection U of subsets of $X \times X$ is an uniform structure, $r \in U$ is called entourages.

Definition 2.1 A pair $(x, y) \in X \times X$ is:

1. proximal if for any $E \in U$, there exists some $n \in \mathbb{Z}^+$, such that $(T^n x, T^n y) \in E$
2. asymptotic if for any $E \in U$, there exists some $n \in \mathbb{Z}^+$, such that $(T^m x, T^m y) \in E$ for any $m > n$.

The lemma of [4] are the same (Check Theorem 2 in [9]).
D is a scrambled set of T if $D \times D/\Delta \subset LYR$.

2.3 Furstenberg Family

For any $A \subset \mathbb{N}$, define the upper density and lower density of A by

$$d_U(A) = \limsup_{n\to\infty} \frac{|A \cap \{0, 1, 2, \ldots, n-1\}|}{n}$$

and

$$d_L(A) = \liminf_{n\to\infty} \frac{|A \cap \{0, 1, 2, \ldots, n-1\}|}{n}$$

respectively. For any sequence $\{E_n\}$ of intervals of positive integers, let $A \subset \mathbb{N}$, we call A has Banach density 1 if

$$\lim_{n\to\infty} \frac{|A \cap E_n|}{|E_n|} = 1.$$

To respect that, we going to use some special collection of Furstenberg family (Ff).Let P a collection of subsets of $\mathbb{Z}^+$, A sub collection of P denoted by F is called (Ff) if for any F_1, F_2 with $F_1 \subset F_2$ and $F_1 \in F$, then $F_2 \in F$. F_{inf} is defined as the family of all infinite subsets of $\mathbb{Z}^+$. F_{pubd} is defined as the family of sets with positive Banach density. F_{cf} is the dual family of F_{inf}. F_{bd1} is the family of all sets with Banach density 1.A pair is F_{bd1}-proximal if $\left\{n \in \mathbb{N} : d(T^n x, T^n y) < \frac{1}{m}\right\}$ for all $m \geq 0$. Also the family of syndetic, thick and piecewise syndetic are denoted as F_s, F_t and F_{ps}. A family $F \subset \mathbb{Z}^+$ is called thick if it contains long runs of positive integers (i.e for every n $\in \mathbb{N}$, there exists some $a_n \in \mathbb{Z}^+$, such that $[a_n, a_n + n] \subset F$). A family $F \subset \mathbb{Z}^+$ is syndetic if there's N $\in \mathbb{N}$, such that $[n, n + N] \cap F \neq \phi$ for every n $\in \mathbb{Z}^+$. A family $F \subset \mathbb{Z}^+$ is piecewise syndetic if its the intersection of thick and syndetic sets. About other classifications, any non-empty collection P of subsets $\mathbb{Z}^+$, it generates a family

$$F(P) = \left\{F \subset \mathbb{Z}^+ : P' \subset F : P' \in P\right\}.$$

For a family F, the dual family denoted as F^* is

$$F^* = \left\{F' \in P : P' \cap F' \neq \phi : P' \in F\right\}.$$

About it, we can see that $F_s^* = F_t$ and $F_{inf}^* = F_{cof}$, where F_{inf}, F_{cof} are the families of all infinite subsets of non integers and co finite subsets of non integers.

For a sequence $\{p_i\}_{i\in\mathbb{N}}$, the finite sum of the sequence denoted by FS has a family called F_{ip} if a subset $F \subset \mathbb{Z}^+$ has a sequence with $p_i\mathbb{N}$ such that FS $\subset F$. For a family F, the difference family of F is defined as

$$\Delta F = \left\{F \subset \mathbb{Z}^+ : \exists F' \in F : \left\{a - b : a, b \in F' : a \geq b\right\} \subset F\right\}.$$

The block family of F, denoted by bF is the family of all the subsets of $\mathbb{Z}^+$, in which there's some F' $\in F$, such that for every finite subset $F'' \subset F'$, then $m + F'' \subset F$ for some m $\in \mathbb{Z}^+$. Some properties are $b(bF) = bF$, $bF_{cof} = F_t$, $bF_s = F_{ps}$ and $bF_{pubd} = F_{pubd}$, $\Delta(F) = \Delta(bF)$. Let (X,T) a DS, that is F-transitive, if for every pair U, V open and non empty subsets of X, the hitting time is an element of the family (i.e $N(U, V) \in F$); also (X,T) is F-recurrent if for every neighborhood U of x, $N(x, U) \in F$, being $x \in \omega_T(x)$; if $\omega_T(x) = X$. (X,T) is a F-transitive-point being $N(x, U) \in F$, for every open non empty U of x. About the definition of Li-York pair, a subset F $\in \mathbb{Z}^+$, if there's a DS, x $\in X$, y is a minimal point and a neighborhood U of y such that (x,y) is proximal and$N(x, U) \in F$, then F is a central set. $F_{central}$ is the family of all the central subsets of $\mathbb{Z}^+$. Based in [5, 6], Let (X,T) a WM, the DS has F_{ip} transitive points and F-Poincare respect transitive (X,T). Let $F \subset \mathbb{Z}^+$ is called Poincare sequence if for every measure preserving system (X,B,μ,T) with $\mu(A) > 0$, there's n $\in F$ with n $\neq 0$, such that $\mu(A \cap T^{-n}A) > 0$. Thus, being (X,T) be a minimal system, then (X,T) is WM iff it has F_{ip}- transitive points. About other theorem, if (X,T) is WM, the set of all the points with proximal pairs is a dense G_δ subset and then it has $F_{central}$ transitive points. Being (X,T) a WM E-system (A DS with invariant measure and full support), as an E-system has positive upper Banach density (F_{pubd}), for every family in a WM E-system, there's a F_{ip} and F_{pubd}. About the work of [2] Being a M-system (A DS of dense minimal points) with generic points, it's implies an E-system, then for every WM-E-system and M-system, there exists a family denoted F_{EM}, where $F_{ip} \cap F_{pubd} \cap F_{ps} \subset F^*_{EM}$, then $F_{central} \cap F_{pubd} \subset F^*_{EM}$. Now being F_1 and F_2 two Furstenberg families, such that $N((x, y), U \times U) \in F_1$ and $N((x, y), X \times X/\Delta_{\frac{1}{\delta}}) \in F_2$ for some $\delta > 0$ and $F_1 \cap F_2 \neq \phi$, then a DS is local chaotic (i.e Li-York, DC and CD) [10, 11]. Also an E-system by F-scattering implies being F_{pd} (i.e being $F_{pd} \in P(\mathbb{Z}^+)$, then F^*_{pd} contains proximal F_{pbd1} (remark: $d_U(F) \geq BD^*(F)$, being $BD^*(F)$ the Banach density. Now as we see, being (X,T) a minimal DS, there's a WM DS iff there's a scattering DS, then a MDS with distal pairs implies equicontinuous points and thus a DS with sensitive maps implies CD then we have the next proposition:

Proposition 2.7 *Let (X,B,μ,T) a MDS-type 1 with DO and sensitive maps, then there's a F-scattering and F-recurrent point with $(F_{ip} \cap F_{ps} \cap F_{pd1})^*$-scattering contains F with dense G_δ subset of $X \times X/cl(\Delta_{\frac{1}{\delta'}})$ for some $\delta' > 0$*

Proof Let (X,B,μ, T) a MDS-type1 with DO, then by property 1, (X,B,μ,T) is an E-system. Let an E-system with F-recurrent point, then it has M-system, then there's a family with F_{ip}-recurrent point and F_{ip}-transitive point. Being an E-system with sensitive map, then for every x $\in X$, there's a open neighborhood U of y, such that there's $\epsilon > 0$, such that $d(T^n x, T^n y) > \epsilon$ for some n $\in F$ and every F $\in F_{ip} \cap F_{ps} \cap F_{pubd}$ by F-scattering and F-transitive (see properties). Let (x,y) $\in X \times X$, and being F_{ip} transitive point is WM in F-recurrent point, by minimal DS, there's a set of proximal pairs, such that F $\in F_{central}$. By lemma [6], $F_{central} \subset F_{ip} \in F_{ps}$, then let $\{z_i\}_{i \in \mathbb{N}}$ a sequence of minimal point dense in X and let the set of points with proximal pairs denoted by P(x), then $\bigcap_{i \in \mathbb{N}} P(z_i)$ is dense in X, then the proximal set is a G_δ. By property 2 and being (X,T) a DS with sensitive map, then the diagonal set has not sensitive map and there's infinite open subsets, then $X \times X/\Delta$ is uncountable and G_δ. Let X being a Hausdorff space, clearly X $\times X$ is also Hausdorff and Δ is closed, then $cl(\Delta)$ is closed for every x $\in X$, then X $\times X/\Delta$ is open, uncountable and dense in X $\in X$, but the system is not asymptotic, then (X,T) has Li-York densely chaos. Thus for MDS-type1 being $X \times X/\Delta$ a G_δ, also $X \times X/cl(\Delta_{\frac{1}{\delta'}})$.

Remark 2 In general a F_{ps}-transitive point does not imply WM. More clearly, we prove also that a MDS-type 1 with DO, dense minimal points and sensitive property implies a CD DS and also a DC, but only if the (X,T) has at least two cyclic points [10] or scrambled sets with ϵ-traces [9]. This result gives us an assumption that for every DS with M-system, F-scattering and without equicontinuous maps, we can links global chaos with local chaos, but we cannot prove an E-system has proximal pairs, under the idea of the support of a invariant measure is a singleton , being (X,T) a subsystem of a DS with the adjoint transformation of T in the dual of C(X).

2.4 Simulations and Some Results

2.4.1 By Point

In this chapter we going to take a case of a DS, where X = $\mathbb{T}^2$, where $\mathbb{T}^2$ is a 2-torus and the map is denoted as $F(x, y) = (1.5x + 0.5y, x + y)$. As we can see, there exist a conjugate relation between this system and the DS with the form $(2x + y, x + y)$, which has positive entropy and it's Devaney chaotic and Li-York chaotic. For built the DS, we going to define a collection of subsets $R_{ij} \times R_{i'j'}$ for $ij, i', j' \in \mathbb{N}_0 \cup [0, n]$ for $n > 0$. Since $n = 3$, then we divide in intervals of length 0.25 each one, from 0 to 1 with (mod 1) and we going to start with the initial states $(x_0\ y_0) = (0.4\ 0)$ and $(x'_0\ y'_0) = (0\ 0.4)$. In the first case we going to define a set action with $\mathbb{Z}^+$ as the greatest set. Some information obtained of the process are the next:

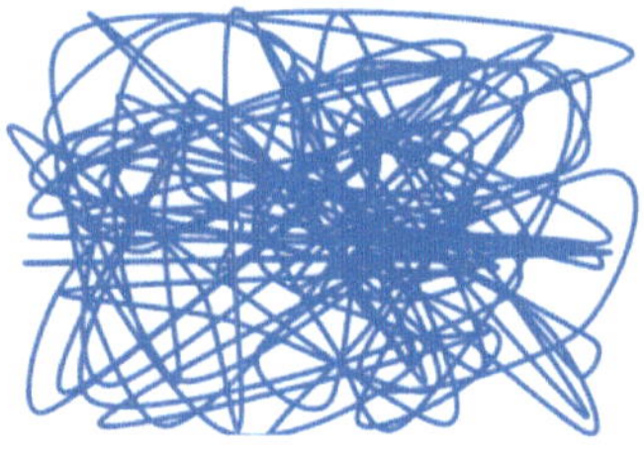

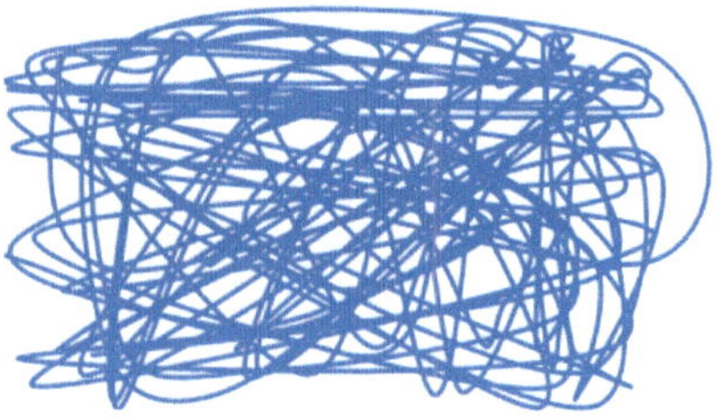

About the map $F(x, y) = (F_0(x, y), F_1(x, y))$, both graphs describes the behaviour of the dynamic to respect the components of $F(x, y)$. The first is about of $(F_0(x, y), F_0(x', y'))$. On the right hand is the relation between the components in $F_1(x, y), F_1(x', y')$.

Now the first step is put each value of an action in a specific region, but also we can take the values of the action and put the numbers in each $R_{ij} \times R_{i'j'}$.

As we can see, the system clearly has a sensitive behaviour and we can prove that choosing one region, where for some $n \in \mathbb{Z}^+$, Either the pair of F_0 or F_1 is outside of some arbitrary region. As clearly the DS is TT and have periodic points (i.e in Figs. 1 and 2, is easy see that there's consecutive values in one region), then there's not equicontinuous points that are transitive points (i.e see the transitive points associated to the change of region), then both components has sensitive behaviour actions in (x x') = (0,4 0)

	0	1	2	3
3	16,81,87,97	2,6,29,30,36,58,92,93,115 117	5,17,20,54,68,71,84,95	13,50,67,77,88,125
2	3,25,39,49,52,53,57,61, 65,70,75,89,114,120	9,14,19,23,41,80,90,94 100,108,109	21,24,37,42,47,56,59,106 112	10,64,76,85,103,124
1	33,40,45,48,69,128,129 132	18,28,34,72,78,79,119, 126,130,	1,7,31,32,51,55,63,73,74 116,123,127	62,99,107,111,113,121, 122,131
0	4,15,46,60,66,110,	0,26,86,104,105,118	8,12,27,35,38,43,91,96 101,102	11,22,44,82,83,98

Fig. 1 F_0

action in
(y y') = (0 0,4)

	0	1	2	3
3	11,25,33,34,40,41,51,57 64,74,96,107,122,123, 128,129,132	42,62,76,89,114,116,130	49,53,70,75,80,82,109, 113,121,124,131	7,18,19,24,31,32,48,52, 55,56,63,69,72,73,78,79 94,104,108,126,127
2	8,22,83	35,91	4,9,15,46,66,97	12,29,38,60,86,101,102, 105,117
1	0,39,103,112	5,10,13,47,67,98	16,61,87,120,	2,23,28,36,45,84,92,100, 106,118
0	26,44,58,68,77,85,90,99 115,125	1,17,27,50,54,59,71,111	3,81,88,110,119	6,14,20,21,30,37,43,65, 93,95

Fig. 2 F_1

in $x, y \in X$. Now as T^2 is perfect, is easy to check that for infinite elements, each region will have infinite elements, under the fact of end the case with every region with values, but that's not possible to prove yet, because we need to study a case where we take a sub region with infinite elements (i.e not necessary uncountable).

The second step is choose a subset F' of $\mathbb{Z}^+$, where the pair would being proximal. As we can see, the elements of action becomes to distribute in different regions. Also we can observe that there's not still a region $R_{i,j} \times R_{i',j'}$, with $i = i'$ and $j = j'$ with respect the results seeing in the last figures , that show the arrivals of elements in different pair of blocks. As every region is an open set and the DS is TT, we can see that there's a element of action that is $F_{central}$ and include the elements of $F_{ip} \cap F_{ps}$. Being $F_{ps} \subset \mathbb{Z}^+$, being (X,T) a AE, then F_{ip} stay on an unique region for any pair of

y	1.3 2.0,2.1 3.0,3.2	0.0,0.3 1.1,1.3 3.0,3.1,3.2	0,0.1,0.2,0.3 1.0,1.2,1.3 3.0,3.1,3.2	
	0.2,0.3 1.3 2.0,2.1,2.3 3.0	1.0		0.2 1.3
	1.0,1.3 3.1	0.3 1.3	0.2 1.3 2.0,2.1,2.3 3.0,3.1	0.2 1.3 2.1,2.3 3.0,3.1
	0.2 1.0 2.3	1.2 2.1 3.0,3.1	0.0,0.2,0.3 1.0,1.3 2.1,2.3 3.0	0.2 1.0,1.3 2.0,2.1,2.3 3.0,3.1,3.2
x'.y'		x		

Fig. 3 Set of regions comparing (x y) with (x' y')

indexes, thus the distribution of elements are more easy to predict. Under the results seeing in this experiment, we can establish the next.

Lemma 2.8 *Let (X,T) being a Devaney chaotic DS and Li-York chaotic DS, then the next properties are valid:*

1. *There exists $n \in F$, such that $F_{ip} \cup F_{ps}$ is not isomorphic to F, which contains proximal cells*
2. *Being P a collection of any quotient space $Q_n \subset \mathbb{Z}^+$, if (X,T) is not distal for some $\mathbb{Z}^+/Q_n$, for any $n \in \mathbb{N}$.*

3 Discussion

Under the results of this work, we can see that if (X,T) is SDIC, then the action can be defined as a subset of $\mathbb{Z}^+$, but not necessarily have a bijection with the set of non negative integers. That means that T being SDIC can have an uncountable set of neighborhoods and actions, such that there's not possible to describe a specific image from an unique region or from one set of action given for studying the behaviour of a DS, unless you had chosen one pair in each component for analyse the results, which is the real difference with a AE DS, where if you choose one pair of points in comparison with others, then there's a set of action where the region doesn't change and the analysis is around this part. If that's the case, then there's a relation between both pairs, such that the system evolve can being represented in terms of one pair or a finite number of pairs. The simulation only had the objective of make see the reader that there's not possible to build a Furstenberg family with $\mathbb{Z}^+$ with the objective to obtain similar results of DS with proximal sets or recurrent sets, unless that we could find the possibility of predict all the possible cells of partition of the iterated process, but we can put indicators that describe when the system has not equicontinuous points as transitive points, assuming that the DS is TT (i.e scattering system, weakly mixing between other type of examples). Now, being the study of one pair and being the space a Hausdorff space, we can think that in closed sets of (X,T), there's bijection between actions sets, but being trivial of these relations, because being (X,T) globally chaotic (Devaney), there's a smallest subset of X to respect the point x, such that if you prove that the subset is composed by at least two points, there's exists a non bijective relation between the actions and the subsets of the DS, unless the subset would being closed, but that is absurd with respect this case, but not impossible at all. The questions that are still without answers had associated if one can really describe in a general form a set of subsets of $\mathbb{Z}^+$, assuming that the system is TT (global) or proximal (local) and if there's a strong relation with the neighborhood (By metric spaces). In the simulation, we only take a finite number of elements (132), but we have not every region with elements, with exception of the projections seeing in the Figs. 1, 2 and 3. Also in the simulation, one can see that including if you have a greater number of blocks, there's still at least two regions with a set of action, which it have a successor number in more than one region, which can be just enough for establish the absence of EA, but not the same for periodic points or nearby points, thus we can say that this system allows the building of M-system and E-system, therefore the existence of elements in $F_{ip} \cap F_{ps} \cap F_{pubd}$, but not necessary contains the greatest element of this set, as we talk in Proposition 2.7. The objective for later will be trying to solve all the dob-ts. At the end the article open the doors to a mixing study between abstraction and experiments focused in find a collection of elements that can describe chaotic behaviours in the most simple form.

References

1. E. Glasner, B. Weiss, Sensitive dependence on initial conditions. Nonlinearity **6** (1993)
2. R.L. Devaney, J.-P. Eckmann, An introduction to chaotic dynamical systems. Phys. Today **40** (1987)
3. W. Huang, X. Ye, Devaney's chaos or 2-scattering implies Li–Yorke's chaos. Topol. Appl. **117**(3), 259–272 (2002)
4. J. Auslander, J.A. Yorke, Interval maps, factors of maps, and chaos. Tohoku Math. J., Second Series, **32**(2), 177-188 (1980)
5. F. Tan, J. Xiong, Chaos via Furstenberg family couple. Topol. Appl. **156**(3), 525–532 (2009)
6. J. Li, Transitive points via Furstenberg family. Topol. Appl. **158**(16), 2221–2231 (2011)
7. J. Li, J. Li, S. Tu, Devaney chaos plus shadowing implies distributional chaos. Chaos: Interdiscip. J. Nonlinear Sci. **26**(9), 93–103 (2016)
8. J. Hantáková, Li-Yorke sensitivity does not imply Li-Yorke chaos. arXiv:1711.10763 (2017)
9. X. Wu, X. Ma, Z. Zhu, T. Lu, Topological ergodic shadowing and chaos on uniform spaces. Int. J. Bifurc. Chaos **28**(03), 1850043 (2018)
10. F. Blanchard, E. Glasner, S. Kolyada, A. Maass, On Li-Yorke pairs. Journal fur die reine und angewandte Mathematik **547**, 51–68 (2002)

11. J. Li, X.D. Ye, Recent development of chaos theory in topological dynamics. Acta Math. Sin., English Series **32**(1), 83–114 (2016)
12. T. Downarowicz, Positive topological entropy implies chaos DC2, in *Proceedings of the American Mathematical Society*, vol. 142, no. 1 (2014), pp. 137–149
13. W. Huang, D. Khilko, S. Kolyada, G. Zhang, Dynamical compactness and sensitivity. J. Differ. Equ. **260**(9), 6800–6827 (2016)
14. E. D'Aniello, T. Steele, Chaotic behaviour of the map x w(x, f). Open Math. **12**(4), 584–592 (2014)
15. M. Brin, G. Stuck, *Introduction to Dynamical Systems* (Cambridge University Press, Cambridge) (2002)
16. J. Li, Chaos and entropy for interval maps. J. Dyn. Differ. Equ. **23**(2), 333–352 (2011)
17. N. C. Bernardes Jr, A. Bonilla, A. Peris, Mean Li-Yorke chaos in Banach spaces. arXiv:1804-03900 (2018)
18. S. Ruette, Transitive sensitive subsystems for interval maps. arXiv:1901-01067 (2019)
19. T. Downarowicz, Positive topological entropy implies chaos DC2, in *Proceedings of the American Mathematical Society*, vol. 142, no. 1 (2014), pp. 137–149

Mosaic Patterns in Reaction-Diffusion Systems

Dalia Ezzeddine and Rabih Sultan

Abstract We study a variety of reaction-diffusion processes that lead to the formation of exotic patterns. 1. We carry out precipitation reactions in gel media, wherein the interdiffusion of the co-precipitates takes place from multiple diffusion sources arranged in a symmetric framework. The precipitation zones are delimited by clear polygonal boundaries in congruence with the spatial distribution of the diffusion pools. 2. A displacement reaction in a solid-gel medium is conducted as a carbonic acid diffusion front invades an agar-calcium hydroxide gel putty. The formation of calcium carbonate yields a diversity of patterns, ranging from mosaic structures to Liesegang bands. 3. A Liesegang experiment precipitating lead chromate from the interdiffusion of lead and chromate ions in 2D yields a pattern of rings exhibiting revert spacing. When the diffusion comes from a constantly fed unstirred source (or reactor, CFUR), the patterns transit to a chaotic regime which is sensitive to the concentrations used and the flow rate.

Keywords Patterns · Chaos · Liesegang · Portlandite · Polygons · Reaction-diffusion

1 Introduction

Complex reaction-diffusion systems display exotic spatio-temporal structures including oscillations and patterns [1]. One class of such systems involves simple precipitation reactions [2], and can give rise to a rich variety of self-organized forms and shapes ranging from precipitate bands to mosaic patterns. The interest in bands (1D) or rings (2D) of precipitate goes back to the time of Liesegang [3], whose name became an icon of periodic patterns in diffusion-precipitation media. In this paper, we extrapolate the study of the Liesegang phenomenon to a number of specific chemical systems, exploring a variety of features with one especially worthy geological analogy. The twin relation between Liesegang patterns and geological banding

D. Ezzeddine · R. Sultan (✉)
Department of Chemistry, American University of Beirut, Beirut, Lebanon
e-mail: rsultan@aub.edu.lb

C. H. Skiadas and Y. Dimotikalis (eds.), *12th Chaotic Modeling and Simulation International Conference*, Springer Proceedings in Complexity,
https://doi.org/10.1007/978-3-030-39515-5_6

(referred to as geochemical self-organization) has received considerable attention since the early discovery of the phenomenon [4, 5], but has notably grown in the second half of the past century [6–12]. In another application, the possibility of the generation of chaotic spatial patterns of $PbCrO_4$ in 2D is presented.

2 Polygonal Boundaries

In the first part of this work, we carry out experiments in a Petri dish of 9.0 cm diameter containing a gel solution of $CoCl_2{\cdot}6H_2O$ (0.050M, 1% agar from Sigma). The mixture is heated to 85 °C with constant stirring for a few minutes until a clear, homogeneous solution is obtained. After gelling, a 1.0 cm-diameter hole is perforated at the center of the gel layer, along with other similar holes (three four, five or six) placed symmetrically and centered between the middle hole and the periphery (i.e. at 2.25 cm from the center). Then a solution of the outer electrolyte containing the co-precipitate ion (0.50 M NaOH or 0.50 M $Na_3PO_4{\cdot}12H_2O$ for the two different systems being studied) is added in the shallow holes, marking the start of the interdiffusion of the two co-precipitates. Figure 1 shows photos of the obtained patterns.

In both systems, we notice a color zonation due to the formation of different precipitates. The characterization of the precipitates of the Co^{2+}–NaOH system (blue-green, beige, dark brown; see frames a–d of Fig. 1) is under way. The blue green precipitate was established to be α-$Co(OH)_2$, as confirmed by the strong peak at $2\theta =$

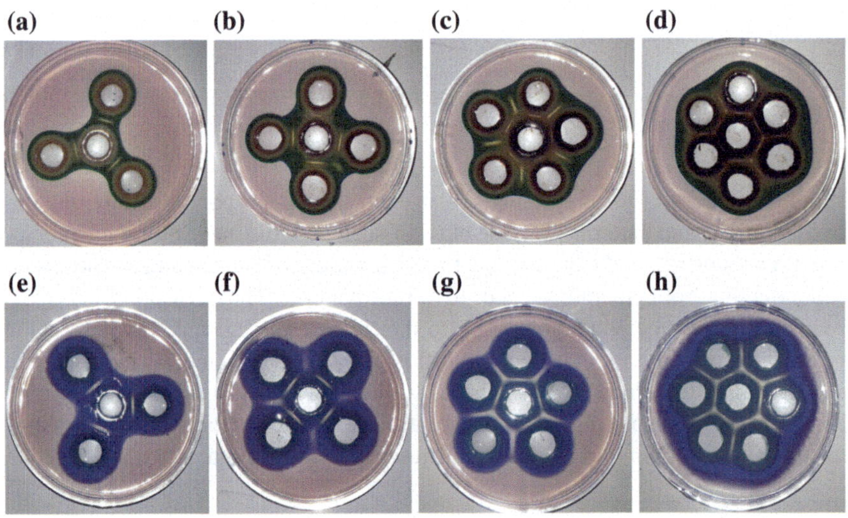

Fig. 1 **a–d** $Co(OH)_2$ mosaic structures with inter-source boundaries in three (**a**), four (**b**) five (**c**) and six (d) diffusion-source systems surrounding a central one. **e–h** Same trend in a cobalt phosphate system. Triangular, square, pentagonal and hexagonal boundaries are apparent

11.7° in the X-ray diffractogram, and in congruence with the results of Batlouni et al. [13]. The dark brown precipitate is believed to be $Co(OH)_3$, and the beige a mixture of the former two. For the phosphate system (see Fig. 1, frames e–h), the dark blue (inner) and the violet-blue (outer) correspond to two cobalt phosphate compounds: $Co_3(PO_4)_2{\cdot}4H_2O$ and its anhydrous form respectively [13, 14].

We most notably observe clear boundaries formed distinctly between the precipitate zones arising from the neighboring diffusion sources: triangular, square, pentagonal and hexagonal for three, four, five and six circular pools respectively surrounding the central source. Those gaps are 'architected' by the depletion in Co^{2+} due to diffusion from the supporting gel toward the precipitated zones.

3 Limestone Zonation in a Portlandite Matrix

In this section we investigate the formation of $CaCO_3$ zones in a portlandite mineral [$Ca(OH)_2$], based on a study published by Rodriguez-Navarro et al. [10]. A portlandite putty is immersed in aqueous carbonic acid, which causes the conversion of $Ca(OH)_2$ to $CaCO_3$, via a displacement reaction in the solid phase. It is this solid displacement reaction which makes the study particularly interesting. In our experiments, the putty is made by merely mixing agar powder with solid $Ca(OH)_2$ in a water-phenolphthalein medium, and heating the mixture to 85 °C to initiate agarose gelation.

3.1 *Experimental Procedure*

A 7.00 g sample of $Ca(OH)_2$ was dissolved in 700 mL of double distilled water with the desired amount of agar powder, yielding 0.135 M $Ca(OH)_2$ solid suspension in 7.0% w/w gel. The mixture was stirred for 40 min until a uniform suspension was obtained. The mixture was then heated to 85 °C after adding 30 ml of phenolphthalein, under continued stirring for around 20 min, then poured into a 3D methacrylate glass cuboid ($9 \times 9 \times 9\,cm^3$), filling it to the rim. The solution was left to gel for eight hours at 18 °C. After gelation, the cube bloc was removed from the cavity then immersed in 250 ml of 0.10 M H_2CO_3, and left uncovered and exposed to air at 18 °C.

3.2 *Result*

The above gel-$Ca(OH)_2$ putty is intended to simulate a portlandite rock exposed to infiltrating carbonated water. The putty is initially all pink (pH = 12.1). The time evolution of the putty is depicted in Fig. 2. Diffusion of H_2CO_3 in the gel causes the displacement reaction equilibrium:

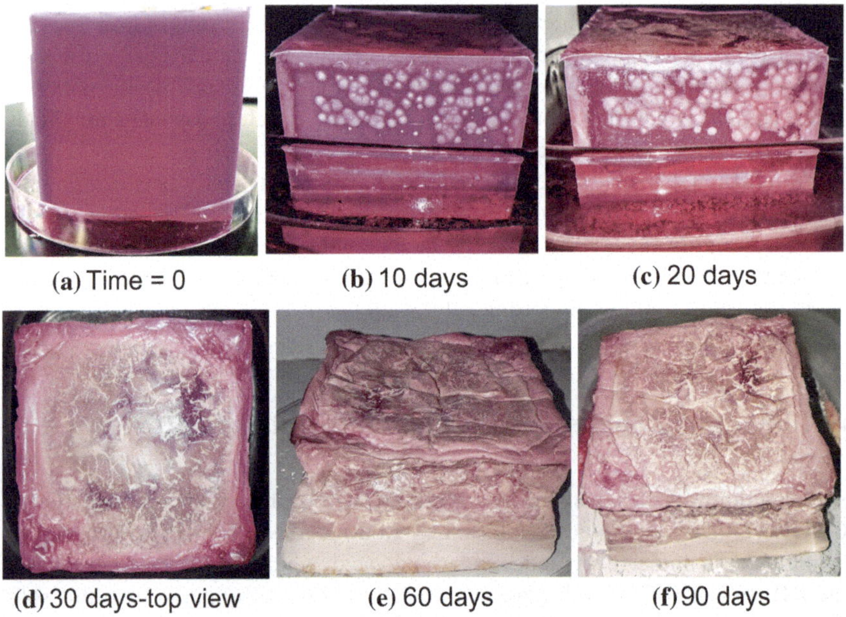

Fig. 2 Diffusion of H_2CO_3 (carbonating water) into a portlandite lime gel putty. White bands and spheroids consist of the calcite mineral ($CaCO_3$) resulting from a carbonation reaction in the solid phase. **a–c** advancement of the diffusion front from the solution into the putty and $CaCO_3$ formation. **d–f** rock evolution after removal from the solution

$$H_2CO_3(aq) + Ca(OH)_2(s) \rightleftharpoons CaCO_3(s) + 2H_2O(l)$$

The carbonation of the putty starts by the formation of a white zone behind the diffusion front (see bottom of the cube in frames 2b and 2c). At later times, white spherulites appear in different regions of the putty (upper zones in frames 2b and 2c.), resembling the orbicular texture of some rocks. The departure from the banded texture (Liesegang bands were rather expected to form) is attributed to the nonhomogeneous distribution of the $Ca(OH)_2$ particles which inevitably aggregate in a uniform sol at the bottom, due to gravity.

In upper zones, the $Ca(OH)_2$ particle distribution in the gel becomes more erratic, thus simulating the texture of a porous rock. Such random distribution disrupts the horizontality of the H_2CO_3 diffusion front, and random percolation becomes dominant. This feature is under present investigation for a more rigorous characterization. Samples from the pink regions and from the white spheroids were taken and subjected to freeze drying, after the gel and the phenolphthalein were washed off with hot double distilled water and the suspension centrifuged at 4000 rpm. The white powders were analyzed by FTIR spectroscopy. The spectra exactly identified $Ca(OH)_2$ (pink regions) and $CaCO_3$ (spheres). Figure 3 exhibits scanning electron micrographs (SEM) of the pink and white regions, with a notable indication of distinctly different crystallinities.

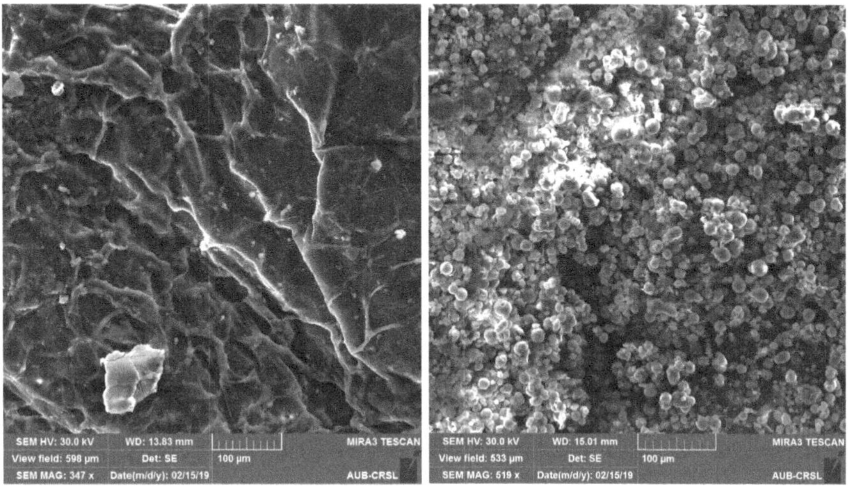

Fig. 3 Scanning electron micrographs of the pink (left) and white (right) regions, delineating between two distinct crystallinities. Left is $Ca(OH)_2$. Right is $CaCO_3$

Frames d–f of Fig. 2 display the evolution of the putty after removal from the bathing solution. A rock like appearance with a mixture of portlandite ($Ca(OH)_2$; pink zones) and calcite ($CaCO_3$; white zones) was attained.

4 Chaos in a 2D $PbCrO_4$ System

Liesegang systems display periodic patterns appearing as bands in a 1D tube, or concentric rings in 2D. In the latter case, diffusion of an outer solution of a co-precipitate ion (say I^-) occurs from a central source into a thin layer of gel containing the other co-precipitate ion (say Pb^{2+}) to yield concentric rings of the sparingly soluble lead iodide, PbI_2. The spacing between the consecutive rings increases with the spatial spread, i.e. as we go away from the circular central junction; the so-called *direct spacing* trend [15, 16]. Likewise, the band width increases with space [16]. In some other systems such as $PbCrO_4$ (studied here), The reverse tendency is observed, a trend known as *revert spacing* [17]. Our aim in the present study is to target patterns that are not periodic, but *aperiodic* in the sense that they do not obey neither the spacing nor the width laws.

4.1 Experimental Procedure

A sample of $Pb(NO_3)_2$ powder was weighed and dissolved in 100 mL of double distilled water to yield 0.0010 M solution of the salt. Difco Bacto agar powder was added to the mixture (1% w/w), and the latter was heated to 85 °C under constant stirring until a homogeneous gel solution was obtained. The gel melt was then poured into a set glass Petri dishes of 15 cm diameter, filling each to the two-thirds in height (around 80 ml of inner solution). The solution was left to gel for 24 h at room temperature.

Afterwards, a 0.20 M K_2CrO_4 solution was delivered at a continuously controlled flow rate in the hollow cavity at the center of the lead nitrate doped gel (see frames a, c, and e of Fig. 4). The experiment was performed while the dish remained permanently exposed to sunlight. This framework is a variant from the traditional Liesegang experiment wherein an initially constant diffusion source aliments the gel. Here, the outer electrolyte is delivered continuously, though at a very slow flow rate, constituting what we call a CFUR (continuously fed unstirred reactor).

4.2 Result

Figure 4 highlights three experiments at three different flow rates: 0.20, 0.067 and 0.025 ml/day. We clearly notice the departure from a periodic pattern toward a gradually increasing chaotic trend, with extremely high sensitivity to initial conditions (here the flow rate). Frames b, d and f of Fig. 4 display the map of bands for a clearer comparison. With these promising starting results, we pursue the study for a more systematic characterization.

5 Conclusions

In this paper, we presented different aspects of diffusion-reaction systems, in a variety of experimental frameworks yielding pattern forming structures with rich dynamical scenarios. In part 1 (Sect. 2) we obtained a panorama of mosaic patterns with multiple diffusion sources. In part 2 (Sect. 3), we simulated the dynamics in a real rock, through an analogy with a laboratory prepared gel putty containing the portlandite mineral. The similarity between the reaction-diffusion gel method and geochemical self-organization is striking. In the last part (Sect. 4) we initiated a study of $PbCrO_4$ precipitate patterns in a 2D CFUR, exploring the transition from periodic to chaotic behavior.

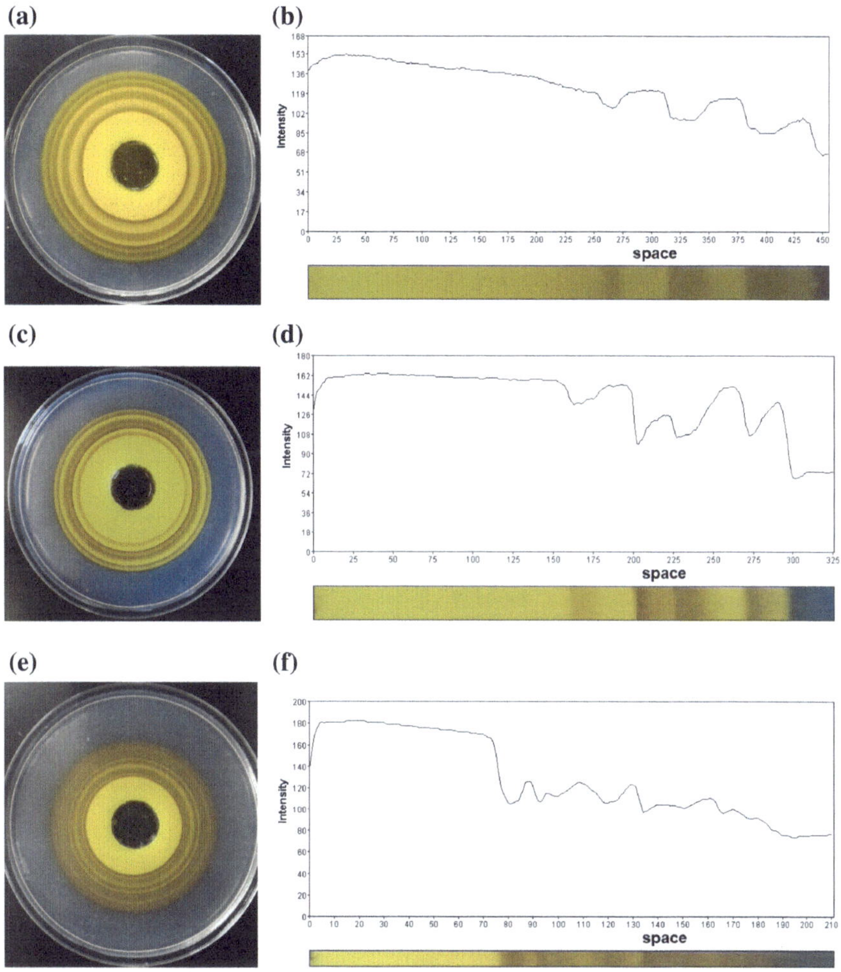

Fig. 4 2D $PbCrO_4$ patterns with the corresponding map of bands. **a**–**b** flow rate = 0.2 ml/day; **c**–**d** flow rate = 0.067 ml/day; **e**–**f** flow rate = 0.025 ml/day. Chaotic trend increasing with decreasing flow rate

Acknowledgements This study was supported by a University Research Board (URB) grant of the American University of Beirut (AUB). X-ray diffraction, IR spectroscopy, freeze drying and SEM were performed at the Central Research Science Lab (CRSL) of the AUB.

References

1. P. Ball, *The Self-Made Tapestry: Pattern Formation in Nature* (Oxford University Press, Oxford, 1999)
2. I. Lagzi (ed.), *Precipitation Patterns in Reaction-Diffusion Systems* (Research Signpost publications, Trivandrum, Kerala, 2011)
3. R.E. Liesegang, Chemische Fernwirkung, *Lieseg. Photograph. Arch.* **37**, 305 (1896); continued in **37**, 331 (1896)
4. R.E. Liesegang, *Geologische Diffusionen* (Steinkopff, Dresden, 1913)
5. R.E. Liesegang, *Die Achate* (Steinkopf, Dresden-Leipsig, 1915)
6. B. Jamtveit, P. Meakin (eds.), *Growth, Dissolution and Pattern Formation in Geosystems* (Kluwer, Dordrecht, 1999)
7. P. Ortoleva, *Geochemical Self-Organization* (Oxford University Press, New York, 1994)
8. A.D. Fowler, I. L'Heureux, Self-organized banded sphalerite and branching galena in the Pine Point ore deposit Northwest Territories. Can. Miner. **34**, 1211–1222 (1996)
9. J.H. Kruhl (ed.), *Fractals, and Dynamic Systems in Geoscience* (Springer, Berlin, 1994)
10. C. Rodriguez-Navarro, O. Cazalla, K. Elert, E. Sebastian, Liesegang pattern development in carbonating traditional lime mortars. Proc. R. Soc. Lond. A **458**, 2261–2273 (2002)
11. R. Sultan, A. Abdel-Rahman, On dynamic self-organization: examples from magmatic and other geochemical systems. Latin Am. J. Solids Struct. (LAJSS) **10**, 59–73 (2013)
12. M. Msharrafieh, M. Al-Ghoul, F. Zaknoun, H. El-Rassy, S. El-Joubeily, R. Sultan, Simulation of geochemical banding I: acidization-precipitation experiments in a ferruginous limestone rock. Chem. Geol. **440**, 42–49 (2016)
13. H. Batlouni, H. El-Rassy, M. Al-Ghoul, Cosynthesis, coexistence, and self-organization of α-and β-Cobalt hydroxide based on diffusion and reaction in organic gels. J. Phys. Chem. A **112**(34), 7755–7757 (2008)
14. T. Karam, H. El-Rassy, F. Zaknoun, Z. Moussa, R. Sultan, Liesegang banding and multiple precipitate formation in cobalt phosphate systems. Chem. Phys. Lett. **525–526**, 54–59 (2012)
15. C.K. Jablczynski, Mémoires présentes à la société chimique. Les anneaux de Liesegang. Bull. Soc. Chim. France **33**, 1592–1602 (1923)
16. M. Droz, J. Magnin, M. Zrinyi, Liesegang patterns: studies on the width law. J. Chem. Phys. **110**(19), 9618–9622 (1999)
17. T. Karam, H. El-Rassy, R. Sultan, Mechanism of revert spacing in a $PbCrO_4$ Liesegang system. J. Phys. Chem. A **115**(14), 2994–2998 (2011)

Modulating the Light-Driven Conductivity in Biosystem

S. Fathizadeh, E. Javan Shour, S. Behnia and F. Nemati

Abstract Biosystems has been proved powerful for use in nanotechnology because of its predictable nanometer-sized structures. One important example is that we can use biosystems to develop the new two- and three-dimensional nanomachines and nanocircuits. To achieve this goal, people hope to control the conductivity properties of system. In this regard, light-regulated circumstances come into play. Photon-induced charge transport can regulate the conductivity properties of system. The role of photoresponse of π electrons of biosystems in light-driven conductivity theoretically studied. We have varied the intensity and frequency of light and investigated their effect on the charge transport properties of system. Therefore, one can modulate the conductivity properties of system via the directly modulation of the light.

1 Introduction

In the recent years, studying the incident light effect on the biological systems has attracted a lot of attentions. Organic materials are easily produced and compatible with biological systems and are easily simulated. Therefore, organic materials are used instead of inorganic materials in the industry [1]. In this regard, Deoxyribonucleic acid (DNA) has been proved powerful for using in nanotechnology because of its predictable nanometer-sized structure [2]. For an example, one can use DNA to develop two- and three-dimensional Nano-machines by designing its nucleotide sequence [2]. To achieve this goal, people hope to control the conductivity properties of the system.

In this paper, a combined model of classical Peyard-Bishop-Dauxois (PBD) [3, 4] and quantum tight-binding (two-leg charge ladder) models [1] is proposed for studying the photo induced charge transfer in DNA (Fig. 1).

12th CHAOS Conference Proceedings, 18–22 June 2019, Chania, Crete, Greece

S. Fathizadeh · E. Javan Shour · S. Behnia (✉) · F. Nemati
Department of Physics, Urmia University of Technology, Urmia, Iran
e-mail: s.behnia@sci.uut.ac.ir

C. H. Skiadas and Y. Dimotikalis (eds.), *12th Chaotic Modeling and Simulation International Conference*, Springer Proceedings in Complexity,
https://doi.org/10.1007/978-3-030-39515-5_7

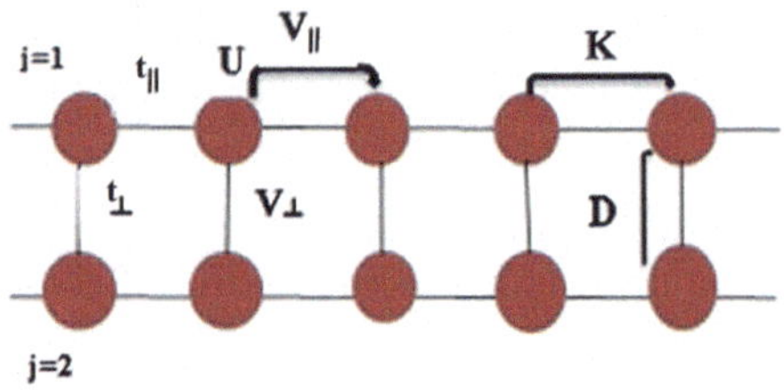

Fig. 1 Schematic view of the new model and applying the incident laser pulse

In this regard, light-regulated circumstances come into play. Photon-induced charge transport can regulate the conductivity properties of system. The role of photo-response of π electrons of bio-systems in light-driven conductivity theoretically studied. We have applied a potential difference to the system via the metal left and right leads and obtained the current-voltage characteristic diagram. We have varied the intensity and frequency of light and investigated their effect on the charge transport properties of system. We can modulate the incident light to gain the appropriate charge transport in our bio-system. Therefore, one can modulate the conductivity properties of system via the directly modulation of the light.

2 Model and Method

We have start with description of the system (Fig. 2) Hamiltonian represented as follows:

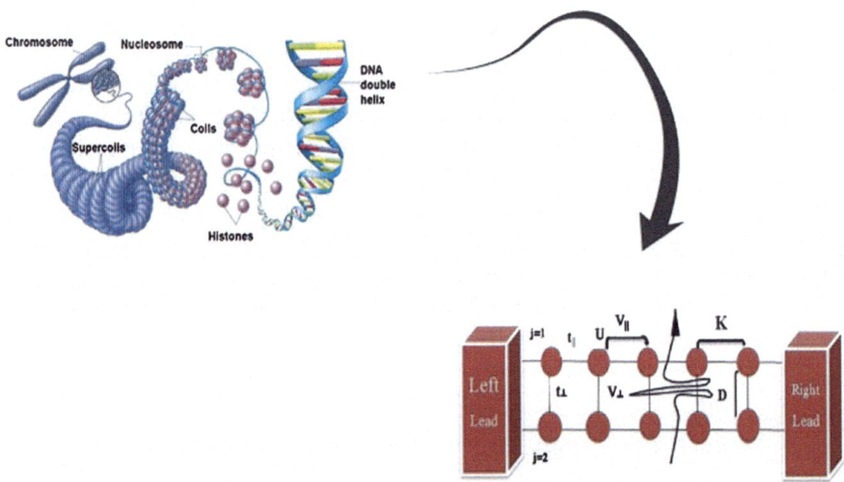

Fig. 2 The schematic illustration of the physical model of light-driven in DNA chain

$$\mathrm{H} = \mathrm{H}_{PBD} + \mathrm{H}_{cl} + \mathrm{H}_{E} + \mathrm{H}_{lead} + \mathrm{H}_{lead-DNA} \tag{1}$$

where H_{PBD} determine the mechanical evolution of base pairs in DNA as follows [3, 5]:

$$\mathrm{H}_{PBD} = \sum_i \frac{1}{2} m \dot{y}_i^2 + V(y_i) + W(y_i + y_{i+1}) \tag{2}$$

Here, m is the mass base pairs, $V(y_i) = D_i\left(e^{-a_i y_i} - 1\right)^2$ is the Morse Potential which indicates the effect of interaction between complementary bases with a_n as the width of the Morse potential and D_n as its depth.

$W(y_i, y_{i+1}) = \frac{K}{2}(1 + \rho \exp(-b(y_i + y_{i+1})))(y_{i+1} - y_i)^2$ is the stacking interaction of neighboring base pairs in the chain, where K is the coupling constant, ρ is the stiffness parameter, and b is the damping coefficient

We have considered Hamiltonian H_{cl} to express the charge dynamic in DNA as following [2]:

$$\begin{aligned} H_{cl} = & \sum_{j=1,2} \sum_i \varepsilon_{j,i} \left(c_{j,i}^{\dagger} c_{j,i} + \mathrm{H.}c. \right) - t_{\parallel} \sum_{j=1,2} \sum_i \left(c_{j,i}^{\dagger} c_{j,i+1} + \mathrm{H.}c. \right) \\ & - t_{\perp} \sum_i \exp(-p y_i) \left(c_{1,i}^{\dagger} c_{2,i} + \mathrm{H.}c. \right) + \frac{U}{4} \sum_{j=1,2} \sum_i n_{j,i}^2 \\ & + V_{\parallel} \sum_{j=1,2} \sum_i n_{j,i} n_{j,i+1} + V_{\perp} \sum_i \exp(-\chi y_i) n_{1,i} n_{2,i} \end{aligned} \tag{3}$$

where, $\varepsilon_{i,j}$ represents the on-site energy on base i of strand j. where $c_{j,i}{}^{\dagger}$ $(c_{j,i})$ indicates the π electron creation (annihilation) operator on base i of strand j. y_i represents the ith H-bond stretching, $t_{\perp}$ represents the intrastrand hopping integral between the nearest-neighboring bases; and $t_{\perp}$ describes the intra-base-pair hopping integral [1]. The exponential function in the $t_{\perp}$ term with the cutoff constant p can describe the exponential tails of the H-bond wave functions during dissociation [1]. $V_{\perp}$ is the Coulomb interaction between the complementary bases, $V_{\parallel}$ is the Coulomb interaction between the adjacent bases and χ is the inverse Debye screening length [1]. Furthermore, the H_E part is expressed as follow [1]:

$$\mathrm{H}_E = -t_{\parallel} \sum_{j=1,2} \sum_i \left[\exp\left(\frac{i e r_0}{\hbar c} A(t) \right) c_{j,i}^{\dagger} c_{j,i+1} + \mathrm{H.}c. \right] \tag{4}$$

where r_0, e, c and

$$A(t) = A_0 \cos(\omega t) \exp\left[-\frac{(t - t_c)^2}{2\tau^2} \right]$$

are the constant distance between the adjacent bases along the strand, elementary electric charge, the light velocity and the vector potential, respectively [1]. A_0, ω, t_c, and τ represents the amplitude, the oscillation frequency, the center, and the half-width of the pulse, respectively [1]. In this case DNA is connected to the left and right metal leads [6], (Fig. 2). The leads are connected to the voltage bias. H_{lead} is known as follows [6]:

$$H_{lead} = \sum_{j=1,2}\sum_{k}(\varepsilon_{Ljk} + eV_b/2)a^{\dagger}_{Ljk}a_{Ljk} + \sum_{j=1,2}\sum_{k}(\varepsilon_{Rjk} - eV_b/2)a^{\dagger}_{Rjk}a_{Rjk} \quad (5)$$

Here, by knowing that, $\beta = L, R$, $\varepsilon_{\beta jk}$ is the on-site energy of lead and $a_{\beta jk}$ ($a_{\beta jk}{}^{\dagger}$) is the creation (annihilation) operator of electron in site (j; k) and V_b is the bias voltage. At the end, we have $H_{lead\text{-}DNA}$, so it is defined as follows [6]:

$$H_{lead-DNA} = \sum_{j=1,2}\sum_{k}\left(t_L a^{\dagger}_{Ljk}c_{j,1} + t_R a^{\dagger}_{Rjk}c_{j,N}\right) \quad (6)$$

In the interaction terms, t_β is the hopping constant from a site k within the lead β to the DNA chain [6].

Finally, in this paper, we have tried to use nonlinear dynamics theory for system analysis, so in this regard, we use ($\dot{a}_n = \frac{-i}{\hbar}[a_n, H]$) Heisenberg approach to determine the dynamics of charges. The evolution equations are formulated as follows:

Then we have

$$\begin{aligned} \ddot{y}_i =& \frac{2a_i D_i}{m}\exp(-a_i y_i)(\exp(-a_i y_i) - 1) + \frac{kb\rho}{2m}\left[\exp(-b(y_i + y_{i-1}))\right. \\ & \left.(y_i - y_{i-1})^2 + \exp(-b(y_{i+1} + y_i))(y_{i+1} - y_i)^2\right] \\ & - \frac{k}{m}[(1 + \rho\exp(-b(y_i + y_{i-1})))(y_i - y_{i-1}) \\ & - (1 + \rho\exp(-b(y_{i+1} + y_i))(y_{i+1} - y_i)) \\ & - \frac{2pt_\perp}{m}\exp(-py_i)\left|\left\langle c^{\dagger}_{1,i}\ c_{2,i}\right\rangle\right| + \frac{V_\perp}{m}\chi\exp^{(-\chi y_i)}\left(\left|\langle n_{1,i}\ n_{2,i}\rangle\right| - \frac{\left|\left\langle c^{t}_{1,i}\ c_{2,i}\right\rangle\right|^2}{2}\right) \end{aligned} \quad (7)$$

$$\begin{aligned} i\hbar\dot{c}_{j,i} =& \varepsilon_{j,i}c_{j,i} + V_{||}c_{j,i}(n_{j,i+1} + n_{j,i-1}) \\ & - t_{||}(c_{j,i+1} + c_{j,i-1}) - t_\perp\exp(-py_i)(c_{2,i} + c_{1,i}) \\ & + V_\perp\exp(-\chi y_i)[n_{1,i}c_{2,i} + n_{2,i}c_{1,i}] \\ & + \frac{U}{4}(n_{j,i} + H.c)c_{j,i} + \sum_k(\delta_{1,i}t_L a_{Ljk} + \delta_{N,i}t_R a_{Rjk}) \end{aligned} \quad (8)$$

$$i\hbar\dot{a}_{Ljk} = (\varepsilon_{Ljk} + eV_b/2)a_{Ljk} + t_L c_{j,1} \quad (9)$$

$$i\hbar\dot{a}_{Rjk} = (\varepsilon_{Rjk} - eV_b/2)a_{Rjk} + t_R c_{j,N} \quad (10)$$

We have used the following parameters values in our model (Tables 1 and 2).

Table 1 The electron hopping constants between different base-pairs [5]

Base-pair sequence	Electron hopping constant
AA, TT	−29
AT	0.5
AG, CT	3
AC, GT	32
TA	2
TG, CA	17
TC, GA	−1
GG, CC	20
GC	−10
CG	−8

Table 2 The on-site energies for two possible base-pairs [5]

Base-pair sequence	ε (eV)
A-T	−4.9
C-G	−4.5

3 Results

To analyze the light transfer mechanism in DNA and investigate the affected parameters, we have tried to obtain the electrical current flowing through DNA. The local electrical current can be defined as $I_i - I_{i-1} = e\frac{dn_i}{dt}$, where $n_i = c_i^\dagger c_i$. So we have the current equation as follows:

$$
\begin{aligned}
I = \frac{ie}{\hbar}\sum_i \sum_{j=1,2} [&t_{\|} \exp(\frac{ier_0}{\hbar c}A(t))(c^\dagger_{j,i+1}c_{j,i} - c^\dagger_{j,i-1}c_{j,i}) \\
&+ t_\perp \exp(-py_i)(c^\dagger_{2,i}c_{1,i} - c^\dagger_{1,i}c_{2,i}) \\
&+ \sum_k (t_L(c^\dagger_{j,1}a_{Ljk} - a^\dagger_{Ljk}c_{j,1}) \\
&+ t_R(-c^\dagger_{j,N}a_{Rjk} + a^\dagger_{Rjk}c_{j,N}))]
\end{aligned}
$$

We have investigated the effect of parameters, such as frequency and amplitude of incident light, gate voltage, and temperature.

3.1 The Incident Light Parameters

The frequency as well as amplitude of incident light are the important factors for studying the light driven conductivity. In this regard, we have varied the frequency and amplitude of light and studied the electrical current through DNA. Figure 3 shows the variation of the current with respect to the incident light frequency. It is appeared that for a certain frequency, the current reaches the maximum value. It can also be expressed at $\omega = 3(\mathrm{THz})$, $\omega = 3.5(\mathrm{THz})$, there is almost no current, so we can obtain a range of light-frequency in which the maximum current flows through the system.

On the other hand, variation of the amplitude of light have the important effect on the DNA charge transport (Fig. 4). We observe that the highest value for the current is at $A_0 = 0.6(hc/er_0)$.

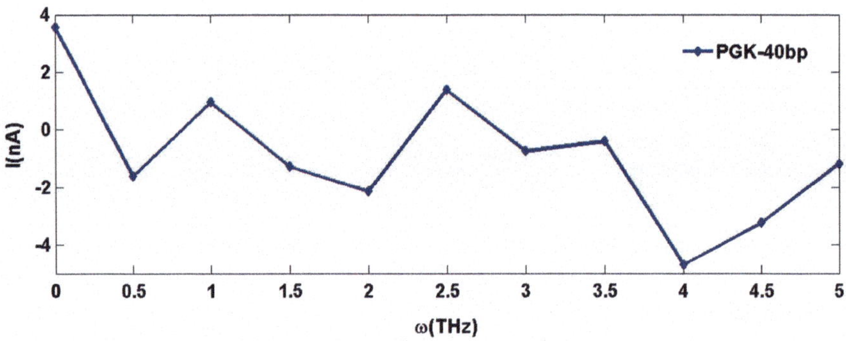

Fig. 3 The electrical current with respect to the radiation light frequency

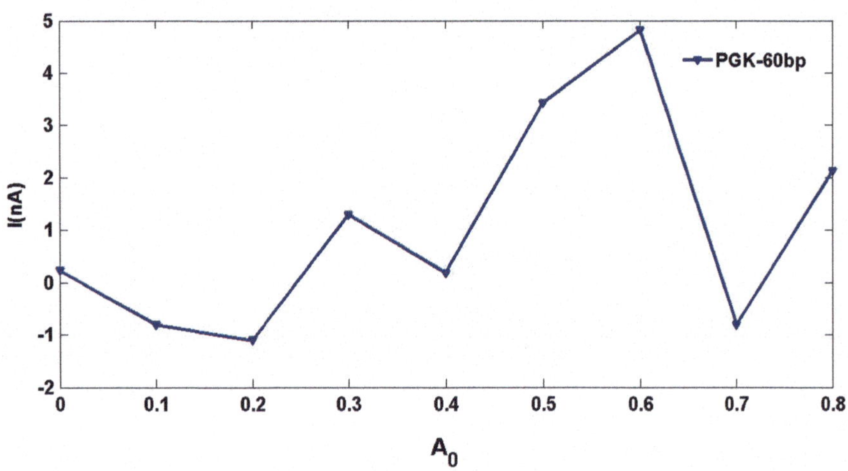

Fig. 4 The electrical current at different amplitudes of light

3.2 The DNA Sequence Effect

The DNA molecule is capable of exhibiting conductivity, semi-conductivity and isolation behavior. We have tried to examine the sequence effect on light driven conductivity in DNA. We have used the DNA sequences in Table 3. Figure 5, shows the time series of the electrical current for different sequences. According to the Fig. 5, the current amplitude increases when the DNA sequence is longer.

3.3 The Gate Voltage Effect

In the current study, we have tried to investigate the effect of an electrical gating on light-driven conductivity of DNA. We put DNA chain in the contact of two metal leads and applied the gate voltage via the leads. We can determine the quasi-ohmic and negative differential resistant (NDR) regions in the I-V characteristic diagram (Fig. 6). It is shown that at a constant frequency, quasi-Ohmic resistance varies by increasing the applied potential. On the other hand, the width of NDR regions is reduced by increasing the voltage.

3.4 The Temperature Effect

The temperature of environment is the effective factor for studying the DNA charge transport. In this regard, we have used a Nosé -Hoover thermostat as a heat source in DNA. So we have connected DNA to a thermostat. The dynamics of the thermostat is determined as [4]:

$$\zeta = \frac{1}{M}\sum_{i} m\dot{y}_i^2 - N\mathrm{K_B}T_0 \tag{12}$$

M is the constant of Nosé–Hoover thermostat that has been set to $M = 1000$, and T_0 is the temperature of thermostat. Also, $\mathrm{K_B}$ is the Boltzmann constant.

We have examined a various temperature range. We have started from 300 K until 355 K (the DNA denaturation temperature). It is shown that at $T = 320(\mathrm{K})$, the electrical current flowing through DNA is maximum (Fig. 7). So we can obtain a range of temperature in which, the maximum current flows through DNA.

Table 3 The DNA sequence used in calculation [5, 6]

Name	Number of base pair	DNA sequence
Ch22	60	AGGGCATCGCTAACGAGGTCGCCGTCCACAGCATCGCTATCGAGGACACCACACCGTCCA
Ch22	40	CAATGCAGTCTATCCACCTGACGGACCCCGACCCGCCTTT
PGK	60	GTCGGACCAGAAACAACGAAGCAGTTCGCTGAGGTGGTTGGACGCGCCAAGACGATTGTG

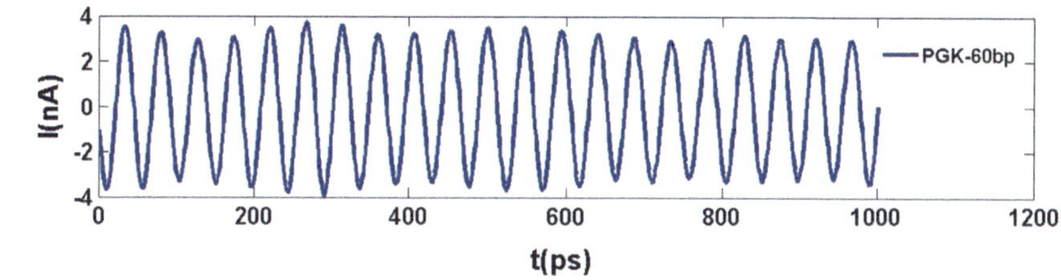

Fig. 5 The electrical current for different sequences

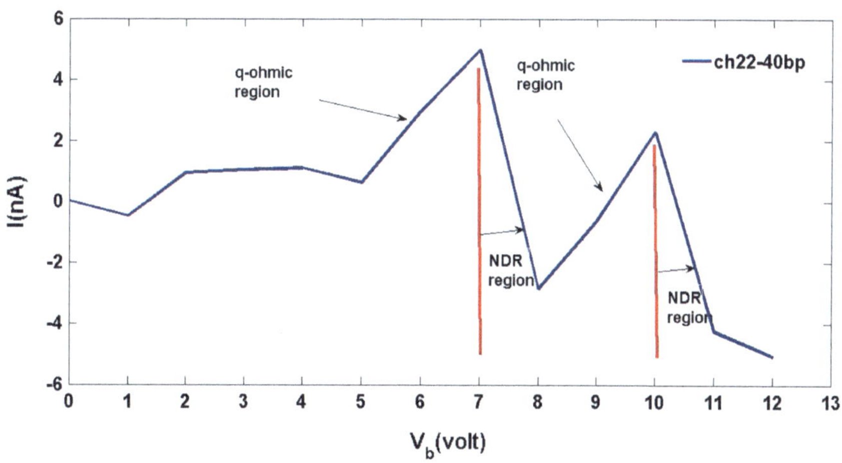

Fig. 6 I-V characteristic diagram for ch22 sequence

4 Conclusions

In the current work, we have tried to study the light driven charge transport in DNA chains. In this regard, we have study the effect of impressive factors to modulate the DNA conductivity. The frequency and amplitude of incident light are important factors to investigate radiation dependent DNA charge transport. We have varied the frequency as well as amplitude of light to determine the range of these parameters in which the maximum electrical current follows through DNA. The DNA sequence

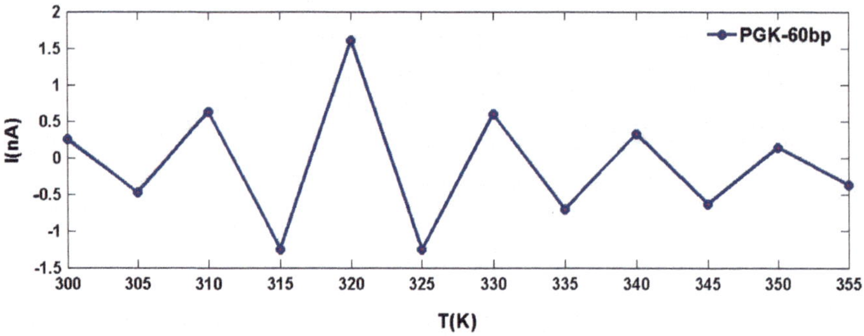

Fig. 7 The electrical current in different temperatures

and temperature of environment are the other factors for modulating the DNA charge transport. We can determine the DNA sequence and the temperature which the current flowing through DNA is maximum. On the other hand, the I-V characteristic diagram shows the quasi-ohmic and NDR regions observed in the previous works.

References

1. L.L. Zhang, S.J. Xie, D.W. Kang, Role of photoresponse of π electrons in light-driven DNA dissociations. Phys. Rev. E **96**(2), 022414 (2017)
2. T.A. Brown, *Gene Cloning and DNA Analysis: An Introduction* (Wiley, 2016)
3. T. Dauxois, M. Peyrard, A.R. Bishop, Phys. Rev. E **47**(1), R44 (1993)
4. S. Behnia, F. Nemati, S. Fathizadeh, Modulation of spin transport in DNA-based nanodevices by temperature gradient: a spin caloritronics approach. Chaos, Solitons Fractals **116**, 8–13 (2018)
5. S. Behnia, S. Fathizadeh, Spintronics in nano scales: an approach from DNA spin polarization. Scientia Iran. **24**(6), 3448–3451 (2017)
6. S. Behnia, S. Fathizadeh, J. Ziaei, Controlling charge current through a DNA based molecular transistor. Phys. Lett. A **381**(1), 36–43 (2017)

Approaches to Estimating the Dynamics of Interacting Populations with Impulse Effects and Uncertainty

Tatiana F. Filippova and Oxana G. Matviychuk

Abstract The problem of estimating reachable sets for nonlinear control systems of Lotka-Volterra type which describe the dynamics of the interaction of predators and their preys under uncertainty conditions is studied. It is assumed that we know only the restrictive set for unknown quantities and there is no additional information (for example, probabilistic, statistical, etc.) on unknown values. Applying the latest results from set-valued analysis we find the external ellipsoidal estimates of corresponding reachable sets in two cases: for systems with classical measurable control and for measure-driven (impulsive control) systems. The models under consideration can describe the behaviour of competing firms, population growth, environmental change, development of individual industries, etc. The results of modeling and numerical simulations based on proposed methods are included to illustrate the main ideas and algorithms.

Keywords Control systems · Nonlinearity of quadratic type · Uncertainty · Impulse control · Ellipsoidal calculus · Tube of trajectories

1 Introduction

The problem of state estimation for the control systems under conditions of uncertainty in initial system states is studied in this paper. The case is investigated here when the probabilistic data of noise and possible errors is not available, and only

12th CHAOS Conference Proceedings, 18–22 June 2019, Chania, Crete, Greece.

T. F. Filippova (✉)
Ural Federal University and Krasovskii Institute of Mathematics and Mechanics,
Ural Branch of Russian Academy of Sciences, Ekaterinburg, Russian Federation
e-mail: ftf@imm.uran.ru

O. G. Matviychuk
Krasovskii Institute of Mathematics and Mechanics, Ural Branch of Russian Academy
of Sciences and Ural Federal University, Ekaterinburg, Russian Federation
e-mail: vog@imm.uran.ru

C. H. Skiadas and Y. Dimotikalis (eds.), *12th Chaotic Modeling and Simulation International Conference*, Springer Proceedings in Complexity,
https://doi.org/10.1007/978-3-030-39515-5_8

some sets restricting unknown quantities are given. Models of this type may be found in many applied problems including physics, economics, biology, ecology, etc. especially when we try to find a mathematical model that takes into account the absence of probabilistic data on uncertain values. For such problems the so-called set-membership estimation approach is used when the mathematical model concerns with a whole set of feasible parameters, consistent with the model structure, with bounded uncertainty, with special types of measurements characterization etc. (see e.g. Kurzhanski and Varaiya [1], Kurzhanski and Filippova [2], Chernousko [3]).

The general schemes to solve such classes of control and estimation problems with set-membership uncertainty are based on the construction and on the analysis of the corresponding reachable sets or their analogs. Effective estimates of reachable sets for models with linear dynamics and for some special classes of nonlinear systems under such set-membership uncertainty (including results basing on differential inclusions theory) were developed earlier. In this paper we study the procedures of upper estimating reachable sets for nonlinear control systems of Lotka-Volterra type. We use here the ideas and results of state estimation theory developed for nonlinear control systems which have a special quadratic dynamical structure (Filippova [4], Filippova and Matviychuk [5], Matviychuk [6]). The case under present study is more complicated than considered before because quadratic terms in differential equations of the control system are arbitrary (not obligatory positive definite as in previous studies). Therefore several new schemes of the problem solution are developed here.

We prove here theoretical results and formulate related numerical algorithms for constructing external ellipsoidal estimates of reachable sets for nonlinear uncertain control systems. Numerical examples and results of related simulations are included to illustrate the basic ideas and results. The approaches and results presented in the paper can be used in applications as a part of the complex solutions of the model predictive control schemes (Allgower and Zheng [7]) for correct modeling and accounting for poorly predicted errors and disturbances.

2 Main Problem Description

First, we give a short list of the most important basic notations. Here R^n denotes the n–dimensional Euclidean space, $compR^n$ is the set of all compact subsets of R^n, $R^{n\times n}$ denotes the set of all $n \times n$–matrices and $x'y = (x, y) = \sum_{i=1}^{n} x_i y_i$ is the usual inner product of $x, y \in R^n$ with prime as a transpose, the Euclidean norm is $\|x\| = (x'x)^{1/2}$.

The symbol $B(a, r)$ denotes the ball in R^n, $B(a, r) = \{x \in R^n : \|x - a\| \leq r\}$, I is the identity $n \times n$-matrix, $diag\{a_1, \ldots, a_n\}$ means the diagonal $n \times n$–matrix with elements a_i at the main diagonal and with zero entries outside this diagonal.

The ellipsoid $E(a, Q)$ in R^n (with a center $a \in R^n$ and a matrix Q) will be denoted as

$$E(a, Q) = \{x \in R^n : (Q^{-1}(x - a), (x - a)) \leq 1\}$$

where the $n \times n$–matrix Q is a symmetric positive definite. The trace of a $n \times n$–matrix Y (the sum of its diagonal elements) will be denoted as $Tr(Y)$.

Consider the dynamical system

$$\dot{x} = Ax + f(x) \cdot d + u(t), \quad x_0 \in X_0, \quad t_0 \leq t \leq T, \tag{1}$$

where $x, d \in R^n$, $\|x\| \leq K$ (the number $K > 0$ is given and is fixed). Here we assume that $f(x)$ is defined as a quadratic form with respect to state variable x, that is

$$f(x) = x'Bx. \tag{2}$$

We assume that the $n \times n$-matrix B is given and is symmetric and positive definite. We assume also that the control functions $u(t)$ in (1) are measurable (in the sense of Lebesgue) functions on $[t_0, T]$ and satisfy the so-called geometrical constraint

$$u(t) \in U, \quad t \in [t_0, T] \ (a.e.), \tag{3}$$

where the set $U \in comp R^m$ is given (here "a.e." means an abbreviation for the expression "almost everywhere" with respect to Lebesgue measure). The matrix A in (1) is assumed to be given.

The bounding set X_0 for initial system states x_0 in (1) is an ellipsoid, $X_0 = E(a_0, Q_0)$, where Q_0 is a symmetric and positive definite matrix and a_0 is the ellipsoid's center.

For any initial state $x_0 \in X_0$ and each admissible control $u(\cdot)$ the related absolutely continuous function $x(t) = x(t; u(\cdot), x_0)$ will denote a solution to the dynamical system (1)–(3). We need to define the following main notion.

Definition 1 For a fixed instant of time t ($t_0 < t \leq T$) the set $X(t)$ at time t ($t_0 < t \leq T$) defined as follows

$$X(t) = \{ x \in R^n : \exists\, x_0 \in X_0, \exists\, u(\cdot) \in U, \text{ such that the equality holds}$$
$$x = x(t) = x(t; u(\cdot), x_0) \}, \quad t_0 < t \leq T,$$

will be called *the reachable set* at time t ($t_0 < t \leq T$) of the system (1)–(3).

The main goal of the paper is to construct the external (with respect to the inclusion of sets) ellipsoidal estimate $E(a^+(t), Q^+(t))$ ($t_0 < t \leq T$) of the reachable set $X(t)$ ($t_0 < t \leq T$), so we need to solve first the following problem.

Problem 1 Find the exact description of the ellipsoid $E(a^+(t), Q^+(t))$ such that the inclusion is true

$$X(t) \subseteq E(a^+(t), Q^+(t)), \quad t_0 < t \leq T.$$

The second goal of the paper consists in concretizing and analyzing the result based on the characteristics of the studied control differential system, which describes the interaction of the predator—prey relationship.

Problem 2 Apply the main estimating approach and the solution of the Problem 1 to find the upper bounds of reachable sets for nonlinear control systems of Lotka-Volterra type.

3 Preliminary Results

We shortly remind in this section some auxiliary results which will be used further. These results concern reachable sets for several special classes of uncertain dynamical systems.

3.1 *Control Systems with Uncertainty*

Consider the control system

$$\dot{x} = Ax + f(x) \cdot d + u(t), \quad x_0 \in X_0 = E(a_0, Q_0), \quad t_0 \leq t \leq T. \tag{4}$$

Here $u(t) \in U = E(\hat{a}, \hat{Q})$, n–vectors $a_0, \hat{a}, d$ are fixed. Suppose that $f(x) = x'Bx$ and Q_0, B and $\hat{Q}$ are positive definite and symmetric $n \times n$–matrices.

Consider the matrix $B^{1/2}Q_0B^{1/2}$ and let k^2 denote its maximal eigenvalue. It is not difficult to find out that k^2 is the smallest number for which we have $X_0 \subseteq E(a_0, k^2B^{-1})$. The following result gives the upper ellipsoidal estimate of the studied set $X(t)$ $(t_0 \leq t \leq T)$.

Theorem 1 (Filippova [4]) *For any $t \in [t_0, T]$ the following inclusion*

$$X(t) \subseteq E(a^+(t), r^+(t)B^{-1}) \tag{5}$$

is true, where $a^+(t)$, $r^+(t)$ satisfy the system of ordinary differential equations

$$\dot{a}^+(t) = Aa^+(t) + ((a^+(t))'Ba^+(t) + r^+(t))d + \hat{a}, \quad t_0 \leq t \leq T, \tag{6}$$

$$\begin{aligned} \dot{r}^+(t) = \max_{\|l\|=1} \Big\{ l'\big(2r^+(t)B^{1/2}(A^0 + 2d \cdot (a^+(t))'B)B^{-1/2} + \\ q^{-1}(r^+(t))B^{1/2}\hat{Q}B^{1/2}\big)l \Big\} + q(r^+(t))r^+(t), \\ q(r) = ((nr)^{-1}\mathrm{Tr}(B\hat{Q}))^{1/2}, \quad a^+(t_0) = a_0, \quad r^+(t_0) = k^2. \end{aligned} \tag{7}$$

Corollary 1 (Filippova [4]) *For sufficiently small $\sigma > 0$ the upper estimate for $X(t_0+\sigma) = X(t_0+\sigma; t_0, X_0)$ is true*

$$X(t_0+\sigma) \subseteq E(a^+(\sigma), Q^+(\sigma)) + o(\sigma)B(0,1), \tag{8}$$

with $\sigma^{-1}o(\sigma) \to 0$ when $\sigma \to +0$ and

$$a^+(\sigma) = a(\sigma) + \sigma\hat{a}, \quad a(\sigma) = a_0 + \sigma(Aa_0 + a_0'Ba_0d + k^2d), \tag{9}$$

$$\begin{gathered} Q^+(\sigma) = (p^{-1}+1)Q(\sigma) + (p+1)\sigma^2\hat{Q}, \\ Q(\sigma) = k^2(I+\sigma R)B^{-1}(I+\sigma R)', \quad R = A + 2d\cdot a_0'B, \end{gathered} \tag{10}$$

where $p > 0$ is the unique root of the equation

$$\sum_{i=1}^{n} \frac{1}{p+\alpha_i} = \frac{n}{p(p+1)}$$

with $\alpha_i \geq 0$ being the roots of the equation $|Q(\sigma) - \alpha\sigma^2\hat{Q}| = 0$ $(i = 1, \ldots, n)$.

Remark 1 The theorem and the corollary formulated above are some adaptations of the corresponding results of the paper Filippova [4] to the problem considered here, therefore their proofs are omitted.

Remark 2 Computational algorithms (together with some numerical examples) basing on Theorem 1 may be found in Filippova and Matviychuk [5].

3.2 Two Quadratic Forms in the Equations of Dynamics of a Control System

Consider the following control system with a more complicated type of nonlinearity,

$$\begin{gathered} \dot{x} = Ax + f^{(1)}(x)\cdot d^{(1)} + f^{(2)}(x)\cdot d^{(2)} + u(t), \quad t_0 \leq t \leq T, \\ u(t) \in U = E(\hat{a}, \hat{Q}), \quad x_0 \in X_0 = E(a_0, Q_0). \end{gathered} \tag{11}$$

Here $d^{(1)}$ and $d^{(2)}$ are n-vectors and $f^{(1)}$, $f^{(2)}$ are quadratic forms,

$$f^{(1)}(x) = x'B^{(1)}x, \quad f^{(2)}(x) = x'B^{(2)}x,$$

with different symmetric and positive definite $n \times n$–matrices $B^{(1)}$, $B^{(2)}$.

Remark 3 It is not difficult to see that it may not be possible to combine these two quadratic functions $f^{(1)}$ $f^{(2)}$ into one positive definite form because the coefficients (the coordinates of vectors $d^{(1)}$ and $d^{(2)}$) may be arbitrary (it may turn out that they are negative or zeros). As a consequence it is not possible to use here the approaches and results of previous investigations and we need to find some new approaches to get estimates of reachable sets of the system under study.

We need also one more auxiliary result.

Denote as k_i^2 $(i = 1, 2)$ the largest eigenvalue of the corresponding matrix $(B^{(i)})^{1/2} Q_0 (B^{(i)})^{1/2}$ $(i = 1, 2)$, therefore we have

$$\max_{z' B^{(1)} z \leq k_1^2} z' B^{(2)} z = k_1^2 \lambda_{12}^2, \quad \max_{z' B^{(2)} z \leq k_2^2} z' B^{(1)} z = k_2^2 \lambda_{21}^2, \tag{12}$$

where λ_{12}^2 and λ_{21}^2 are maximal eigenvalues of matrices $(B^{(1)})^{-1/2} B^{(2)} (B^{(1)})^{-1/2}$ and $(B^{(2)})^{-1/2} B^{(1)} (B^{(2)})^{-1/2}$, respectively.

Lemma 1 *The following upper estimate is true*

$$X_0 \subseteq E(a_0, k_1^2 (B^{(1)})^{-1}) \bigcap E(a_0, k_2^2 (B^{(2)})^{-1}). \tag{13}$$

Proof The proof is a consequence of the fact that each of two inclusions

$$X_0 \subseteq E(a_0, k_i^2 (B^{(i)})^{-1}), \quad i = 1, 2,$$

follows directly from the definition of parameters k_1 and k_2.

The following theorem gives the upper estimate for the considered case.

Theorem 2 (Filippova and Matviychuk [5]) *For any $\sigma > 0$ and for $X(t_0 + \sigma) = X(t_0 + \sigma, t_0, X_0)$ the following upper estimate is valid*

$$X(t_0 + \sigma) \subseteq E(a^{(1)}(\sigma), Q^{(1)}(\sigma)) \bigcap E(a^{(2)}(\sigma), Q^{(2)}(\sigma)) + o(\sigma) B(0, 1), \tag{14}$$

here $\sigma^{-1} o(\sigma) \to 0$ when $\sigma \to +0$ and

$$\begin{aligned}
a^{(1)}(\sigma) &= a(\sigma) + \sigma k_1^2 \lambda_{12}^2 d^{(2)} + \sigma \hat{a}, \\
a^{(2)}(\sigma) &= a(\sigma) + \sigma k_2^2 \lambda_{21}^2 d^{(1)} + \sigma \hat{a}, \\
a(\sigma) &= (I + \sigma A) a + \sigma a' B^{(1)} a d^{(1)} + \sigma a' B^{(2)} a d^{(2)}, \\
Q^{(1)}(\sigma) &= (p_1^{-1} + 1)(I + \sigma R) k_1^2 (B^{(1)})^{-1} (I + \sigma R)' + \\
&\quad + (p_1 + 1) \sigma^2 (\|d^{(2)}\|^2 k_1^4 \lambda_{12}^4 \cdot I + \hat{Q}), \\
Q^{(2)}(\sigma) &= (p_2^{-1} + 1)(I + \sigma R) k_2^2 (B^{(2)})^{-1} (I + \sigma R)' + \\
&\quad + (p_2 + 1) \sigma^2 (\|d^{(1)}\|^2 k_2^4 \lambda_{21}^4 \cdot I + \hat{Q}), \\
R &= A + 2 d^{(1)} a' B^{(1)} + 2 d^{(2)} a' B^{(2)},
\end{aligned} \tag{15}$$

with the unique positive solutions p_1 and p_2 of related algebraic equations

$$\sum_{i=1}^{n} \frac{1}{p_1 + \alpha_i} = \frac{n}{p_1 (p_1 + 1)}, \quad \sum_{i=1}^{n} \frac{1}{p_2 + \beta_i} = \frac{n}{p_2 (p_2 + 1)} \tag{16}$$

where $\alpha_i, \beta_i \geq 0$ $(i = 1, \ldots, n)$ *are the roots of the following equations*

$$\begin{aligned} \det((I + \sigma R)k_1^2(B^{(1)})^{-1}(I + \sigma R)' - \alpha\sigma^2\|d^{(2)}\|^2 k_1^4 \lambda_{12}^4 \cdot I) = 0, \\ \det((I + \sigma R)k_2^2(B^{(2)})^{-1}(I + \sigma R)' - \beta\sigma^2\|d^{(1)}\|^2 k_2^4 \lambda_{21}^4 \cdot I) = 0. \end{aligned} \tag{17}$$

Proof The proof follows the main lines of the proof of Theorem 2 from the paper Filippova and Matviychuk [8] (the result was announced also in Filippova and Matviychuk [5]). However we consider here a more complicated case of initial conditions for the system under study. Namely, we assume here that the initial position of the system is formed by the intersection of two ellipsoids, this complicates to some extent the whole scheme of reasoning.

4 Main Results

4.1 Lotka-Volterra Control Systems Under Uncertainty Conditions

Consider the following Lotka-Volterra system which describes the classical ecological predator-prey (or parasite-host) model with additional control functions (Bayen and Rapaport [9], Bonneuil and Mullers [10], Lotka [11], Murray [12], Prostyakov [13]):

$$\begin{cases} \dot{x}_1(t) = \quad ax_1 - bx_1x_2 + u_1, \\ \dot{x}_2(t) = -cx_2 + dx_1x_2 + u_2, \end{cases} \quad x(t_0) = x_0, \quad t_0 \leq t \leq T. \tag{18}$$

Here we assume that numbers $a, b, c, d > 0$ are given and initial vectors x_0 are unknown but bounded, that is we have the inclusion $x_0 \in X_0$, where X_0 is a given compact subset of R^2. This assumption may be interpreted for example in such a way that we do not know exactly the initial states (or amounts) of predators and prey. We assume also that controls $u(t)$ in (18) are taken measurable in Lebesgue on $[t_0, T]$, controls should satisfy also the inclusion

$$u(t) \in U, \quad \text{a.e. } t \in [t_0, T], \tag{19}$$

where $U \in comp R^2$. The choice of a control can influence, in particular, the rate of change in amounts of predators and prey.

Let us first construct the state space transformation $z = Zx$ with the nondegenerate matrix Z of the state space R^2 (here we have $x, z \in R^2$)

$$Z = \frac{1}{\sqrt{2}} \begin{pmatrix} 1 & -1 \\ 1 & 1 \end{pmatrix}. \tag{20}$$

We will have new equations (in the space of new coordinates)

$$\begin{cases} \dot{z}_1(t) = & Az_1 - Cz_2 + B(z_1^2 - z_2^2) + w_1, \\ \dot{z}_2(t) = & -Cz_1 + Az_2 + D(z_1^2 - z_2^2) + w_2, \end{cases} \quad t_0 \le t \le T, \tag{21}$$

with $A = (a - c)/2$, $B = \sqrt{2}(d - b)/4$, $C = (a + c)/2$, $D = \sqrt{2}(d + b)/4$. Now we have new constraints $z_0 \in \mathcal{Z}_0 = E(Za_0, ZQ_0Z')$ and $w \in \mathcal{W} = E(Z\hat{a}, Z\hat{Q}Z')$.

For small $\varepsilon > 0$ define

$$f_\varepsilon^{(1)}(z) = z_1^2 + \varepsilon^2 z_2^2, \quad f_\varepsilon^{(2)}(z) = \varepsilon^2 z_1^2 + z_2^2,$$

$$d^{(1)} = (B, D)', \quad d^{(2)} = -d^{(1)}.$$

Therefore we come to the following system

$$\begin{gathered} \dot{z} = A^* z + f_\varepsilon^{(1)}(z) \cdot d^{(1)} + f_\varepsilon^{(2)}(z) \cdot d^{(2)} + w(t), \\ z_0 \in \mathcal{Z}_0, \quad w \in \mathcal{W}, \quad t_0 \le t \le T, \end{gathered} \tag{22}$$

with

$$A^* = \begin{pmatrix} A & -C \\ -C & A \end{pmatrix}$$

and with functions $f_\varepsilon^{(1)}(z)$ and $f_\varepsilon^{(2)}(z)$ being the positive definite quadratic forms with matrices $B_\varepsilon^{(1)} = diag\{1, \varepsilon^2\}$ and $B_\varepsilon^{(2)} = diag\{\varepsilon^2, 1\}$, respectively.

We find now the external ellipsoidal estimates of reachable set $Z(t)$ of the system (22). We apply for this purpose Theorem 2 (see also Matviychuk [6]).

Theorem 3 *Let $\sigma > 0$ and let $X(t_0 + \sigma) = X(t_0 + \sigma, t_0, X_0)$ be the reachable set of the system* (18). *The following upper estimate holds true*

$$\begin{gathered} X(t_0 + \sigma) \subseteq E(a^{(1)}(\sigma), Q^{(1)}(\sigma)) \bigcap E(a^{(2)}(\sigma), Q^{(2)}(\sigma)) + o(\sigma)B(0, 1), \\ \sigma^{-1} o(\sigma) \to 0 \quad when\ \sigma \to +0, \end{gathered} \tag{23}$$

where

$$\begin{aligned} a^{(1)}(\sigma) &= Z^{-1}(a(\sigma) + \sigma k_1^2 \lambda_{12}^2 d^{(2)}) + \sigma\hat{a}, \\ a^{(2)}(\sigma) &= Z^{-1}(a(\sigma) + \sigma k_2^2 \lambda_{21}^2 d^{(1)}) + \sigma\hat{a}, \\ a(\sigma) &= (I + \sigma ZAZ')Za_0 + \sigma(Za_0)' B_\varepsilon^{(1)} Za_0 d^{(1)} + \sigma(Za_0)' B_\varepsilon^{(2)} Za_0 d^{(2)}, \\ Q^{(1)}(\sigma) &= Z^{-1}\big((p_1^{-1} + 1)(I + \sigma R) k_1^2 (B_\varepsilon^{(1)})^{-1} (I + \sigma R)' + \\ &+ (p_1 + 1)\sigma^2 (\|d^{(2)}\|^2 k_1^4 \lambda_{12}^4 \cdot I + Z\hat{Q}Z')\big)(Z^{-1})', \\ Q^{(2)}(\sigma) &= Z^{-1}\big((p_2^{-1} + 1)(I + \sigma R) k_2^2 (B_\varepsilon^{(2)})^{-1} (I + \sigma R)' + \\ &+ (p_2 + 1)\sigma^2 (\|d^{(1)}\|^2 k_2^4 \lambda_{21}^4 \cdot I + Z\hat{Q}Z')\big)(Z^{-1})', \\ R &= ZAZ' + 2d^{(1)}(Za_0)' B_\varepsilon^{(1)} + 2d^{(2)}(Za_0)' B_\varepsilon^{(2)}, \end{aligned}$$

where $B_\varepsilon^{(1)} = diag\{1, \varepsilon^2\}$, $B_\varepsilon^{(2)} = diag\{\varepsilon^2, 1\}$, k_1^2, k_2^2, λ_{12}^2 *and* λ_{21}^2 *are the maximal eigenvalue of the matrices*

$$(B_\varepsilon^{(1)})^{1/2} Z Q_0 Z' (B_\varepsilon^{(1)})^{1/2}, \quad (B_\varepsilon^{(2)})^{1/2} Z Q_0 Z' (B_\varepsilon^{(2)})^{1/2},$$
$$(B_\varepsilon^{(1)})^{-1/2} B_\varepsilon^{(2)} (B_\varepsilon^{(1)})^{-1/2}, \quad (B_\varepsilon^{(2)})^{-1/2} B_\varepsilon^{(1)} (B_\varepsilon^{(2)})^{-1/2},$$

respectively, numbers p_1, p_2 *are the unique positive solutions of related algebraic equations*

$$\sum_{i=1}^{n} \frac{1}{p_1 + \alpha_i} = \frac{n}{p_1(p_1+1)}, \quad \sum_{i=1}^{n} \frac{1}{p_2 + \beta_i} = \frac{n}{p_2(p_2+1)}$$

with $\alpha_i, \beta_i \geq 0$ $(i = 1, \ldots, n)$ *being the roots of the following equations*

$$\det\big((I+\sigma R) k_1^2 (B_\varepsilon^{(1)})^{-1} (I+\sigma R)' - \alpha\sigma^2 (||d^{(2)}||^2 k_1^4 \lambda_{12}^4 \cdot I + Z\hat{Q}Z')\big) = 0,$$
$$\det\big((I+\sigma R) k_2^2 (B_\varepsilon^{(2)})^{-1} (I+\sigma R)' - \beta\sigma^2 (||d^{(1)}||^2 k_2^4 \lambda_{21}^4 \cdot I + Z\hat{Q}Z')\big) = 0.$$

Proof The above formulas for ellipsoidal estimates of reachable set of the system under study specify the main relations of the Theorem 2, they may be easily verified by direct calculations.

Remark 4 The relations of Theorem 3 look very complicated, since they are obtained using the common scheme and are based on general results of Theorem 2. In the following we will partially simplify the above ellipsoidal estimate.

We transform further the system (18) and replace variables in the same way as it was done in [14, p. 44–45]. After such replacement we come to the following system (which is equivalent to (18))

$$\begin{cases} \dot{v_1} = \quad v_1 - v_1 v_2 + u_1, \\ \dot{v_2} = -\alpha v_2 + v_1 v_2 + u_2, \end{cases} \quad v(t_0) = v_0, \quad t_0 \leq t \leq T. \tag{24}$$

Here $\alpha > 0$ is given parameter and we assume as before that initial state vectors v_0 are unknown but bounded, with a bound

$$v_0 \in B(0, 1) = \{v = \{v_1, v_2\} : v_1^2 + v_2^2 \leq 1\}.$$

Control functions $u(t)$ are Lebesgue measurable on $[t_0, T]$, they satisfy the following constraint (with $r > 0$)

$$u(t) \in B(0, r), \quad \text{for a.e. } t \in [t_0, T]. \tag{25}$$

Replace here ones again the variables as $w_1 = v_1 + v_2, w_2 = v_1 - v_2$ and consider the system

$$\begin{aligned} \dot{w} &= Dw + d^{(1)} f_1(w) + d^{(2)} f_2(w), \\ w(t_0) &= w_0 \in B(0, r), \quad r = 2^{-1/2}, \quad t_0 \le t \le T, \end{aligned} \tag{26}$$

with $f_1(w) = 2w_1^2 + w_2^2$, $f_2(w) = w_1^2 + 2w_2^2$,

$$D = \frac{1}{2}\begin{pmatrix} (1-\alpha) & (1+\alpha) \\ (1+\alpha) & (1-\alpha) \end{pmatrix} \tag{27}$$

and with related trajectory tube denoted as $W(\cdot) = W(\cdot, t_0, X_0)$.

Theorem 4 *For any $\sigma > 0$ and for $W(t_0 + \sigma) = W(t_0 + \sigma, t_0, X_0)$ the upper estimate is valid*

$$W(t_0 + \sigma) \subseteq E(a^{(1)}(\sigma), Q^{(1)}(\sigma)) \bigcap E(a^{(2)}(\sigma), Q^{(2)}(\sigma)) + o(\sigma)B(0, 1), \tag{28}$$

where $\sigma^{-1}o(\sigma) \to 0$ when $\sigma \to +0$ and parameters of the estimating ellipsoids $E(a^{(1)}(\sigma), Q^{(1)}(\sigma))$ and $E(a^{(2)}(\sigma), Q^{(2)}(\sigma))$ are defined in Theorem 1 with the following simplifications

$$k_1^2 = k_2^2 = 2/r^2, \quad \lambda_{12}^2 = \lambda_{21}^2 = 2, \quad a = \hat{a} = 0,$$

$$d^{(1)} = -d^{(2)} = -(0, 1)', \; \hat{Q} = r^2 I,$$

$$B^{(1)} = \begin{pmatrix} 2 & 0 \\ 0 & 1 \end{pmatrix}, \quad B^{(2)} = \begin{pmatrix} 1 & 0 \\ 0 & 2 \end{pmatrix}.$$

Proof The proof of this upper estimate for the reachable set follows the lines of reasoning used in constructing the modified system and it follows directly from results of Theorems 1–2.

4.2 Iterative Algorithm and Results of Numerical Simulations

The iterative algorithm can be built on the basis of Theorems 3 and 4 to calculate the external ellipsoidal tube which will estimate the solution tube $X(t)$ on the whole time interval $t \in [t_0, T]$.

Algorithm. We divide the interval $[t_0, T]$ into sub-intervals $[t_i, t_{i+1}]$ with $t_i = t_0 + ih$ $(i = 1, \ldots, m)$ and $\sigma = (T - t_0)/m$, $t_m = T$.

- Taken $\mathcal{X}_0 = E(a_0, Q_0)$, find ellipsoids $E(a^{(1)}(\sigma), Q^{(1)}(\sigma))$ and $E(a^{(2)}(\sigma), Q^{(2)}(\sigma))$ basing on results of Theorem 3.

- Find the smallest ([1, 3]) ellipsoid $E(a_1, Q_1)$ containing the intersection

$$E(a^{(1)}(\sigma), Q^{(1)}(\sigma)) \bigcap E(a^{(2)}(\sigma), Q^{(2)}(\sigma)) \subseteq E(a_1, Q_1).$$

- Go to the next subsegment $[t_1, t_2]$ and take $E(a_1, Q_1)$ as the initial ellipsoid at the instant t_1.
- Continue these iterations to the end of the interval. So we will get the external ellipsoidal estimate $E(a^+(t), Q^+(t))$ of the whole tube $X(t)$ with accuracy which goes to zero when $m \to \infty$.

Example 1 Consider the Lotka-Volterra control system:

$$\begin{cases} \dot{x_1}(t) = \quad x_1 - x_1x_2 + u_1, \\ \dot{x_2}(t) = -x_2 + x_1x_2 + u_2, \end{cases} \\ x_0 \in X_0, \quad t_0 \leq t \leq T. \tag{29}$$

Here we take $t_0 = 0$, $T = 1$, $X_0 = B(0, 1)$ and $U = B(0, 0.1)$.

The trajectory tube $X(t)$ and some reachable sets are shown in Figs. 1 and 2. The tube $X(t)$ (Fig. 1) is found approximately by using results of [15], we will consider this tube as a reference object for comparison with other approximately found objects. Ellipsoidal estimates for the reachable set $X(0.01)$ which were found on the basis of results of Theorems 3 and 4 are shown in Fig. 3. Several steps of the main iterative Algorithm are given in the Fig. 4 where the external ellipsoidal estimating tube $E(a^+(t), Q^+(t))$ for the reachable set $X(t)$ is shown.

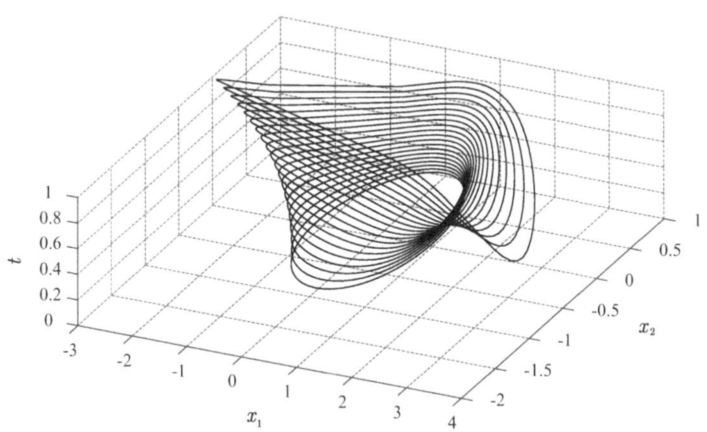

Fig. 1 The trajectory tube $X(t)$ of the system (29)

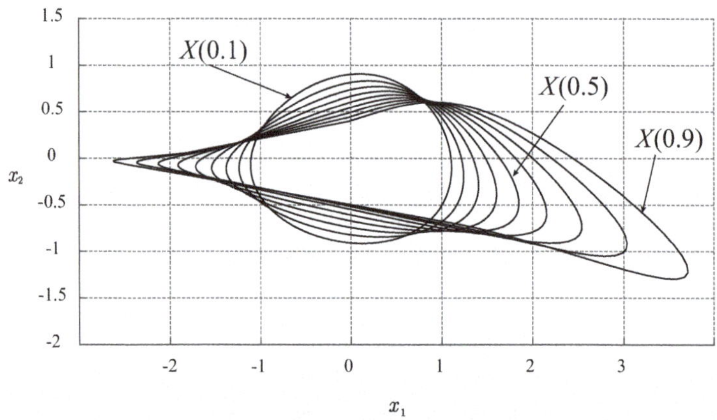

Fig. 2 Reachable sets $X(t)$ of the system (29) at the moments $t = 0.1; 0.2; \ldots; 0.9$

Fig. 3 Ellipsoidal estimate for the reachable set $X(0.01)$

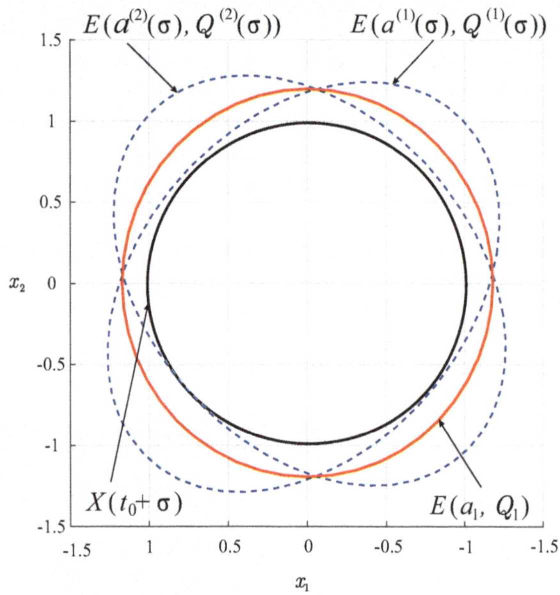

4.3 *Generalization of the Approach for the System with Impulsive Controls*

The main goal in this section is to try to apply the previously proposed ideas for estimating the states of systems described by ordinary differential equations to systems with more complicated dynamics.

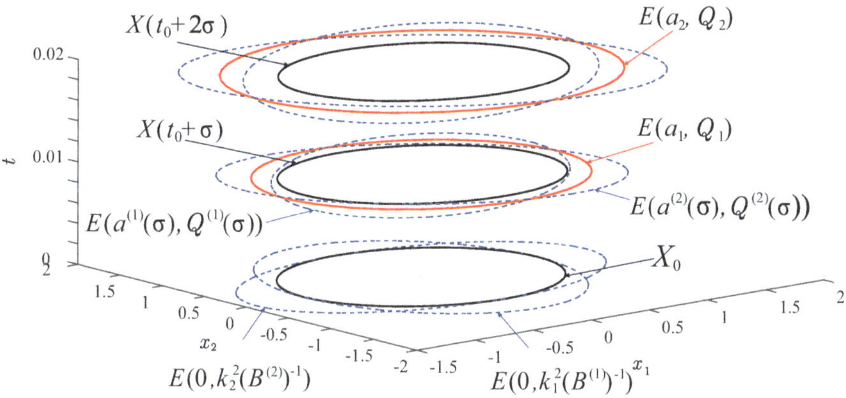

Fig. 4 Steps of the Algorithm of ellipsoidal estimating process for the trajectory tube $X(t)$ ($\sigma = 0.01$)

Consider the problem of estimating reachable sets for impulsive control system with nonlinearity and uncertainty

$$dx(t) = \big(Ax(t) + x'Bx \cdot d + u(t)\big)dt + Cdv(t). \tag{30}$$

Here $t_0 \leq t \leq T$, $x \in R^n$, B is a symmetric positive definite matrix; d, C are n-vectors. The impulsive function $v : [t_0, T] \to R$ is of bounded variation on $[t_0, T]$, monotonically increasing and right-continuous, with bounded variation defined as

$$\mathrm{Var}_{t\in[t_0,T]}\, v(t) = \sup_{\{t_i\}} \sum_{i=1}^{k} |v(t_i) - v(t_{i-1})| \leq \mu,$$

where $t_i : t_0 \leq t_1 \leq \ldots \leq t_k = T$, $\mu > 0$.

We assume also that $X_0 = E(a, k^2B^{-1})$ $(k \neq 0)$, $u(t) \in U = E(\hat{a}, \hat{Q})$.

Let us introduce a new time variable (Filippova and Matviychuk [8])

$$\eta(t) = t + \int\limits_{t_0}^{t} dv(s),$$

and a new state coordinate $\tau(\eta) = \inf\{t \mid \eta(t) \geq \eta\}$ and consider the following differential inclusion

$$\frac{d}{d\eta}\begin{pmatrix} z \\ \tau \end{pmatrix} \in H(\tau, z),$$
$$z(t_0) = x_0 \in X_0, \quad \tau(t_0) = t_0, \quad t_0 \le \eta \le T + \mu, \tag{31}$$
$$H(\tau, z) = \bigcup_{0 \le \nu \le 1} \left\{ \nu \begin{pmatrix} C \\ 0 \end{pmatrix} + (1 - \nu) \begin{pmatrix} Az + z'Bz \cdot d + E(\hat{a}, \hat{Q}) \\ 1 \end{pmatrix} \right\}.$$

Denote as $w = \{z, \tau\}$ the extended state vector of the system (31). The reachable set of the system (31) is denoted further as $W(\eta) = W(\eta; t_0, w_0, A, X_0 \times \{t_0\})(t_0 \le \eta \le T + \mu)$.

The following result explains the reason for constructing an auxiliary differential inclusion (31).

Theorem 5 *The reachable set $X(T)$ is the projection of $W(T + \mu)$ at the subspace of variable z: $X(T) = \pi_z W(T + \mu)$.*

Proof The proof of the theorem uses the the concrete structure of the control system and follows the general scheme of reasoning presented in ([8]).

Therefore, the related estimation procedure and the numerical algorithm basing on a combination of results of Theorems 3–5 may be also reformulated and applied to produce external ellipsoidal estimates also for the reachable sets of the impulsive control system (30).

5 Conclusions

The paper deals with the problems of state estimation for dynamical control systems with unknown but bounded initial vector state.

The case of the system nonlinearity generated by several quadratic functions in related differential equations is studied. We formulated the main problem as the problem of the motion of set-valued states with a nonlinear dynamics and under assumptions of uncertainty of a set-membership kind.

Basing on recent results of ellipsoidal calculus we present the modified state estimation approach which is based on the special structure of nonlinearity and uncertainty of the dynamical system and allows to construct the external ellipsoidal estimates of studied reachable sets numerically.

Acknowledgements The research was supported by the Russian Foundation for Basic Researches (RFBR) under Projects No. 18-01-00544-a and No. 16-29-04191-ofi-m.

References

1. A.B. Kurzhanski, P. Varaiya, *Dynamics and Control of Trajectory Tubes* (Theory and Computation. Springer, New York, 2014)
2. A.B. Kurzhanski, T.F. Filippova, On the theory of trajectory tubes—a mathematical formalism for uncertain dynamics, viability and control, in *Advances in Nonlinear Dynamics and Control: a Report from Russia, Progress in Systems and Control Theory*, ed. by A.B. Kurzhanski (Birkhauser, Boston, 1993), vol. 17, pp. 22–188
3. F.L. Chernousko, *State Estimation for Dynamic Systems* (CRC Press, Boca Raton, 1994)
4. T.F. Filippova, Differential equations of ellipsoidal state estimates for bilinear-quadratic control systems under uncertainty. J. Chaotic Model. Simul. (CMSIM) **1**, 85–93 (2017)
5. T.F. Filippova, O.G. Matviychuk, Algorithms of estimating reachable sets of nonlinear control systems with uncertainty. in *Proceedings of the 7th Chaotic Modeling and Simulation International Conference (Lisbon, Portugal: 7–10 June, 2014). Published by ISAST: International Society for the Advancement of Science and Technology*, Portugal, ed. by Christos H. Skiadas (2014), pp. 115–124
6. O.G. Matviychuk, Estimates of the reachable set of nonlinear control system under uncertainty. AIP Conf. Proc. **2025**(100005), 1–9 (2018)
7. F. Allgower, A. Zheng (eds.), *Nonlinear Model Predictive Control* (Birkhauser, Basel, 2000)
8. T.F. Filippova, O.G. Matviychuk, Algorithms of estimating reachable sets of nonlinear control systems with uncertainty. J. Chaotic Model. Simul. (CMSIM) **3**, 205–214 (2015)
9. T. Bayen, A. Rapaport, Minimal time crisis versus minimum time to reach a viability kernel: a case study in the prey-predator model. Optim. Control. Appl. Methods, Wiley **40**(2), 330–350 (2019)
10. N. Bonneuil, K. Mullers, Viable populations in a prey-predator system. J. Math. Biol. **35**, 261–293 (1997)
11. A.J. Lotka, *Elements of Physical Biology* (Williams and Wilkins, Baltimore, 1925)
12. J.D. Murray, *Mathematical Biology* (Springer, New York, 1993)
13. P. Prostyakov, Construction of the reachability set of the Lotka-Volterra system. Differ. Equ. **42**(3), 391–399 (2006)
14. A.S. Bratus, A.S. Novozhilov, A.P. Platonov, *Dynamical Systems and Models of Biology* (Fizmatlit, Moscow, 2010)
15. A.R. Matviychuk, O.G. Matviychuk, A method of approximate reachable set construction on the plane for a bilinear control system with uncertainty. AIP Conf. Proc. **2025**(100004), 1–6 (2018)

Tricritical Directed Percolation with Long-Range Spreading

Michal Hnatič, Tomáš Lučivjanský and Lukáš Mižišin

Abstract The effect of turbulent mixing on the model of tricritical directed percolation is studied. The turbulent advecting velocity field is modeled by means of the Kraichnan's rapid change ensemble. The non-equilibrium critical phenomena of one-component reaction-diffusion systems are investigated employing field theoretic renormalization group technique. We give a brief overview of the field-theoretic approach to the problem including renormalization group analysis. The renormalization procedure is performed in the framework of double (ε, y)-expansion scheme, where the ε is the deviation from the upper critical dimension $d_c = 3$ and the y is the exponent describing scaling behavior of velocity fluctuations. The corresponding asymptotic behavior is analyzed in leading order of perturbation theory.

Keywords Tricritical directed percolation · Long-range interaction · Field-theoretic renormalization group

1 Introduction

The directed percolation (DP) has been studied for many decades [1, 2] as one of the simplest non-equilibrium model of statistical physics. In the literature DP is known under other names such as simple epidemic process, an epidemic with recovery,

12th CHAOS Conference Proceedings, 18–22 June 2019, Chania, Crete, Greece.

M. Hnatič · L. Mižišin (✉)
Bogoliubov laboratory of Theoretical Physics, JINR, Joliot-Curie 6, 141980 Dubna, Moscow Region, Russia
e-mail: mizisin@theor.jinr.ru

Institute of Experimental Physics, Slovak Academy of Sciences, Watsonova 47, 040 01 Košice, Slovakia

M. Hnatič · T. Lučivjanský
Faculty of Science, P. J. Šafarik University, Šrobarova 2, 041 54 Košice, Slovakia

C. H. Skiadas and Y. Dimotikalis (eds.), *12th Chaotic Modeling and Simulation International Conference*, Springer Proceedings in Complexity,
https://doi.org/10.1007/978-3-030-39515-5_9

the Gribov process, the stochastic version of Schlögl's first reaction or Reggeon field theory [1, 3–7]. The model is applicable in different research areas in physics, biology, chemistry and even sociology.

The percolation process exhibits non-equilibrium second order phase transition between absorbing and active state [1]. The corresponding universal properties are well-known today due to numerous studies and simulations of lattice models [2, 6, 7]. On the other hand, modification of percolation spreading, quenched disorder, additional symmetry or broken rapidity reversal symmetry might result into a new model with different universal properties. One of the universal classes that is directly related to DP is tricritical directed percolation (TDP) [4]. TDP exhibits a tricritical point which separates continuous DP transition from a first order transition. The model is related to the fundamental model of tricriticality in critical statics [11], i.e. the ϕ^6 model, where the action takes the following form $\mathcal{S} = (\partial\phi)/2 + g_2\phi^2 + g_4\phi^4 + g_6\phi^6$. If parameter g_4 is positive, then the next term ϕ^6 is irrelevant and one obtains mean-field theory with a second order transition located at the $g_2 = 0$. For negative g_4, the system has the first order transition at critical value of the g_2, which depends on g_4 and g_6. As a result tricritical point $g_2 = g_4 = 0$ appears.

Our aim is to analyze the asymptotic properties of TDP process in a turbulent medium and to study the effects of turbulent stirring and mixing on the tricritical behavior. It was shown, that the flow can radically change the usual critical behavior, for example ϕ^4 model to complex behavior described by new nonequilibrium universality classes [8, 9]. In this paper, the turbulent mixing is described by a Gaussian velocity ensemble with prescribed statistics, which is known in the literature as the rapid-change model [10]. The model of passive scalar field advection by such velocity ensembles is relatively simple but offers anomalous scaling of the turbulent heat or mass transport (see the paper [10]).

The model is defined in the next section and we introduce higher order term in a construction of dynamical response functional. Finally, powerful machinery of the field–theoretic renormalization group is employed in order to determine universal properties.

2 Tricritical Directed Percolation

Renormalization group (RG) approach to stochastic dynamics is based on a proper construction of dynamic equation for slow-modes in a system. In reaction-diffusion problems such quantity is a density $\psi \equiv \psi(t, \mathbf{x})$ of active particles. Straightforward way for a construction of TDP consists in an inclusion of a higher order effective coupling into the Langevin equation of DP process [5]

$$\partial_t \psi = D\nabla^2\psi - D\left(\tau + \frac{u}{2}\psi + \frac{g'}{6}\psi^2\right)\psi + \zeta, \qquad (1)$$

where $\partial_t = \partial/\partial t$ is the time derivative, ∇^2 is the Laplace operator, ζ is the random noise that has to vanish in absorbing state, i.e. for a state where there are no active particles $\psi(t, \mathbf{x}) = 0$. The Gaussian random noise $\zeta(t, \mathbf{x})$ is fully characterized by its correlations

$$\langle \zeta(t, \mathbf{x})\zeta(t', \mathbf{x}')\rangle = Dg''\psi(t, \mathbf{x})\delta^d(\mathbf{x} - \mathbf{x}')\delta(t - t'). \tag{2}$$

In this notation the g', g'', u are positive constants, D is a diffusion constant, τ is a deviation from threshold value and d denotes space dimension. In the mean field approximation, the tricritical point which separates the continuous DP transition from a discontinuous one is thus found at $\tau = u = 0$.

3 Field-Theoretic Formulation

Dynamic response functional $\mathcal{S}$ can be derived in a standard fashion [11] from the Langevin equation (1) with noise correlation (2), and it takes the form [12]

$$\mathcal{S}\left[\tilde{\psi}, \psi\right] = \tilde{\psi}\left(\partial_t - D_0\nabla^2 + D_0\tau_0\right)\psi + \frac{D_0\tilde{\psi}}{2}\left(u_0\psi + \frac{g_0'}{3}\psi^2 - g_0''\tilde{\psi}\right)\psi, \tag{3}$$

where $\tilde{\psi} \equiv \tilde{\psi}(t, \mathbf{x})$ denotes auxiliary response field conjugated to the particle density $\psi(t, \mathbf{x})$. We use condensed notation in which integrals over the spatial and time variables are implicitly assumed. For example, the first term on the right hand side actually stands for $\int dt \int d^d\mathbf{x}\tilde{\psi}(t, \mathbf{x})\partial_t\psi(t, \mathbf{x})$. In the action (3), we have reserved subscript "0" for unrenormalized (bare) quantities.

In order to take advantage of field-theoretic methods and renormalization group, as a first step canonical dimensions have to be calculated. The main problem now is a general observation that models that exhibit a transition to an absorbing state contains in a corresponding action functional $\mathcal{S}$ superfluous variable. This has to be removed by rescaling transformation before actual analysis [4]. Otherwise it is not possible to determine them unambiguously. This redundant variable can be removed by the following transformation

$$\psi \to a\psi, \quad \tilde{\psi} \to a^{-1}\tilde{\psi}, \quad u \to a^{-1}u, \quad g' \to a^{-2}g', \quad g'' \to ag'', \tag{4}$$

which keeps the action $\mathcal{S}$ invariant. Using these properties, we may set $a = g''^{-1}$ (i.e. $g' \to g'/g''$ and $g'' \to 1$) to fix the redundancy. It can be done only if the parameters are finite positive quantities in the studied part of phase diagram. Finally, the dynamical response functional takes form

$$\mathcal{S}\left[\tilde{\psi}, \psi\right] = \tilde{\psi}\left(\partial_t - D_0\nabla^2 + D_0\tau_0\right)\psi + \frac{D_0}{2}\left(u_0\tilde{\psi}\psi^2 + \frac{g_0}{3}\tilde{\psi}\psi^3 - \tilde{\psi}^2\psi\right), \tag{5}$$

as minimal field theoretic model for TDP universality class with new coupling constant g_0.

Coupling with velocity field $\mathbf{v} = \{v_i(t, \mathbf{x})\}$ is introduced by replacement

$$\partial_t \to \nabla_t = \partial_t + v_i \partial_i \tag{6}$$

where $\partial_i = \partial/\partial x_i$ is i-th component of spatial gradient and ∇_t is known as Lagrangian or convective derivative. It is assumed that the velocity field corresponds to the incompressible flow, i.e. $\partial_i v_i = 0$. The model for velocity field is called rapid–change model with zero mean value and correlation function is chosen in the following form

$$\begin{aligned} \langle v_i(t, \mathbf{x}) v_j(t', \mathbf{x}')\rangle &= \delta(t - t') D_{ij}(\mathbf{x} - \mathbf{x}') \\ D_{ij}(\mathbf{x} - \mathbf{x}') = D_0 w_0 &\int\limits_{|\mathbf{k}|>m} \frac{d^d\mathbf{k}}{(2\pi)^d} P_{ij}(\mathbf{k}) \frac{1}{|\mathbf{k}|^{d+y}} \exp\{i\mathbf{k} \cdot (\mathbf{x} - \mathbf{x}')\} \end{aligned} \tag{7}$$

where $P_{ij}(\mathbf{k}) = \delta_{ij} - k_i k_j / k^2$ is transverse projector, ξ is free parameter. The cutoff in the integral at $|\mathbf{k}| = m$, where $m = 1/L$ is the reciprocal of the integral turbulence scale L, provides IR regularization. The cutoff is the simplest way for the practical calculation of Feynman diagrams and it is also used for computation of "pure" TDP diagrams.

Finally, the full action functional with included velocity field takes the form

$$\begin{aligned} \mathcal{S}\left[\tilde{\psi}, \psi, \mathbf{v}\right] &= \tilde{\psi}\left(\partial_t + \mathbf{v} \cdot \nabla - D_0 \nabla^2 + \tau_0 D_0\right) \psi \\ &+ \frac{D_0 u_0}{2} \tilde{\psi} \psi^2 + \frac{D_0 g_0}{6} \tilde{\psi} \psi^3 - \frac{D_0}{2} \tilde{\psi}^2 \psi + \frac{1}{2} v_i D_{ij}^{-1} v_j, \end{aligned} \tag{8}$$

where the D_{ij}^{-1} is the kernel of the inverse linear operation for the function $D_{ij}(\mathbf{x} - \mathbf{x}')$. The new model contains two coupling constants g_0, w_0 and the double expansion parameters ε, y.

4 RG Analysis

A starting point for the field-theoretical RG approach is an analysis of canonical dimensions. As a rule, dynamical models of type (8) exhibit two independent scales. This means that to each quantity Q two independent canonical dimensions are assigned, i.e., the momentum dimension $d_k[Q]$ and the frequency dimension $d_\omega[Q]$ determined from the standard normalization conditions

$$\begin{aligned} d_k[k] &= -d_k[x] = 1, \qquad & d_k[\omega] &= d_k[t] = 0, \\ d_\omega[k] &= d_\omega[x] = 0, \qquad & d_\omega[\omega] &= -d_\omega[t] = 1, \end{aligned} \tag{9}$$

Table 1 The canonical dimension fields and parameters of full action (8)

Q	ψ	$\tilde{\psi}$	$\mathbf{v}$	D_0	τ_0	u_0	g_0	w_0
$d_\omega[Q]$	0	0	1	1	0	0	0	0
$d_k[Q]$	$d-2$	2	-1	-2	2	$4-d$	2ε	y
$d[Q]$	$d-2$	2	1	0	2	$4-d$	2ε	y

and from the requirement that each term in the action must be dimensionless [11]. The total canonical dimension is $d[Q] = d_k[Q] + 2d_\omega[Q]$ and the canonical dimensions for the model with action functional (8) are summarized in Table 1. The upper critical dimension for given model at which $d[g] = 0$ has the following value

$$d_c = 3 \qquad \varepsilon = d_c - d = 3 - d. \tag{10}$$

The elimination of IR irrelevant term leads to an ultraviolet (UV) renormalizable theory at and below d_c. The total canonical dimension of a 1-irreducible Green functions (1PI) is given by the relation

$$d[\Gamma] = d_k[\Gamma] + 2d_\omega[\Gamma] = d + 2 - n_\psi d[\psi] - n_{\tilde{\psi}} d[\tilde{\psi}] - n_{\mathbf{v}} d[\mathbf{v}], \tag{11}$$

where n_ψ and $n_{\tilde{\psi}}$ are numbers of fields ψ and $\tilde{\psi}$, respectively. The formal degree of the UV divergence is $\delta[\Gamma] = d[\Gamma]|_{\varepsilon=0}$ in logarithmic theory. Power counting shows that primitive divergences appear in the Green functions $\Gamma_{1,1}$, $\Gamma_{1,2}$, $\Gamma_{1,3}$ and $\Gamma_{2,1}$.

The calculation of a given Green function follows the standard Feynman diagrammatic rules and all diagrammatic elements are easily derived from the action (8). The propagator of TDP can be read off from Gaussian part of model (5) and in the frequency-momentum representation reads

$$\langle \psi \tilde{\psi} \rangle_0 = \langle \tilde{\psi} \psi \rangle_0^* = \frac{1}{-i\omega + D_0(k^2 + \tau_0)}, \tag{12}$$

or in the time–momentum representation as follows

$$\langle \psi \tilde{\psi} \rangle_0 = \theta(t) \exp(-D_0[k^2 + \tau_0]t), \quad \langle \tilde{\psi} \psi \rangle_0 = \theta(-t) \exp(D_0[k^2 + \tau_0]t), \tag{13}$$

where $\theta(t)$ is Heaviside step function and velocity propagator (7). The interaction part gives rise to the four vertices $\tilde{\psi}^2\psi$, $\tilde{\psi}\psi^2$, $\tilde{\psi}\psi^3$ and $\tilde{\psi}\psi\mathbf{v}$. All graphic elements of the perturbation theory are depicted in Fig. 1.

As the dimensional analysis shows all primitively divergent Green functions are already contained in the action (8). The renormalized action for the renormalized fields can be written in the following form

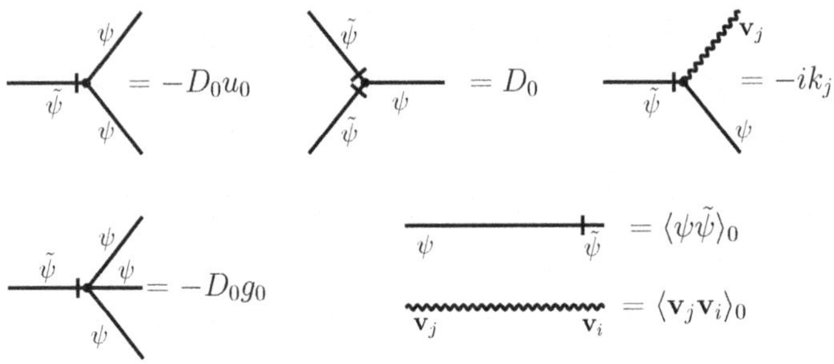

Fig. 1 The graphical representation of diagrammatic elements for full action (8)

$$
\begin{aligned}
\mathcal{S}_{\mathcal{R}} = \tilde{\psi}_R \left(Z_1 \partial_t + Z_1 \mathbf{v} \cdot \nabla - Z_2 D \nabla^2 + Z_3^{\tau} D \tau + Z_3^{u^2} D u^2 \right) \psi_R + \\
+ Z_4 \frac{Du}{2} \mu^{\varepsilon} \tilde{\psi}_R \psi_R^2 + \frac{Dg}{6} \mu^{2\varepsilon} Z_5 \tilde{\psi}_R \psi_R^3 - \frac{D}{2} Z_6 \tilde{\psi}_R^2 \psi_R + \frac{1}{2} v_i D_{ij}^{-1} v_j
\end{aligned}
\tag{14}
$$

with statistical weight $\exp\{-\mathcal{S}\}$ and the μ is the renormalization mass. In he action, the last term is not renormalized, the amplitude factor $D_0 w_0$ is expressed as $D w \mu^y$ in renormalized parameters. The renormalized fields and parameters are obtained from the following multiplicative renormalization

$$
\begin{aligned}
\psi = Z_\psi \psi_R, \quad \tilde{\psi} = Z_{\tilde{\psi}} \tilde{\psi}_R, \quad \tau_0 = Z_\tau \tau + u^2 \bar{Z}, \quad w_0 = \mu^{\xi} w Z_w, \\
D_0 = Z_D D, \quad u_0 = \mu^{\varepsilon} Z_u u, \quad g_0 = \mu^{2\varepsilon} g Z_g, \quad Z_w = Z_D^{-1},
\end{aligned}
\tag{15}
$$

and the velocity field is not renormalized therefore $Z_v = 1$. Relations between renormalization constants directly follow

$$
\begin{aligned}
Z_1 &= Z_\psi Z_{\tilde{\psi}}, & Z_\psi &= Z_1 Z_2 Z_6^{-1}, \\
Z_2 &= Z_\psi Z_{\tilde{\psi}} Z_D, & Z_{\tilde{\psi}} &= Z_2^{-1} Z_6, \\
Z_3^{\tau} &= Z_\psi Z_{\tilde{\psi}} Z_D Z_\tau, & Z_D &= Z_1^{-1 Z_2}, \\
Z_3^{u^2} &= Z_\psi Z_{\tilde{\psi}} Z_D \bar{Z}, & \bar{Z} &= Z_2^{-1} Z_3^{u^2}, \\
Z_4 &= Z_\psi^2 Z_{\tilde{\psi}} Z_D Z_u, & Z_\tau &= Z_2^{-1} Z_3^{\tau}, \\
Z_5 &= Z_\psi^3 Z_{\tilde{\psi}} Z_D Z_g, & Z_u &= Z_1^{-1} Z_2^{-2} Z_4 Z_6, \\
Z_6 &= Z_\psi Z_{\tilde{\psi}}^2 Z_D, & Z_g &= Z_1^{-2} Z_2^{-3} Z_5 Z_6^2.
\end{aligned}
\tag{16}
$$

Relevant Feynman diagrams are displayed in Figs. 2, 3, 4 and 5 for the full model (8) and are calculated in the minimal subtraction scheme [11]. However, one loop diagrams do not possess divergence in the TDP model and for that reason at least two-loop approximation is needed. In that case of TDP model, two-loop diagrams contain only superficial divergent part with a pole $1/\varepsilon$ and the coupling constant g.

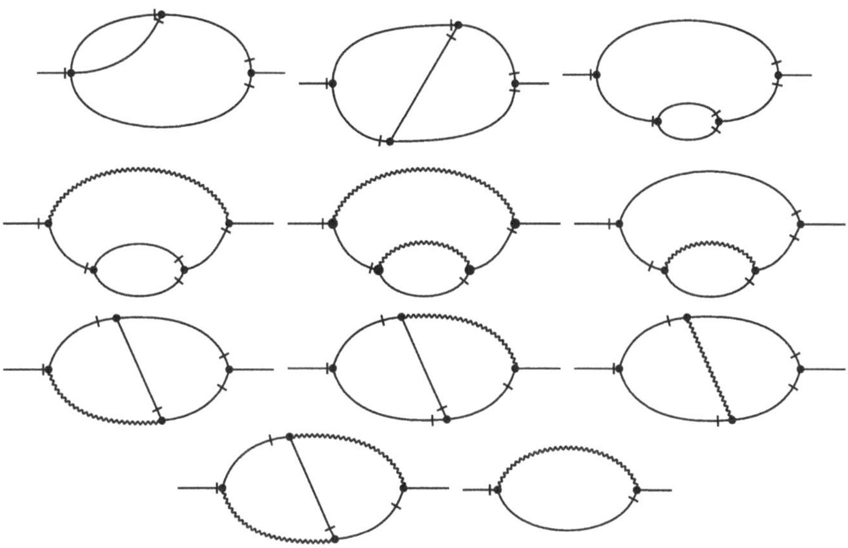

Fig. 2 Feynman diagrams that contribute to 1PI Green function $\Gamma_{1,1}$

On the other hand, the full model (8) contains one-loop diagram with divergent part (see Fig. 2 last diagrams) and then a diagrams containing velocity propagator might contain a combination of poles ε and y.

The self energy diagrams are shown on Fig. 2. The first one contributes to the RG constants Z_1, Z_2 and Z_3^τ. The rest of them yield corrections to Z_3^u which are proportional to u^2. It is necessary to distinguish contribution to Z_3 as

$$\tau_0 = Z_\tau \tau = Z_2^{-1}(Z_3^\tau \tau + Z_3^u u^2). \tag{17}$$

The element proportional to u^2 is generated by diagrams with two vertices of the $\tilde{\psi}\psi^2$ while diagrams with three and more such vertices will not be superficially divergent owing to their dimension. At last, the diagrams with velocity field also contribute to $Z_3^{u^2}$ and Z_2.

Feynman diagrams that correspond to the interaction part are shown in Figs. 3, 4, and 5. It can be easily seen that calculation of divergent part of Feynman diagrams are reduced to a computation of three different integrals for "pure" TDP model. The last diagrams in Fig. 4 (and the first diagrams in the third row in Fig. 5), which contain

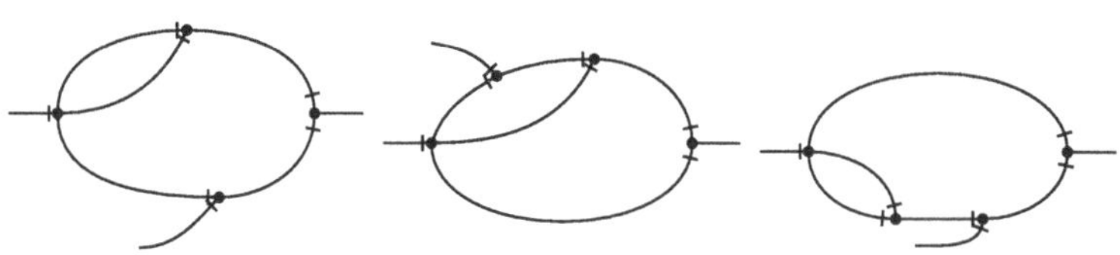

Fig. 3 Feynman diagrams that contribute to 1PI Green functions $\Gamma_{2,1}$ and renormalization constant Z_6

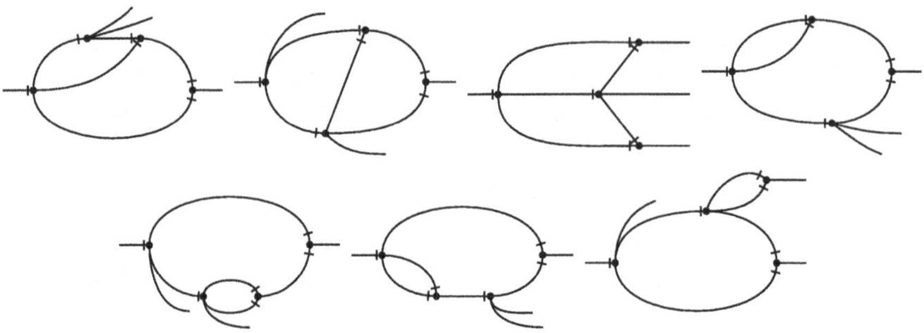

Fig. 4 Feynman diagrams that contribute to 1PI Green functions $\Gamma_{1,3}$ and renormalization constant Z_5

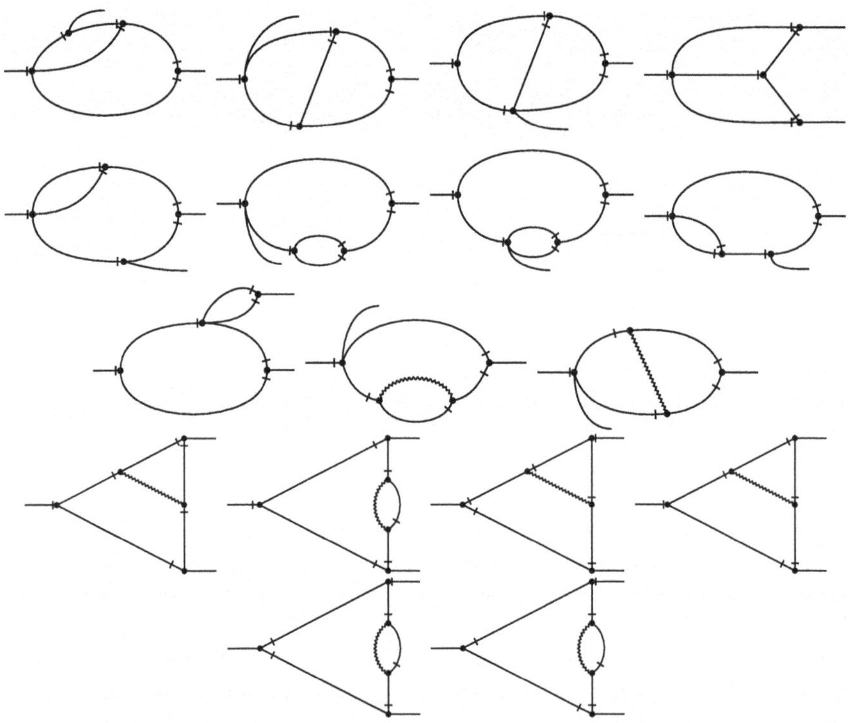

Fig. 5 Feynman diagrams that contribute to 1PI Green functions $\Gamma_{1,2}$ and renormalization constant Z_4

insertion of divergent one-loop diagrams, are finite in dimensional regularization. The 1PI Green function $\Gamma_{2,1}$ and $\Gamma_{1,3}$ do not contain diagrams with velocity propagator. The calculation is still in progress for diagrams containing velocity propagator. Further, RG constants are presented for pure TDP model

$$
\begin{aligned}
&Z_1 = 1 - \frac{1}{24}\frac{\sqrt{3}}{\varepsilon}\frac{g}{(4\pi)^3} + O(g^2), && Z_2 = 1 - \frac{1}{24}\frac{7\sqrt{3}}{9\varepsilon}\frac{g}{(4\pi)^3} + O(g^2),\\
&Z_3^{\tau} = 1 + 0g + O(g^2), && Z_3^{u^2} = \frac{1}{(4\pi)^2}\frac{1}{16}\frac{4\pi - 3\sqrt{3}}{6\pi\varepsilon} + O(g^1),\\
&Z_4 = 1 + \frac{1}{24}\frac{20\pi - 12\sqrt{3}}{\varepsilon}\frac{g}{(4\pi)^3} + O(g^2), && Z_5 = 1 + \frac{1}{24}\frac{36\pi - 18\sqrt{3}}{\varepsilon}\frac{g}{(4\pi)^3} + O(g^2),\\
&Z_6 = 1 + 0g + O(g^2).
\end{aligned}
\tag{18}
$$

The RG constants Z_1, Z_3^{τ}, Z_5, Z_6 are valid for full model.

After calculation of all contribution of Feynman diagrams to RG constants Z_i, the renormalization group equation will be used to explore the scaling properties of TDP. In particular, the unrenormalized (bare) Green function $\Gamma_{n_{\tilde{\psi}},n_{\psi}}$ must be independent of μ, i.e.

$$
\mu\partial_{\mu}\Big|_0 \Gamma_{n_{\tilde{\psi}},n_{\psi},n_v} = 0 \tag{19}
$$

where $\partial_{\mu}|_0$ denotes μ derivatives at fixed bare parameters. Passing from bare to renormalized parameters and fields, the relevant Green function can be rewritten to

$$
\left[D_{RG} - n_{\psi}\gamma_{\psi} - n_{\tilde{\psi}}\gamma_{\tilde{\psi}}\right]\Gamma^R_{n_{\tilde{\psi}},n_{\psi}} = 0. \tag{20}
$$

Here, the RG operator D_{RG} is given by the formula

$$
D_{RG} \equiv \mu\partial_{\mu} + \beta_g\partial_g + \beta_w\partial_w - \gamma_D D\partial_D - (\gamma_3^{\tau}\tau + u^2\bar{\gamma})\partial_{\tau} - \gamma_u u\partial_u, \tag{21}
$$

where $\beta_{e_i} = \mu\partial_{\mu}\big|_0 e_i$ for $e_i \in \{g, w\}$ and anomalous dimension of the quantity Q is defined as follows

$$
\gamma_Q \equiv \mu\partial_{\mu}\big|_0 \ln Z_Q. \tag{22}
$$

The form of RG functions depends on the constants Z_i and calculation RG constants $i = 2, 3, 4$ are still in progress.

Our results for "pure" TDP process agree with results [13, 14]. In our case we use different IR regulator in calculation of Feynman diagrams, but universal quantities (critical exponents) have to be independent of the precise form of IR regularization.

Acknowledgements The work was supported by VEGA Grant No. 1/0345/17 of the Ministry of Education, Science, Research and Sport of the Slovak Republic and the grant of the Slovak Research and Development Agency under the contract No. APVV-16-0186.

References

1. U. Täuber, *Critical Dynamics: a Field Theory Approach to Equilibrium and Non-Equilibrium Scaling Behavior* (Cambridge University Press, New York, 2014)
2. H. Hinrichsen, Adv. Phys. **49**(7), 815–958 (2000)

3. H.K. Janssen, U.C. Täuber, Ann. Phys. **315**, 147 (2004)
4. H.K. Janssen, M. Müller, O. Stenull, Phys. Rev. E **70**, 026114 (2004)
5. H.K. Janssen, J. Phys.: Condens. Matter **17**, S1973–S1993 (2005)
6. S. Lübeck, J. Stat. Phys. **123**(1), 193–221 (2005)
7. P. Grassberger, J. Stat. Mech. **P01004**, (2006)
8. N.V. Antonov, M. Hnatich, J. Honkonen, J. Phys. A: Math. Gen. **39**, 7867 (2006)
9. N.V. Antonov, A.A. Ignatieva, J. Phys. A: Math. Gen. **39**, 13593 (2006)
10. G. Falkovich, K. Gawędzki, M. Vergassola, Rev. Mod. Phys. **73**, 913 (2001)
11. A.N. Vasil'ev, *The Field Theoretic Renormalization Group in Critical Behavior Theory and Stochastic Dynamics* (Chapman & Hall/CRC, Boca Raton, 2004)
12. P. C. Martin, E. D. Siggia, H. A. Rose, Phys. Rev. A **8**, 423 (1973); C. De Dominicis, J. Phys. (Paris), Colloq. 37, C1-247 (1976); H. K. Janssen, Z. Phys. B **23**, 377 (1976); R. Bausch, H. K. Janssen, H. Wagner, ibid. **24**, 113 (1976)
13. T. Ohtsuki, T. Keyes, Phys. Rev. A **35**, 2697 (1987)
14. T. Ohtsuki, T. Keyes, Phys. Rev. A **36**, 4434 (1987)

Anisotropic MHD Turbulence Near Two Spatial Dimensions: General Field Theoretic Renormalization Group Analysis

Eva Jurčišinová, Marian Jurčišin and Richard Remecký

Abstract Using the so-called double-expansion approach we perform a general field theoretic renormalization group analysis of the MHD turbulence with the presence of the weak small-scale uniaxial anisotropy valid for all spatial dimensions $d \geq 2$. The ultraviolet divergent Green's functions are identified and the renormalization of the model is performed in the first order of the corresponding perturbation theory. The explicit form of all renormalization constants is found and the corresponding anomalous dimensions are determined. The explicit form of the system of all renormalization group β functions is found, which can be directly used for the analysis of all possible scaling regimes of the model deep inside of the inertial range.

Keywords MHD turbulence · Two-dimensional turbulence · Field theory · Renormalization group

---

12th CHAOS Conference Proceedings, 18–22 June 2019, Chania, Crete, Greece

---

E. Jurčišinová · M. Jurčišin
Institute of Experimental Physics, Slovak Academy of Sciences,
Watsonova 47, 040 01 Košice, Slovakia
e-mail: jurcisine@saske.sk

M. Jurčišin
e-mail: jurcisin@saske.sk

R. Remecký (✉)
Bogoliubov Laboratory of Theoretical Physics,
Joint Institute for Nuclear Research, 141 980 Dubna,
Moscow Region, Russian Federation
e-mail: remecky@saske.sk

C. H. Skiadas and Y. Dimotikalis (eds.), *12th Chaotic Modeling and Simulation International Conference*, Springer Proceedings in Complexity,
https://doi.org/10.1007/978-3-030-39515-5_10

1 Introduction

The renormalization group technique represents an effective and powerful method for systematic investigation of a self-similar scaling behavior [1–3]. It was widely used in the theory of critical phenomena to explain the origin of the critical scaling as well as to calculate corresponding universal quantities (e.g., critical exponents) [4].

During a few last decades the field theoretic renormalization group (RG) technique has also been used intensively for the investigation of the universal properties of various turbulent systems starting from simple models of passive quantities advected by various turbulent environments with Gaussian statistic of the velocity field up to such complex systems as, e.g., the magnetohydrodynamic (MHD) turbulence. In this respect, in Adzhemyan et al. [5] the standard ε-expansion perturbative approach was used for the first time for the investigation of the infrared properties of the anisotropically driven MHD turbulence in spatial dimensions $d > 2$. There, the influence of the presence of the anisotropy on the stability of the scaling regimes of the MHD turbulence was investigated and it was shown that the presence of the small-scale anisotropy increases the role of the Lorentz force especially in the case of the so-called kinetic regime. However, it is well-known that the field-theoretic model of the MHD turbulence contains additional divergences in two spatial dimensions [6, 7]. As was shown by Jurcisin and Stehlik [7] in the framework of the isotropic MHD turbulence, these additional divergences can have nontrivial impact on the stability of the scaling regimes of the MHD turbulence even in the three-dimensional case. At the same time, as was shown by Hnatich et al. [8], the taking into account the existence of the additional divergences in two spatial dimensions in the Navier-Stokes turbulence with the presence of the small-scale anisotropy significantly enlarges the region of the stability of the scaling regime in the parametric space of the model.

Using all these facts, it is clear that the existence of additional divergences in two spatial dimensions must also have a nontrivial impact on the properties of the scaling regimes of the MHD turbulence in the case when the presence of the small-scale anisotropy in the system is considered. In this respect, in the present paper we shall start with the systematic investigation of the properties of the anisotropic MHD turbulence near two spatial dimensions, namely, we shall give a self-contained formulation of the model which is necessary for the consistent investigation of the MHD turbulence for spatial dimensions $d \geq 2$ with the presence of the weak small-scale uniaxial anisotropy.

As was shown by Honkonen and Nalimov [9], to investigate consistently turbulent systems driven by the stochastic Navier-Stokes equation near two spatial dimensions it is necessary to take into account additional divergences of the correlation functions that appear for $d = 2$. There, the correct self-consistent approach was proposed based on a double-expansion scheme, which takes into account these new divergencies, in the framework of which a new parameter δ is introduced defined by the relation $2\delta = d - 2$, which controls deviation from spatial dimension $d = 2$.

In what follows, we shall use the aforementioned double-expansion regularization scheme to begin with the systematic investigation of the field theoretic model of the anisotropic MHD turbulence near two spatial dimensions, namely, we shall define consistently the multiplicatively renormalizable field theoretic model of the MHD turbulence driven by the random forces with the presence of the uniaxial anisotropy for spatial dimensions $d \geq 2$ and the corresponding renormalization of the model will be present in the leading order of the perturbation theory, i.e., in the one-loop approximation. At the same time, a subsequent detailed analysis of all stable scaling regimes of the model will be given elsewhere.

The paper is organized as follows. In Sect. 2 the MHD turbulence with the presence of the small-scale anisotropy is formulated. The field theoretic formulation of the model is present in Sect. 3. The renormalization of the model is performed in Sect. 4. Finally, a brief discussion of the obtained results is given in Sect. 5.

2 The model of MHD turbulence with small-scale uniaxial anisotropy

In what follows, we are interested in fully developed MHD turbulence described by the following system of stochastic equations

$$\partial_t \mathbf{v} + (\mathbf{v}\cdot\partial)\mathbf{v} = \nu_0 \Delta \mathbf{v} + (\mathbf{b}\cdot\partial)\mathbf{b} - \partial\mathcal{P} + \mathbf{f}^{\mathbf{v}}, \tag{1}$$

$$\partial_t \mathbf{b} + (\mathbf{v}\cdot\partial)\mathbf{b} = \nu_0 u_0 \Delta \mathbf{b} + (\mathbf{b}\cdot\partial)\mathbf{v} + \mathbf{f}^{\mathbf{b}}, \tag{2}$$

where $\mathbf{v} \equiv \mathbf{v}(x)$ and $\mathbf{b} \equiv \mathbf{b}(x)$ $[x \equiv (\mathbf{x}, t)]$ are the transverse velocity and magnetic fields, ν_0 is the molecular viscosity (the subscript 0 always denotes bare parameters of the unrenormalized theory), $\nu_0 u_0$ represents the magnetic diffusivity, u_0 is the inverse magnetic Prandtl number, and $\mathcal{P}$ is the pressure. Eq. (1) is the Navier-Stokes equation with the presence of the Lorentz force and Eq. (2), which drives the dynamics of the magnetic field, follows from the Maxwell equations for a continuous medium. Due to the fact that we suppose incompressibility of the studied turbulent system the velocity field as well as magnetic field are divergenceless: $\partial \cdot \mathbf{v} = 0$ and $\partial \cdot \mathbf{b} = 0$.

To maintain the steady state of the dissipative system described by Eqs. (1) and (2) it is necessary to pump continuously energy into the system at large scales. The corresponding kinetic and magnetic infrared energy pumping into the system at large scales is performed through the incompressible random forces $\mathbf{f}^{\mathbf{v}} \equiv \mathbf{f}^{\mathbf{v}}(x)$ and $\mathbf{f}^{\mathbf{b}} \equiv \mathbf{f}^{\mathbf{b}}(x)$, for which we supposed Gaussian statistics with zero mean and the following specific form of the correlators

$$\langle f_i^{\mathbf{v}}(\mathbf{x}_1, t_1) f_j^{\mathbf{v}}(\mathbf{x}_2, t_2)\rangle = \delta(t_1 - t_2) \int \frac{\mathrm{d}\mathbf{k}}{(2\pi)^d} \mathcal{D}_{ij}^{\mathbf{v}}(\mathbf{k}) e^{\mathrm{i}\mathbf{k}\cdot(\mathbf{x}_1 - \mathbf{x}_2)}, \tag{3}$$

$$\langle f_i^{\mathbf{b}}(\mathbf{x}_1, t_1) f_j^{\mathbf{b}}(\mathbf{x}_2, t_2)\rangle = \delta(t_1 - t_2) \int \frac{\mathrm{d}\mathbf{k}}{(2\pi)^d} \mathcal{D}_{ij}^{\mathbf{b}}(\mathbf{k}) e^{\mathrm{i}\mathbf{k}\cdot(\mathbf{x}_1 - \mathbf{x}_2)}, \tag{4}$$

where

$$\mathcal{D}^{\mathbf{v}}_{ij}(\mathbf{k}) = g_{v10}\nu_0^3 u_0 k^{2-2\delta-2\varepsilon}[(1+\alpha_1\xi_k^2)P_{ij}(\mathbf{k}) + \alpha_2 R_{ij}(\mathbf{k})] + \\ + g_{v20}\nu_0^3 u_0 k^2[(1+\alpha_{30}\xi_k^2)P_{ij}(\mathbf{k}) + (\alpha_{40}+\alpha_{50}\xi_k^2)R_{ij}(\mathbf{k})], \quad (5)$$

$$\mathcal{D}^{\mathbf{b}}_{ij}(\mathbf{k}) = g_{b10}\nu_0^3 u_0^2 k^{2-2\delta-2a\varepsilon}[(1+\sigma_1\xi_k^2)P_{ij}(\mathbf{k}) + \sigma_2 R_{ij}(\mathbf{k})] + \\ + g_{b20}\nu_0^3 u_0^2 k^2[(1+\sigma_{30}\xi_k^2)P_{ij}(\mathbf{k}) + (\sigma_{40}+\sigma_{50}\xi_k^2)R_{ij}(\mathbf{k})]. \quad (6)$$

The correlators (3) and (4) are taken in the form that, on one hand, simulate the small scale uniaxial anisotropy of the system (strictly speaking, the uniaxial anisotropic energy pumping into the system) and, on the other hand, allows one to investigate the scaling properties of the model in spatial dimensions $d \geq 2$ using the field theoretic RG technique (see the next section). In correlators (5) and (6), $\mathbf{k}$ is the wave vector (momentum), $k = |\mathbf{k}|$, the parameter ε determines the powerlike falloff of the long-range correlations with physical value $\varepsilon = 2$, δ represents the deviation of the spatial dimension from $d = 2$: $\delta = (d-2)/2$, the parameter a controls the form of the magnetic forcing in comparison to the form of the kinetic energy forcing, g_{v10}, g_{v20}, g_{b10}, and g_{b20} play the role of the model coupling constants, parameters α_i and σ_i $(i = 1, 2)$ are parameters that control the form of the anisotropic energy pumping into the system, and α_{i0} and σ_{i0} $(i = 3, 4, 5)$ are another anisotropy parameters needed for correct renormalization of the model near spatial dimension $d = 2$. In addition, $P_{ij}(k)$ and $R_{ij}(k)$ are the isotropic and uniaxial anisotropic transverse projectors which have the following explicit form

$$P_{ij}(\mathbf{k}) = \delta_{ij} - \frac{k_i k_j}{k^2}, \quad R_{ij}(\mathbf{k}) \equiv P_{is}(\mathbf{k}) n_s n_t P_{tj}(\mathbf{k}) = \left(n_i - \xi_k \frac{k_i}{k}\right)\left(n_j - \xi_k \frac{k_j}{k}\right), \quad (7)$$

where

$$\xi_k = \frac{\mathbf{k}\cdot\mathbf{n}}{k}, \quad (8)$$

and the unit vector $\mathbf{n}$ defines the direction of the uniaxial anisotropy.

3 The field theoretic formulation of the model

According to the well-known theorem [10–13] the stochastic problem given by Eqs. (1) and (2) with correlators of random forces (3) and (4) can be rewritten into the field theoretic model with double set of fields $\Phi = \{\mathbf{v}, \mathbf{b}, \mathbf{v}', \mathbf{b}'\}$ described by the following action functional

$$\mathcal{S}(\Phi) = \frac{1}{2}\int \mathrm{d}x_1 x_2 \left[v'_i(x_1)\langle f^{\mathbf{v}}_i(x_1) f^{\mathbf{v}}_j(x_2)\rangle v'_j(x_2) + b'_i(x_1)\langle f^{\mathbf{b}}_i(x_1) f^{\mathbf{b}}_j(x_2)\rangle b'_j(x_2)\right] \\ + \int \mathrm{d}x \Big\{\mathbf{v}'\cdot[-\partial_t\mathbf{v} + \nu_0\Delta\mathbf{v} - (\mathbf{v}\cdot\partial)\mathbf{v} - (\mathbf{b}\cdot\partial)\mathbf{b}]$$

$$
\begin{aligned}
&+ \mathbf{b}' \cdot [-\partial_t \mathbf{b} + \nu_0 u_0 \Delta \mathbf{b} + (\mathbf{b} \cdot \partial)\mathbf{v} - (\mathbf{v} \cdot \partial)\mathbf{b}] \\
&+ \nu_0 \mathbf{v}' \cdot \left[\chi_{10}(\mathbf{n} \cdot \partial)^2 \mathbf{v} + \chi_{20} \mathbf{n} \Delta(\mathbf{n} \cdot \mathbf{v}) + \chi_{30} \mathbf{n}(\mathbf{n} \cdot \partial)^2 (\mathbf{n} \cdot \mathbf{v})\right] \\
&+ \nu_0 u_0 \mathbf{b}' \cdot \left[\tau_{10}(\mathbf{n} \cdot \partial)^2 \mathbf{b} + \tau_{20} \mathbf{n} \Delta(\mathbf{n} \cdot \mathbf{b}) + \tau_{30} \mathbf{n}(\mathbf{n} \cdot \partial)^2 (\mathbf{n} \cdot \mathbf{b})\right] \\
&+ \mathbf{v}' \cdot \left[\lambda_{10} \mathbf{b}(\mathbf{n} \cdot \partial)(\mathbf{n} \cdot \mathbf{b}) + \lambda_{20} \mathbf{n}(\mathbf{n} \cdot \partial)\mathbf{b}^2 + \lambda_{30} \mathbf{n}(\mathbf{b} \cdot \partial)(\mathbf{n} \cdot \mathbf{b}) \right. \\
&\qquad \left. \left. + \lambda_{40} \mathbf{n}(\mathbf{n} \cdot \partial)(\mathbf{n} \cdot \mathbf{b})^2\right]\right\}, \qquad (9)
\end{aligned}
$$

where the last four lines represent added new anisotropy terms with corresponding bare parameters χ_{i0}, τ_{i0}, and λ_{j0} ($i = 1, 2, 3$ and $j = 1, 2, 3, 4$), which are necessary to make the field theoretic model multiplicative renormalizable.

The formulation of the studied stochastic problem in the form of the field theoretic model with action functional (9) allows one to perform the perturbative analysis of the model using the standard Feynman diagrammatic technique with well defined propagators and interaction vertices.

In what follows, we always suppose that the anisotropy of the system is weak, i.e., we consider only small deviations from the full isotropy of the model. Using this assumption, the quadratic part of the action leads to the set of bare propagators of the model, which have the following form in the frequency-momentum representation and in the weak anisotropy limit (only linear parts of the propagators with respect to the anisotropy parameters are taken into account)

$$
\begin{aligned}
\Delta_{ij}^{v'v}(\omega_k, \mathbf{k}) &= \frac{1}{\mathrm{i}\omega_k + \nu_0 k^2} \\
&\times \left\{\left[1 - \frac{\chi_{10}\nu_0(\mathbf{n} \cdot \mathbf{k})^2}{\mathrm{i}\omega_k + \nu_0 k^2}\right] P_{ij}(\mathbf{k}) + \frac{\nu_0[\chi_{20}\mathbf{k}^2 + \chi_{30}(\mathbf{n} \cdot \mathbf{k})^2]}{\mathrm{i}\omega_k + \nu_0 k^2} R_{ij}(\mathbf{k})\right\} \qquad (10) \\
\Delta_{ij}^{b'b}(\omega_k, \mathbf{k}) &= \frac{1}{\mathrm{i}\omega_k + \nu_0 u_0 k^2} \\
&\times \left\{\left[1 - \frac{\tau_{10}\nu_0 u_0(\mathbf{n} \cdot \mathbf{k})^2}{\mathrm{i}\omega_k + \nu_0 u_0 k^2}\right] P_{ij}(\mathbf{k}) + \frac{\nu_0 u_0[\tau_{20}\mathbf{k}^2 + \tau_{30}(\mathbf{n} \cdot \mathbf{k})^2]}{\mathrm{i}\omega_k + \nu_0 u_0 k^2} R_{ij}(\mathbf{k})\right\} \qquad (11) \\
\Delta_{ij}^{vv}(\omega_k, \mathbf{k}) &= \frac{u_0 \nu_0^3 k^2}{|\mathrm{i}\omega_k + \nu_0 k^2|^2} \left[g_{v10} k^{-2\delta - 2\varepsilon} A_v + g_{v20} B_v\right], \qquad (12) \\
\Delta_{ij}^{bb}(\omega_k, \mathbf{k}) &= \frac{u_0^2 \nu_0^3 k^2}{|\mathrm{i}\omega_k + \nu_0 u_0 k^2|^2} \left[g_{b10} k^{-2\delta - 2\varepsilon} A_b + g_{b20} B_b\right], \qquad (13)
\end{aligned}
$$

where quantities A_v, A_b, B_v, and B_b are given as follows

$$
\begin{aligned}
A_v &= \left[1 + \alpha_1 \xi_k^2 - \nu_0 \chi_{10} (\mathbf{n} \cdot \mathbf{k})^2 C\right] P_{ij}(\mathbf{k}) \\
&\quad + \left[\alpha_2 - \nu_0 (\chi_{20} k^2 + \chi_{30} (\mathbf{n} \cdot \mathbf{k})^2) C\right] R_{ij}(\mathbf{k}) \qquad (14) \\
B_v &= \left[1 + \alpha_{30} \xi_k^2 - \nu_0 \chi_{10} (\mathbf{n} \cdot \mathbf{k})^2 C\right] P_{ij}(\mathbf{k}) \\
&\quad + \left[\alpha_{40} + \alpha_{50} \xi_k^2 - \nu_0 (\chi_{20} k^2 + \chi_{30} (\mathbf{n} \cdot \mathbf{k})^2) C\right] R_{ij}(\mathbf{k}) \qquad (15) \\
A_b &= \left[1 + \sigma_1 \xi_k^2 - \nu_0 u_0 \tau_{10} (\mathbf{n} \cdot \mathbf{k})^2 D\right] P_{ij}(\mathbf{k})
\end{aligned}
$$

$$+ \left[\sigma_2 - \nu_0 u_0 (\tau_{20} k^2 + \tau_{30} (\mathbf{n}\cdot\mathbf{k})^2) D\right] R_{ij}(\mathbf{k}) \tag{16}$$

$$B_b = \left[1 + \sigma_{30}\xi_k^2 - \nu_0 u_0 \tau_{10} (\mathbf{n}\cdot\mathbf{k})^2 D\right] P_{ij}(\mathbf{k})$$
$$+ \left[\sigma_{40} + \sigma_{50}\xi_k^2 - \nu_0 u_0 (\tau_{20} k^2 + \tau_{30} (\mathbf{n}\cdot\mathbf{k})^2) D\right] R_{ij}(\mathbf{k}), \tag{17}$$

and

$$C = \frac{1}{\mathrm{i}\omega_k + \nu_0 k^2} + \frac{1}{-\mathrm{i}\omega_k + \nu_0 k^2}, \tag{18}$$

$$D = \frac{1}{\mathrm{i}\omega_k + \nu_0 u_0 k^2} + \frac{1}{-\mathrm{i}\omega_k + \nu_0 u_0 k^2}. \tag{19}$$

On the other hand, from the interaction part of the action functional (9) the following interaction vertices are defined

$$\mathcal{V}_{jsm}(k) = \mathrm{i}(k_s \delta_{jm} + k_m \delta_{js}), \tag{20}$$

$$\mathcal{U}_{jsm}(k) = \mathrm{i}(k_s \delta_{jm} - k_m \delta_{js}), \tag{21}$$

$$\mathcal{W}_{jsm}(k) = \mathrm{i}(k_s \delta_{jm} + k_m \delta_{js}) + \mathrm{i}\lambda_{10}(n_s \delta_{jm} + n_m \delta_{js})(\mathbf{n}\cdot\mathbf{k}) +$$
$$+ \mathrm{i}2\lambda_{20} n_j \delta_{sm} (\mathbf{n}\cdot\mathbf{k}) + \mathrm{i}\lambda_{30}(k_s n_m + k_m n_s) n_j +$$
$$+ \mathrm{i}2\lambda_{40} n_s n_m n_j (\mathbf{n}\cdot\mathbf{k}). \tag{22}$$

The standard graphical representation of the propagators and the interaction vertices of the model can be seen in Fig. 1 and Fig. 2, respectively. Let us also note that the momentum $\mathbf{k}$ enters every vertex via the corresponding auxiliary field $\mathbf{v}'$ and $\mathbf{b}'$, respectively.

4 The multiplicative renormalization of the model and the RG functions

The dimensional analysis of the model shows that, in the case with spatial dimensions $d \geq 2$, the superficial ultraviolet (UV) divergences exist in the following one-irreducible Green's functions: $\langle v_i' v_j \rangle_{1-ir}$, $\langle b_i' b_j \rangle_{1-ir}$, $\langle v_i v_j \rangle_{1-ir}$, $\langle b_i b_j \rangle_{1-ir}$, and

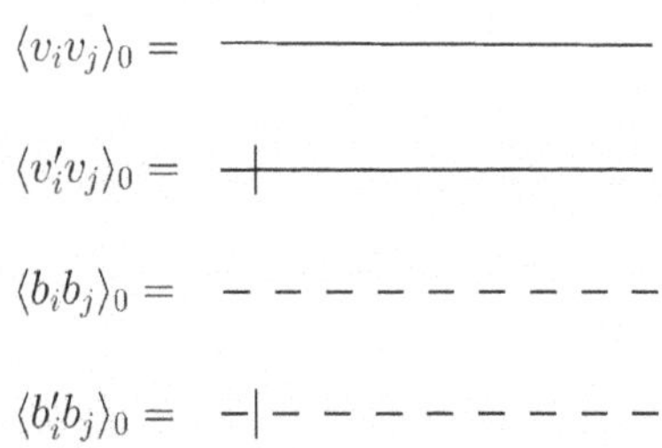

Fig. 1 Graphical representation of the propagators of the model

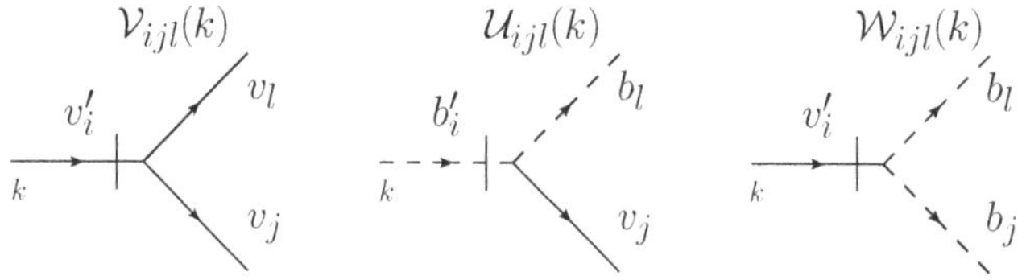

Fig. 2 Graphical representation of the interaction vertices $\mathcal{V}_{jsm}, \mathcal{U}_{jsm}$, and $\mathcal{W}_{jsm}$

$\langle v_i' b_j b_l \rangle_{1-ir}$. In general, in our uniaxial anisotropic case, all these divergences can be removed multiplicatively with the following form of the counter terms, which must be added to the basic action to make the corresponding Green's functions UV finite:

$$
\begin{aligned}
\Delta S = \int \mathrm{d}x \big\{ & \nu \mathbf{v}' \cdot [(Z_1 - 1)\Delta \mathbf{v} + \chi_1 (Z_2 - 1)(\mathbf{n}\cdot\partial)^2 \mathbf{v} + \chi_2 (Z_3 - 1)\mathbf{n}\Delta(\mathbf{n}\cdot\mathbf{v}) \\
& + \chi_3 (Z_4 - 1)\mathbf{n}(\mathbf{n}\cdot\partial)^2(\mathbf{n}\cdot\mathbf{v})] + u\nu \mathbf{b}' \cdot [(Z_5 - 1)\Delta\mathbf{b} + \tau_1 (Z_6 - 1)(\mathbf{n}\cdot\partial)^2\mathbf{b} \\
& + \tau_2 (Z_7 - 1)\mathbf{n}\Delta(\mathbf{n}\cdot\mathbf{b}) + \tau_3 (Z_8 - 1)\mathbf{n}(\mathbf{n}\cdot\partial)^2(\mathbf{n}\cdot\mathbf{b})] \\
& + u\nu^3 g_{v2}\mu^{-2\delta}\mathbf{v}' \cdot [(1 - Z_9)\Delta\mathbf{v}' + \alpha_3 (1 - Z_{10})(\mathbf{n}\cdot\partial)^2\mathbf{v}' \\
& + \alpha_4 (1 - Z_{11})\mathbf{n}\Delta(\mathbf{n}\cdot\mathbf{v}') + \alpha_5 (1 - Z_{12})\mathbf{n}(\mathbf{n}\cdot\partial)^2(\mathbf{n}\cdot\mathbf{v}')]/2 \\
& + u^2\nu^3 g_{b2}\mu^{-2\delta}\mathbf{b}' \cdot [(1 - Z_{13})\Delta\mathbf{b}' + \sigma_3 (1 - Z_{14})(\mathbf{n}\cdot\partial)^2\mathbf{b}' \\
& + \sigma_4 (1 - Z_{15})\mathbf{n}\Delta(\mathbf{n}\cdot\mathbf{b}') + \sigma_5 (1 - Z_{16})\mathbf{n}(\mathbf{n}\cdot\partial)^2(\mathbf{n}\cdot\mathbf{b}')]/2 \\
& + \mathbf{v}' \cdot [(Z_{17} - 1)(\mathbf{b}\cdot\partial)\mathbf{b} + \lambda_1 (Z_{18} - 1)\mathbf{b}(\mathbf{n}\cdot\partial)(\mathbf{n}\cdot\mathbf{b}) \\
& + \lambda_2 (Z_{19} - 1)\mathbf{n}(\mathbf{n}\cdot\partial)\mathbf{b}^2 + \lambda_3 (Z_{20} - 1)\mathbf{n}(\mathbf{b}\cdot\partial)(\mathbf{n}\cdot\mathbf{b}) \\
& + \lambda_4 (Z_{21} - 1)\mathbf{n}(\mathbf{n}\cdot\partial)(\mathbf{n}\cdot\mathbf{b})^2]\big\}, \qquad (23)
\end{aligned}
$$

where, all quantities without subscript 0 are renormalized counterparts of the unrenormalized ones in the action (9), μ is the renormalization mass (a scale-setting parameter), and Z_i, $i = 1, \ldots, 21$ are the independent set of the renormalization constants of the model.

Thus, the model is multiplicatively renormalized by the renormalization of the bare parameters in the following general form

$$
g_{v10} = g_{v1}\mu^{2\varepsilon} Z_{g_{v1}}, \quad g_{v20} = g_{v2}\mu^{-2\delta} Z_{g_{v2}}, \quad g_{b10} = g_{b1}\mu^{2a\varepsilon} Z_{g_{b1}}, \qquad (24)
$$
$$
g_{b20} = g_{b2}\mu^{-2\delta} Z_{g_{b2}}, \quad u_0 = u Z_u, \quad \nu_0 = \nu Z_\nu, \quad \chi_{i0} = \chi_i Z_{\chi_i} \qquad (25)
$$
$$
\tau_{i0} = \tau_i Z_{\tau_i}, \quad \alpha_{i+2,0} = \alpha_{i+2} Z_{\alpha_{i+2}}, \quad \sigma_{i+2,0} = \sigma_{i+2} Z_{\sigma_{i+2}} \qquad (26)
$$
$$
\lambda_{j0} = \lambda_j Z_{\lambda_j}, \qquad (27)
$$

for $i = 1, 2, 3$, and $j = 1, 2, 3, 4$. At the same time, the renormalization constants in Eqs. (24)–(27) are directly related to the renormalization constants introduced in Eq. (23) as follows

$$Z_\nu = Z_1, \quad Z_{\chi_i} = Z_1^{-1} Z_{i+1}, \quad Z_u = Z_1^{-1} Z_5, \quad Z_{\tau_i} = Z_5^{-1} Z_{5+i}, \tag{28}$$

$$Z_{g_{v2}} = Z_1^{-2} Z_5^{-1} Z_9, \quad Z_{\alpha_{2+i}} = Z_9^{-1} Z_{9+i}, \quad Z_{g_{b2}} = Z_1^{-1} Z_5^{-2} Z_{13} Z_{17}, \tag{29}$$

$$Z_{\sigma_{2+i}} = Z_{13}^{-1} Z_{13+i}, \quad Z_{\mathbf{b}} = Z_{\mathbf{b}'}^{-1} = Z_{17}^{1/2}, \quad Z_{\lambda_j} = Z_{17}^{-1} Z_{17+j}, \tag{30}$$

$$Z_{g_{v1}} = Z_1^{-2} Z_5^{-1}, \quad Z_{g_{b1}} = Z_1^{-1} Z_5^{-2} Z_{17} \tag{31}$$

where again $i = 1, 2, 3$, and $j = 1, 2, 3, 4$.

In the framework of the one-loop approximation, the renormalization constants Z_i, ($i = 1, \ldots, 21$) are determined by the calculation of 13 different Feynman diagrams: two diagrams for $\langle v'v \rangle$, two diagrams for $\langle v'v' \rangle$, two diagrams for $\langle b'b \rangle$, one diagram for $\langle b'b' \rangle$, and six diagrams for the interaction vertex $\langle v'bb \rangle$. All one-loop Feynman diagrams that contribute to the renormalization of the propagators are explicitly shown in Fig. 3. On the other hand, all one-loop Feynman diagrams that contribute to the renormalization of the interaction vertex $\mathcal{W}_{jsm}$ are shown in Fig. 4.

In the framework of the minimal subtraction (MS) scheme, the one-loop expressions for the renormalization constants have the following general form

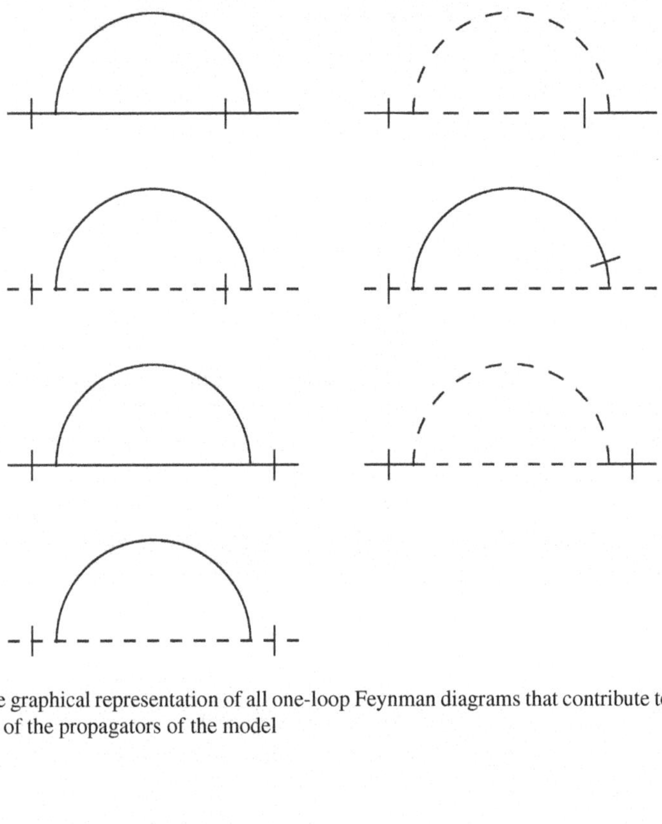

Fig. 3 The graphical representation of all one-loop Feynman diagrams that contribute to the renormalization of the propagators of the model

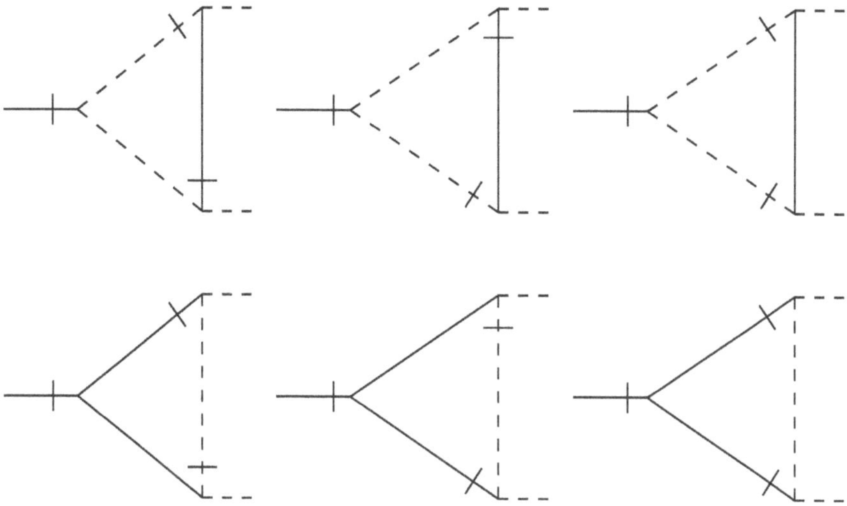

Fig. 4 The graphical representation of all one-loop Feynman diagrams that contribute to the renormalization of the interaction vertex $\mathcal{W}_{jsm}$

$$Z_i = 1 + \frac{S_d}{(2\pi)^d}\left(\frac{g_{v1}}{2\varepsilon}F_{i1} + \frac{g_{v2}}{-2\delta}F_{i2} + \frac{g_{b1}}{2a\varepsilon}F_{i3} + \frac{g_{b2}}{-2\delta}F_{i4}\right), \quad i = 1, \ldots, 8, \quad (32)$$

$$Z_i = 1 + \frac{S_d}{(2\pi)^d}\left(\frac{g_{v1}^2}{4\varepsilon + 2\delta}\frac{F_{i1}}{g_{v2}} + \frac{g_{v1}}{2\varepsilon}F_{i2} + \frac{g_{v2}}{-2\delta}F_{i3} + \frac{g_{b1}^2}{4a\varepsilon + 2\delta}\frac{F_{i4}}{g_{v2}}\right.$$
$$\left. + \frac{g_{b1}g_{b2}}{2a\varepsilon}\frac{F_{i5}}{g_{v2}} + \frac{g_{b2}^2}{-2\delta}\frac{F_{i6}}{g_{v2}}\right), \quad i = 9, \ldots, 12, \quad (33)$$

$$Z_i = 1 + \frac{S_d}{(2\pi)^d}\left(\frac{g_{v1}g_{b1}}{2\varepsilon(1+a) + 2\delta}\frac{F_{i1}}{g_{b2}} + \frac{g_{v1}}{2\varepsilon}F_{i2} + \frac{g_{v2}g_{b1}}{2a\varepsilon}\frac{F_{i3}}{g_{b2}} + \frac{g_{v2}}{-2\delta}F_{i4}\right),$$
$$i = 13, \ldots, 16 \quad (34)$$

$$Z_i = 1 + \frac{S_d}{(2\pi)^d}\left(\frac{g_{v1}}{2\varepsilon}F_{i1} + \frac{g_{v2}}{-2\delta}F_{i2} + \frac{g_{b1}}{2a\varepsilon}F_{i3} + \frac{g_{b2}}{-2\delta}F_{i4}\right), \quad i = 17, \ldots, 21, (35)$$

where the coefficients F_{ij} are rather huge functions of the model parameters and therefore we shall not show all of them here. However, as an example, let us show the explicit form of the functions $F_{1i}, i = 1, \ldots, 4$ for the renormalization constant Z_1. The explicit form of all functions F_{ij} will be given elsewhere. Thus, the explicit form of functions $F_{1i}, i = 1, \ldots, 4$ is the following

$$\frac{F_{11}}{u} = \frac{1-d}{4(2+d)} - \frac{d+1}{4d(4+d)}\alpha_1 - \frac{d-2}{4d(2+d)(4+d)}\alpha_2 + \frac{d^2+3d+3}{2d(2+d)(4+d)}\chi_1 + \frac{3d-2}{4d(2+d)(4+d)}\chi_2 + \frac{3(3d+2)}{4d(2+d)(4+d)(6+d)}\chi_3, \tag{36}$$

$$\frac{F_{12}}{u} = \frac{1-d}{4(2+d)} - \frac{d+1}{4d(4+d)}\alpha_3 - \frac{d-2}{4d(2+d)(4+d)}\alpha_4 - \frac{3}{4d(2+d)(4+d)(6+d)}\alpha_5 + \frac{d^2+3d+3}{2d(2+d)(4+d)}\chi_1 + \frac{3d-2}{4d(2+d)(4+d)}\chi_2 + \frac{3(3d+2)}{4d(2+d)(4+d)(6+d)}\chi_3, \tag{37}$$

$$F_{13} = -\frac{d^2+d-4}{4d(2+d)} - \frac{d^2+5d+2}{4d(2+d)(4+d)}\sigma_1 - \frac{d+6}{4d(2+d)(4+d)}\sigma_2 + \frac{d^2+5d+1}{2d(2+d)(4+d)}\tau_1 + \frac{d+10}{4d(2+d)(4+d)}\tau_2 + \frac{3(d+14)}{4d(2+d)(4+d)(6+d)}\tau_3, \tag{38}$$

and

$$F_{14} = -\frac{d^2+d-4}{4d(2+d)} - \frac{d^2+5d+2}{4d(2+d)(4+d)}\sigma_3 - \frac{d+6}{4d(2+d)(4+d)}\sigma_4 - \frac{3(d+8)}{4d(2+d)(4+d)(6+d)}\sigma_5 + \frac{d^2+5d+1}{2d(2+d)(4+d)}\tau_1 + \frac{d+10}{4d(2+d)(4+d)}\tau_2 + \frac{3(d+14)}{4d(2+d)(4+d)(6+d)}\tau_3. \tag{39}$$

All possible infrared (IR) scaling regimes of the model are driven by the IR stable fixed points of the RG equations, the coordinates of which are given by the zero points of the system of the so-called RG beta functions:

$$\beta_C(C_*) = 0, \quad C = \{g_{v1}, g_{v2}, g_{b1}, g_{b2}, u, \alpha_i, \sigma_i, \chi_j, \tau_j, \lambda_l\}, \tag{40}$$

for $i = 3, 4, 5$, $j = 1, 2, 3$, $l = 1, \ldots, 4$, and the star denotes the fixed point values of the parameters. At the same time, a fixed point C_* is IR stable when the real parts of all eigenvalues of the matrix

$$\Omega_{ij} = \left(\frac{\partial \beta_{C_i}}{\partial_{C_j}}\right)_{C=C_*} \tag{41}$$

are positive.

The system of all β functions of the present model are given as follows (note that, in general, the β function for an invariant charge g is defined as $\beta_g = \mu\partial g/\partial\mu$)

$$\beta_{g_{v1}} = g_{v1}(-2\varepsilon + 2\gamma_1 + \gamma_5), \quad \beta_{g_{b1}} = g_{b1}(-2a\varepsilon + \gamma_1 + 2\gamma_5 - \gamma_{17}), \tag{42}$$
$$\beta_{g_{v2}} = g_{v2}(2\delta + 2\gamma_1 + \gamma_5 - \gamma_9), \quad \beta_{g_{b2}} = g_{b2}(2\delta + \gamma_1 + 2\gamma_5 - \gamma_{17} - \gamma_{13}), \tag{43}$$
$$\beta_u = u(\gamma_1 - \gamma_5), \quad \beta_{\chi_i} = \chi_i(\gamma_1 - \gamma_{i+1}), \quad \beta_{\tau_i} = \tau_i(\gamma_5 - \gamma_{i+5}), \tag{44}$$
$$\beta_{\alpha_{2+i}} = \alpha_{2+i}(\gamma_9 - \gamma_{9+i}), \quad \beta_{\sigma_{2+i}} = \sigma_{2+i}(\gamma_{13} - \gamma_{13+i}), \tag{45}$$
$$\beta_{\lambda_j} = \lambda_j(\gamma_{17} - \gamma_{17+j}), \tag{46}$$

where $i = 1, 2, 3$, $j = 1, \ldots, 4$, and the so-called anomalous dimensions γ_i, $(i = 1, \ldots, 21)$ are defined directly through the corresponding renormalization constants Z_i as follows

$$\gamma_i = \mu \frac{\partial Z_i}{\partial \mu}. \tag{47}$$

Thus, now we have all necessary tools for the analysis of the influence of the weak small scale anisotropy on the stability of the scaling regimes of the MHD turbulence for all spatial dimensions $d \geq 2$. However, the corresponding detailed analysis of all possible scaling regimes of the model as well as of the influence of the presence of the small-scale uniaxial anisotropy on their stability will be given elsewhere.

5 Conclusion

In the present paper we have provided the field theoretic formulation of the MHD turbulence for spatial dimensions $d \geq 2$ with the presence of the weak small scale anisotropy. We have performed the corresponding general RG analysis of the model, the explicit form of all renormalization constants of the model was found, and the system of all RG β functions of the model is determined. The results of the present paper will be used for detailed analysis of all IR scaling regimes of the model in near future.

Acknowledgements The work was supported by the VEGA grants No. 2/0065/17 and No. 2/0058/19, by the grant APVV-17-0020, and by the realization of the project ITMS No. 26220120029, based on the supporting operational Research and Development Program financed from the European Regional Development Fund.

References

1. J. Zinn-Justin, *Quantum Field Theory and Critical Phenomena* (Clarendon, Oxford, 1989)
2. A.N. Vasilev. *Quantum-Field Renormalization Group in the Theory of Critical Phenomena and Stochastic Dynamics*, Gordon & Breach, London (2004)
3. J.C. Collins, *Renormalization: an Introduction to Renormalization, the Renormalization Group, and the Operator Product Expansion* (Cambridge University Press, Cambridge, U.K., 1984)
4. D.J. Amit, V. Martin-Mayor, *Field Theory, Renormalization Group, and Critical Phenomena* (Graphs to Computers, World Scientific, Singapore, 2005)

5. L.Ts. Adzhemyan, M. Hnatich, D. Horváth, M. Stehlik, Infrared properties of an anisotropically driven MHD turbulence. Int. J. Modern Phys. B, **9**(26), 3401–3419 (1995)
6. M. Hnatich, J. Honkonen, M. Jurcisin, Stochastic magnetohydrodynamic turbulence in space dimensions $d \geq 2$. Phys. Rev. E **64**(5), 056411 (2001)
7. M. Jurcisin, M. Stehlik, D-dimensional developed MHD turbulence: double expansion model. J. Phys. A: Math. General **39**(25), 8035–8050 (2006)
8. M. Hnatich, E. Jonyova, M. Jurcisin, M. Stehlik, Stability of scaling regimes in $d \geq 2$ developed turbulence with weak anisotropy. Phys. Rev. E. **64**(1), 016312 (2001)
9. J. Honkonen, M.Y. Nalimov, Two-parameter expansion in the renormalization-group analysis of turbulence. Zeitschrift für Physik B: Condensed Matter **99**(2), 297–303 (1996)
10. P.C. Martin, E.D. Siggia, H.A. Rose, Statistical dynamics of classical systems. Phys. Rev. A **8**(1), 423–437 (1973)
11. C. De Dominicis, Techniques de renormalization de la théorie des champs et dynamique des phénoménes critiques. J. de Phys.: Colloq. C1 **37**(1), C1-247–C1-253 (1976)
12. H.K. Janssen, On a lagrangean for classical field dynamics and renormalization group calculations of dynamical critical properties. Zeitschrift für Physik B **23**, 377–380 (1976)
13. R. Bausch, H.K. Janssen, H. Wagner, Renormalized field theory of critical dynamics. Zeitschrift für Physik B **24**, 113–127 (1976)

Intermittency of Chaos Functions and the Belousov-Zhabotinsky Reaction

Shunji Kawamoto

Abstract Intermittent time series generated by the one-dimensional (1-D) solvable chaos map consisting of time-dependent chaos functions are firstly presented, without the accumulation of round-off error caused by numerical iterations. Then, the 1-D map is applied for deriving a 2-D solvable chaos map corresponding to the Belousov-Zhabotinsky (BZ) reaction, which is known to have chemical waves in time. Finally, discrete limit cycles with chaotic dynamics and the pattern formation depending on the system parameter are obtained numerically by considering the bifurcation diagram, and are discussed on the basis of the numerical result for chemical cells of the BZ reaction, as one of non-equilibrium open systems.

Keywords Intermittency · Chaos function · 2-D solvable chaos map · Belousov-Zhabotinsky reaction · Bifurcation diagram · Limit cycle · Pattern formation · Non-equilibrium open system

1 Introduction

It is well known that many papers and books have been published on nonlinear dynamics, such as soliton, chaos and fractals [1–3], and have been focused on theoretical, numerical and experimental studies, as nonlinear science [4]. In particular, it has been considered that the one-dimensional (1-D) nonlinear difference equations possess a rich spectrum of dynamic behavior as chaos in the field of biological, physical, chemical and social sciences. Then, the chaos theory has been widely extended to medicine, optics and human sciences [5].

In the meantime, an intermittent chaos of the 3-D system has been shown numerically at the transition from regular to irregular chaotic motion by controlling the variable parameter of the system [6]. After that, the intermittent transition was explained

12th CHAOS Conference Proceedings, 18–22 June 2019, Chania, Crete, Greece.

S. Kawamoto (✉)
Osaka Prefecture University, Sakai, Osaka, Japan
e-mail: kawamoto@eis.osakafu-u.ac.jp

C. H. Skiadas and Y. Dimotikalis (eds.), *12th Chaotic Modeling and Simulation International Conference*, Springer Proceedings in Complexity,
https://doi.org/10.1007/978-3-030-39515-5_11

in terms of 1-D chaotic maps or nonlinear difference equations [7]. At the same time, a discrete dissipative dynamical system for the transition is presented, and a connection with the widely observed 1/f noise phenomena is suggested [8]. Later, statistical properties of the intermittent chaos have been investigated for a simple 1-D map approximating the 3-D system and in coupled dynamical systems [9, 10]. The intermittency in a variety of different systems and types has been discussed and classified [11].

On the other hand, the propagation of waves observed in a chemical reaction system has been reported [12, 13], and the so-called Brusselator, which is a simplified reaction model, has been considered as a non-equilibrium open system [14, 15]. In addition, the data obtained in the experiment on the Belousov-Zhabotinsky (BZ) reaction [16] have been analyzed by constructing the phase portrait and the Poincaré section [17, 18], and after assumptions and rescales, a 2-D model of nonlinear differential equations has been derived for the interaction between local reactions and diffusion as chaotic dynamics [19, 20].

Recently, time-dependent chaos functions and its application to engineering have been proposed, and 1-D, 2-D and 3-D solvable chaos maps are derived from the chaos solutions [21, 22]. Furthermore, the chaos functions are introduced for solvable maps corresponding to the logistic map and the FitzHugh-Nagumo model, which explain population growth and neural phenomena, respectively [23, 24].

This paper presents firstly that intermittent time series of different types are generated on the basis of the time-dependent chaos functions without the accumulation of round-off error caused by numerical iterations, in Sect. 2. In Sect. 3, the BZ reaction model is introduced, and the 1-D map obtained from the chaos functions is applied for deriving 2-D solvable maps corresponding to the BZ reaction model. Section 4 is devoted to the numerical calculation for finding bifurcation diagrams and discrete limit cycles with chaotic dynamics, and the MATLAB program is given in Appendix. Finally, conclusions are summarized in the last section.

2 Intermittency of Chaos Functions

For a time-dependent chaos function;

$$x_n(t) = \cos(2^n t), \tag{1}$$

$$t \neq \pm m\pi/2^l \tag{2}$$

with finite positive integers $\{l, m\}$, we have a 1-D solvable chaos map in discrete time t_i satisfying (2) as

$$\begin{aligned} x_{n+1}(t_{i+1}) &= \cos^2(2^n t_i) - \sin^2(2^n t_i) \\ &= 2\cos^2(2^n t_i) - 1 \end{aligned}$$

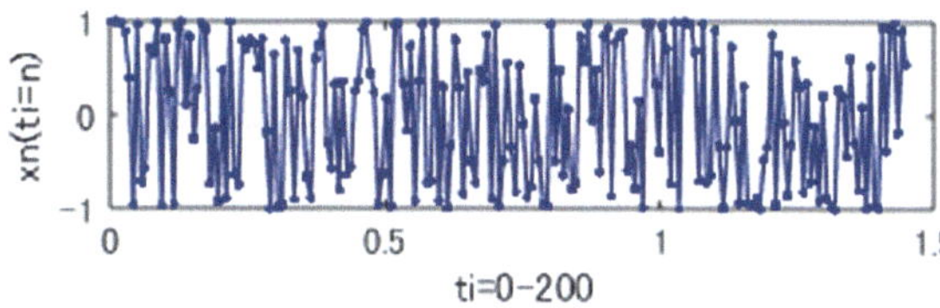

Fig. 1 Chaotic time series of $x_n(t_i) = \cos(2^n t_i)$ with an initial value $x_0(t_0) = 1.0$ and the time step $\Delta t = \pi/431 \approx 0.0073$

$$= 2x_n^2(t_i) - 1, \tag{3}$$

where the passage from a point $x_n(t_i)$ to the next one $x_{n+1}(t_{i+1})$ with the time step $\Delta t \equiv t_{i+1} - t_i$ can be considered as a mapping. Then, it is known that the chaos solution (1) with (2) and the 1-D map (3) are extended to the 2-D solvable chaos maps and the 2-D chaotic maps [21, 22]. In Fig. 1, the numerical result of (1) with (2) is illustrated as regular chaotic time series without the accumulation of round-off error caused by numerical iterations [23, 25].

Similarly, from the following function consisting of (1) with (2) as

$$x_n(t_i) = \frac{a}{b + \cos(2^n t_i)}, \tag{4}$$

where $\{a \neq 0, b\}$ are real constants, we obtain numerically chaotic time series as shown in Fig. 2a–e without the accumulation of round-off error, which indicate intermittent time series, particularly (c) and (d) show spiky time series observed in neuroscience. Then, we find a 1-D solvable chaos map from (4);

$$x_{n+1}(t_{i+1}) = \frac{a x_n^2(t_i)}{(2b^2 + b - 1)x_n^2(t_i) - 4abx_n(t_i) + 2a^2}, \tag{5}$$

here it is important to note that the form of the 1-D map (5) plays a keyrole for finding a 2-D map corresponding to the BZ reaction model shown in Sect. 3.

3 The BZ Reaction Model and 2-D Solvable Chaos Maps

As is known, the propagation of waves in chemical reaction systems have been observed, and the activity is explained as chemical oscillations [12, 13]. Then, a simplified reaction model, the so-called Brusselator has been discussed as a non-equilibrium system [14, 15], and the data obtained in the experiment on the BZ reaction [16] are analyzed by constructing the phase portrait and the Poincaré section [17, 18]. After assumptions and rescales, a two-variable model of nonlinear differential equations is derived for the interaction between local reactions and diffusion to describe the complex reaction dynamics;

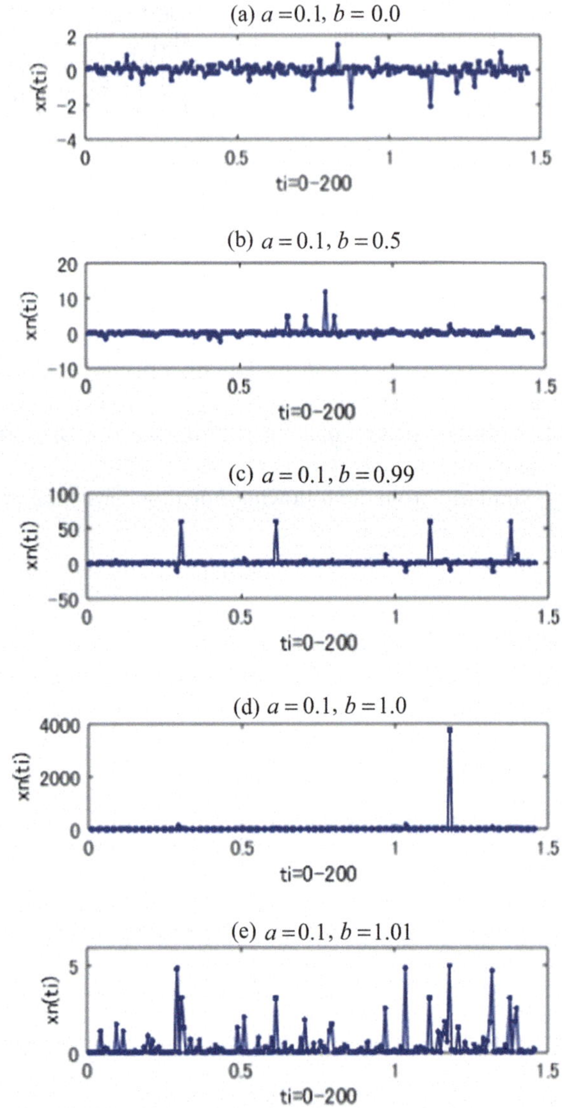

Fig. 2 Chaotic and intermittent time series of $x_n(t_i) = a/(b + \cos(2^n t_i))$ with constants $\{a \neq 0, b\}$ and an initial value $x_0(t_0) = a/(b + 1.0)$

$$\varepsilon \frac{du}{dt} = u(1 - u) + \frac{qv(p - u)}{p + u}, \tag{6}$$

$$\frac{dv}{dt} = u - v, \tag{7}$$

where $u(t)$ and $v(t)$ are dimensionless variables, and $\{\varepsilon, p, q\}$ are scaling parameters, respectively [19, 20].

Moreover, it is interesting to note that a simple model for fish and nutrients interplay is given by

$$\frac{dx}{dt} = x(1-x) - \frac{xy}{x+a}, \tag{8}$$

$$\frac{dy}{dt} = b\frac{xy}{x+h} - y - \frac{y^2}{y^2+c}, \tag{9}$$

which present the interaction of different plankton biomass with real constants $\{a, b, c, h\}$ [26], and has a similar form presented by (5) to the BZ reaction model (6) and (7). Later, it is reported that chemical waves in the BZ reaction can transport objects along a desired path to a desired location [27].

Then, we find 2-D chaotic maps corresponding to the BZ reaction model (6) and (7), which has chemical waves in time. Firstly, from the time-dependent chaos functions given by

$$x_n(t_i) = \sin^2(2^n t_i), \tag{10}$$

$$y_n(t_i) = \frac{a}{b + \cos(2^n t_i)}, \tag{11}$$

we have a condition

$$x_n(t_i) + (\frac{a}{y_n(t_i)} - b)^2 = 1, \tag{12}$$

and the 2-D solvable chaos maps are derived from (10), (11) and (12) as

$$x_{n+1}(t_{i+1}) = 4x_n(t_i)(1 - x_n(t_i)) \tag{13a}$$

$$= x_n(t_i)(1 - x_n(t_i)) + 3x_n(t_i)(1 - x_n(t_i)) \tag{13b}$$

$$= x_n(t_i)(1 - x_n(t_i)) + 3x_n(t_i)(\frac{a}{y_n(t_i)} - b)^2, \tag{13c}$$

$$y_{n+1}(t_{i+1}) = \frac{a}{b + (\cos^2(2^n t_i) - \sin^2(2^n t_i))} = \frac{a}{(1+b) - 2x_n(t_i)}, \tag{14}$$

where $x_n(t_i)$ and $y_n(t_i)$ are chaos solutions (10) and (11) of the discrete time t_i satisfying (2). In addition, the Eq. (13c) is derived by using the condition (12) in (13b), and we have three 2-D chaos maps; (13a) and (14), (13b) and (14), (13c) and (14), which have chaos solutions (10) and (11) with the condition (12), respectively. In Fig. 3a–d, time series of $x_n(t_i)$ and $y_n(t_i)$ are shown in (a) and (b), and the orbit

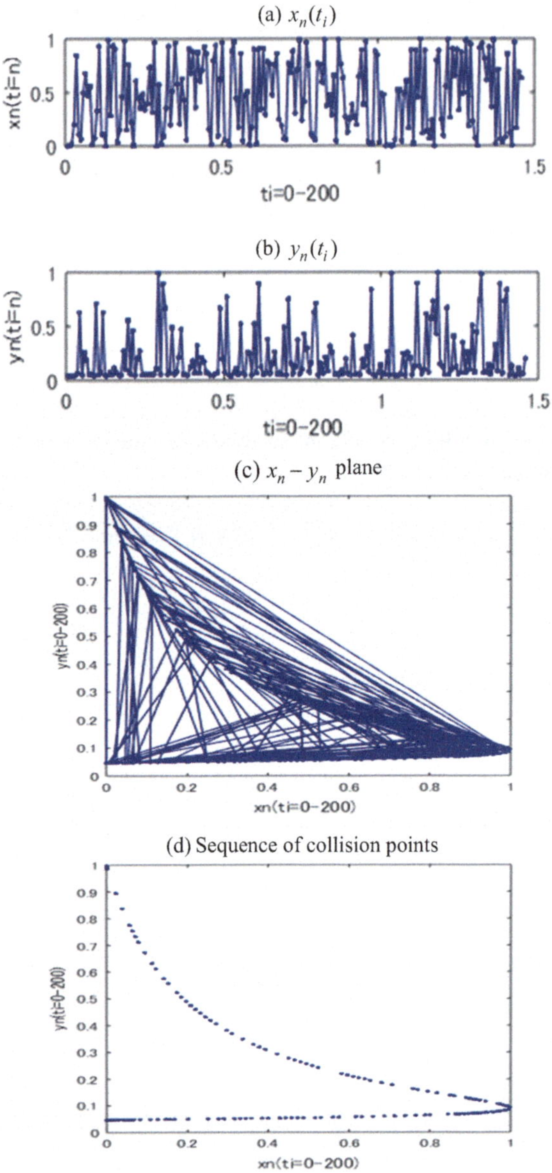

Fig. 3 Chaotic and intermittent time series of $x_n(t_i)$ and $y_n(t_i)$ in **a** and **b**, respectively, the orbit of one chemical cell in **c**, and the sequence of collision points in **d**, with an initial point $(x_0, y_0) = (0.0, a/(1+b))$, $a = 0.1$ and $b = 1.1$

of one chemical cell with an initial point $(x_0, y_0) = (0.0, a/(1+b))$, $a = 0.1$ and $b = 1.1$ is illustrated in (c). The sequence of collision points is presented in (d), which gives the discrete curve (12) obtained from the chaos solutions (10) and (11). Here, it should be noticed that Fig. 3a–d is calculated without the accumulation of round-off error for $x_n(t_i)$ and $y_n(t_i)$ [23, 25], and a MATLAB program for Fig. 3 is given in Appendix.

Thus, we obtain the following three 2-D chaotic maps by introducing a system parameter α, $0 \le \alpha \le 4.0$ as

Case 1:

$$x_{n+1}(t_{i+1}) = \alpha x_n(t_i)(1 - x_n(t_i)), \tag{15}$$

$$y_{n+1}(t_{i+1}) = \frac{a}{(1+b) - 2x_n(t_i)}, \tag{16}$$

Case 2:

$$x_{n+1}(t_{i+1}) = x_n(t_i)(1 - x_n(t_i)) + (\alpha - 1)x_n(t_i)(1 - x_n(t_i)), \tag{17}$$

$$y_{n+1}(t_{i+1}) = \frac{a}{(1+b) - 2x_n(t_i)}, \tag{18}$$

Case 3:

$$x_{n+1}(t_{i+1}) = x_n(t_i)(1 - x_n(t_i)) + (\alpha - 1)x_n(t_i)(\frac{a}{y_n(t_i)} - b)^2, \tag{19}$$

$$y_{n+1}(t_{i+1}) = \frac{a}{(1+b) - 2x_n(t_i)}, \tag{20}$$

where three 2-D chaotic maps have chaos solutions (10) and (11) at $\alpha = 4.0$, and the solutions $x_n(t_i)$ and $y_n(t_i)$ are unstable numerically for $\alpha > 4.0$. In Sect. 4, we obtain bifurcation diagrams for Cases 1, 2 and 3, and find discrete limit cycles, which may correspond to chemical waves observed in the BZ reaction [12, 13].

4 Discrete Limit Cycles with Chaotic Dynamics

The three 2-D chaotic maps, that is, Cases 1, 2 and 3 are derived with a system parameter α in Sect. 3, and we obtain and discuss the bifurcation diagrams in order to find discrete limit cycles for the 2-D chaotic maps in this Section.

Bifurcation diagrams calculated for Cases 1, 2 and 3 are illustrated in Fig. 4a, b, and have the first bifurcation point arising at $\alpha = 3.0$ in the interval $0 \le \alpha \le 4.0$. Here, we carry out 200 iterations for the 2-D map, and drop the first 150 iterations to show the remaining 50 subsequent values of $x_n(t_i)$ and $y_n(t_i)$ at each value of α. Cases 1 and 2 have the same diagrams as shown in Fig. 4a, and Case 3 given by (19) and (20) has a different chaotic region from Cases 1 and 2 in the diagram. However, as shown numerically in Fig. 5a–d for the orbit and the sequence of collision points, Cases 1 and 2 have no limit cycles, and Case 3 has discrete and stable limit cycles with chaotic dynamics of $x_n(t_i)$ and $y_n(t_i)$ on the $x_n - y_n$ plane in the interval $3.0 \le \alpha \le 3.5$. It is found that the sequence of points shows discrete and stable

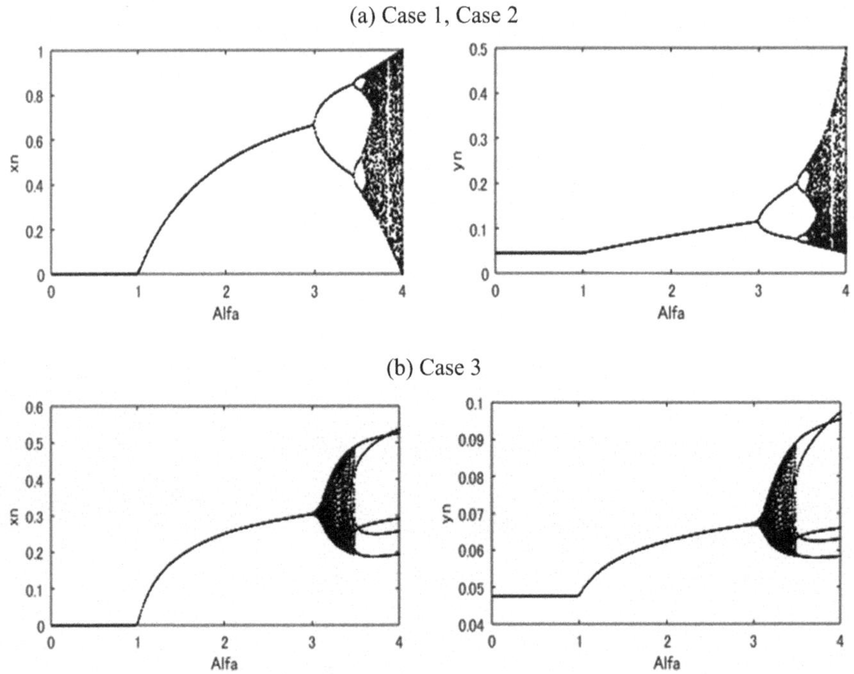

Fig. 4 Bifurcation diagrams of $x_n(t_i)$ and $y_n(t_i)$ for **a** Case 1: (15) and (16), Case 2: (17) and (18), and **b** Case 3: (19) and (20)

limit cycles as presented in Fig. 5, and the orbit converges to the limit cycle from the initial point $(x_0, y_0) = (0.3, 0.07)$.

Thus, the sequence of collision points forms different patterns of limit cycle as the system parameter α increases. Therefore, the pattern formation of discrete limit cycles depends on the system parameter α, and moreover may depend on initial, external and boundary conditions for Case 3, as the waves observed originally in chemical reaction systems.

5 Conclusions

We have presented in this paper that intermittent time series of different types are numerically calculated by introducing chaos functions, without the accumulation of round-off error caused by the numerical iterations. Next, 2-D chaotic maps with a system parameter α, which correspond to the BZ reaction model, are derived on the basis of the chaos functions, and the bifurcation diagrams are calculated. Then, discrete limit cycles with chaotic dynamics and the pattern formation depending on the parameter α, are numerically obtained from the 2-D map (19) and (20). The result

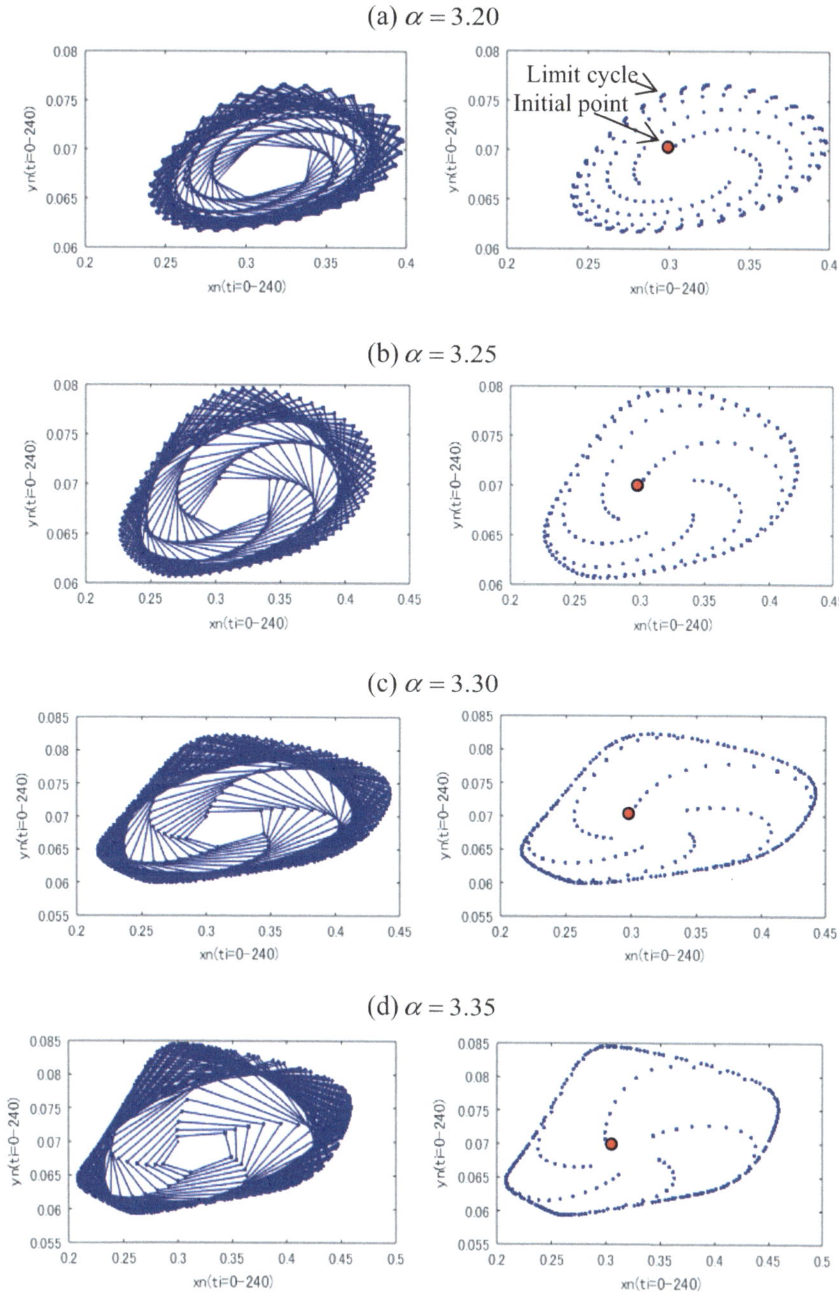

Fig. 5 Discrete and chaotic dynamics of limit cycles with initial point $(x_0, y_0) = (0.3, 0.07)$; **a** $\alpha = 3.20$, **b** $\alpha = 3.25$, **c** $\alpha = 3.30$, **d** $\alpha = 3.35$ for Case 3 given by (19) and (20)

would be helpful for explaining the chaotic dynamics observed in the BZ reaction, as one of non-equilibrium open systems.

Appendix

```
% MATLAB program for Fig. 3 by S. Kawamoto
% initial conditions
T = zeros(1,200);
ILN1 = zeros(200,200);
ILN2 = zeros(200,200);
X = zeros(200,200);
Y = zeros(200,200);
XX = zeros(1,200);
YY = zeros(1,200);
L = 1;
PR = 431;
A = 0.1;
B = 1.1;
% 2-D chaos solution on the Xn-Yn plane
for I = 1:200, T(I) = I.*L.*pi./PR; end
for I = 1:200

  for N = 1
    ILN1(I,N) = mod(2^N.*I.*L,PR);
    X(I,N) = (sin(ILN1(I,N).*pi./PR)).^2;
    ILN2(I,N) = mod(2^N.*I.*L,2*PR);
    Y(I,N) = A/(B + cos(ILN2(I,N).*pi./PR));
  end
  for N = 2:I
    ILN1(I,N) = mod(2.*ILN1(I,N-1),PR);
    X(I,N) = (sin(ILN1(I,N).*pi./PR)).^2;
    ILN2(I,N) = mod(2.*ILN2(I,N-1),2*PR);
    Y(I,N) = A/(B + cos(ILN2(I,N).*pi./PR));
  end

end
for I = 1:200

  XX(I) = X(I,I);
  YY(I) = Y(I,I);

end
% figures Fig. 3a–d
figure('Position',[100 100 350 100])
plot(T,XX,'−b.','MarkerFaceColor','b','MakerSize',7);
```

```
xlabel('ti = 0–200'); ylabel('xn(ti = n)')
figure('Position',[100 100 350 100])
plot(T,YY,'−b.','MarkerFaceColor','b','MakerSize',7);
xlabel('ti = 0–200'); ylabel('yn(ti = n)')
figure('Position',[100 100 350 350])
plot(XX,YY,'−b.','MarkerFaceColor','b','MakerSize',7);
xlabel('xn(ti = 0–200)'); ylabel('yn(ti = 0–200)')
figure('Position',[100 100 350 350])
plot(XX,YY,'b.','MarkerFaceColor','b','MakerSize',7);
xlabel('xn(ti = 0–200)'); ylabel('yn(ti = 0–200)')
```

References

1. M.J. Ablowitz, H. Segur, *Solitons and the Inverse Scattering Transform* (SIAM, Philadelphia, 1981)
2. F.C. Moon, *Chaotic and Fractal Dynamics* (Wiley, New York, 1992)
3. H. Peitgen, H. Jurgens, D. Saupe, *Chaos and Fractals—New Frontiers of Science* (Springer, New York, 1992)
4. A. Scott, *Nonlinear Science* (Routledge, London, 2005)
5. C. H. Skiadas, C. Skiadas (eds.). *Handbook of Application of Chaos Theory* (Chapman and Hall/CRC Press, 2016)
6. P. Manneville, Y. Pomeau, Intermittency and the Lorenz model. Phys. Lett. **75A**, 1–2 (1979)
7. Y. Pomeau, P. Manneville, Intermittent transition to turbulence in dissipative dynamical systems. Commun. Math. Phys. **74**, 189–197 (1980)
8. P. Manneville, Intermittency, self-similarity and 1/f spectrum in dissipative dynamical systems. J. Phys. **41**, 1235–1243 (1980)
9. T. Kohyama, Y. Aizawa, Theory of the intermittent chaos—1/f spectrum and the Pareto-Zipf law. Prog. Theor. Phys. **71**, 917–929 (1984)
10. H. Fujisaka, T. Yamada, A new intermittency in coupled dynamical systems. Prog. Theor. Phys. **74**, 918–921 (1985)
11. E. del Rio, S. Elaskar. The intermittency route to chaos. *Handbook of Application of Chaos Theory*, ed. by C.H. Skiadas, C. Skiadas (Chapman and Hall/CRC Press, 2016), pp. 3–20
12. A.N. Zaikin, A.M. Zhabotinsky, Concentration wave propagation in two-dimensional liquid-phase self-oscillating system. Nature **225**, 535–537 (1970)
13. A.T. Winfree, Spiral waves of chemical activity. Science **175**, 634–636 (1972)
14. R.J. Field, R.M. Noyes, Explanation of spatial band propagation in the Belousov reaction. Nature **237**, 390–392 (1972)
15. G. Nicolis, I. Prigogine, *Self-Organization in Non-Equilibrium Systems* (Wiley, New York, 1977)
16. A.M. Zhabotinsky, Belousov-Zhabotinsky reaction. Scholarpedia **2**(9), 1435 (2007)
17. J.-C. Roux, R.H. Simoyi, H.L. Swinney, Observation of a strange attractor. Physica **8D**, 257–266 (1983)
18. F. Argoul, A. Arneodo, P. Richetti, J.C. Roux, H.L. Swinney, Chemical chaos: from hints to confirmation. Acc. Chem. Res. **20**, 436–442 (1987)
19. J.J. Tyson, P.C. Fife, Target patterns in a realistic model of the Belousov-Zhabotinskii reaction. J. Chem. Phys. **73**, 2224–2237 (1980)
20. J.A. Vastano, T. Russo, H.L. Swinney, Bifurcation to spatially induced chaos in a reaction-diffusion system. Phys. D **46**, 23–42 (1990)

21. S. Kawamoto, 2-D and 3-D solvable chaos maps. Chaotic Model. Simul. (CMSIM) **1**, 107–118 (2017)
22. S. Kawamoto, Chaotic time series by time-discretization of periodic functions and its application to engineering. Chaotic Model. Simul. (CMSIM) **2**, 193–204 (2017)
23. S. Kawamoto, The FitzHugh-Nagumo model and 2-D solvable chaos maps. Chaotic Model. Simul. (CMSIM) **3**, 269–283 (2018)
24. S. Kawamoto. The FitzHugh-Nagumo model and spatiotemporal fractal sets based on time-dependent chaos functions. in *11th Chaotic Modeling and Simulation International Conference*, Springer Proceedings in Complexity, ed. by C.H. Skiadas, I. Lubashevsky (Springer Nature Switzerland AG, 2019), pp. 159–173
25. S. Kawamoto, T. Horiuchi, Algorithm for exact long time chaotic series and its appication to cryptosystems. Int. J. Bifurcat. Chaos **14**(10), 3607–3611 (2004)
26. M. Scheffer, Fish and nutrients interplay determines algal biomass: a minimal model. Oikos **62**, 271–282 (1991)
27. T. Ichino, T. Asahi, H. Kitahata, N. Magome, K. Agladge, K. Yoshikawa, Microfreight delivered by chemical waves. J. Phys. Chem. C **112**, 3032–3035 (2008)

Reaction-Diffusion Systems and Propagation of Limit Cycles with Chaotic Dynamics

Shunji Kawamoto

Abstract Travelling wave solutions to reaction-diffusion systems are considered from the standpoint based on chaos functions. Firstly, the Fisher-KPP equation, which describes a model for the propagation of gene as nonlinear dynamics, is introduced and is transformed into a two-dimensional (2-D) system of nonlinear differential equations. Then, 2-D solvable chaos maps for the 2-D system are derived from chaos functions, and the bifurcation diagrams are numerically calculated to find a system parameter for limit cycles with discrete and chaotic properties. Finally, the chaotic dynamics are discussed by presenting the so-called entrainment and synchronization, and by illustrating the propagation of limit cycles as travelling waves on a phase plane corresponding to the original plane.

Keywords Reaction-diffusion system · Fisher-KPP equation · Chaos function · Bifurcation diagram · Limit cycle · Entrainment · Synchronization · Travelling wave

1 Introduction

Nonlinear dynamic phenomena, such as soliton, chaos and fractals have been of interest in connection with a wide variety of different property in physics, engineering, biology and economics, as nonlinear science, and are usually described by nonlinear differential equations or nonlinear difference equations [1–4]. For chaotic dynamics, the one-dimensional (1-D) nonlinear difference equations are known to possess a rich spectrum of dynamical behavior as chaos, and the chaos theory has been applied to living systems, medicine, optics and human sciences [5, 6].

12th *CHAOS Conference Proceedings, 18–22 June 2019, Chania, Crete, Greece*

S. Kawamoto (✉)
Osaka Prefecture University, Sakai, Osaka, Japan
e-mail: kawamoto@eis.osakafu-u.ac.jp

C. H. Skiadas and Y. Dimotikalis (eds.), *12th Chaotic Modeling and Simulation International Conference*, Springer Proceedings in Complexity,
https://doi.org/10.1007/978-3-030-39515-5_12

On the other hand, travelling wave solutions to reaction-diffusion systems have been discussed extensively, and one of simple examples is the Fisher-Kolmogorov, Petrovsky and Piskounov (F-KPP) equation, which is used in population dynamics and genetics [7, 8]. Originally, the Fisher equation is proposed as a model for the propagation of gene, and the foundation for analytical study of reaction-diffusion systems has been presented by analyzing the F-KPP equation. Later, the effect of boundaries on convection in a shallow layer of fluid heated from below has been considered [9, 10], and entire travelling front solutions to a scalar reaction-diffusion equation have been studied, which can be applied to the F-KPP equation [11]. Here, it is interesting to note that the F-KPP equation includes a simple reaction term, and is related to the logistic function and the well-discussed logistic map for population growth [12, 13].

Recently, time-dependent chaos functions have been proposed, and 1-D, 2-D and 3-D solvable chaos maps are derived [14]. In addition, its application to engineering has been presented [15], and the solvable chaos maps corresponding to the logistic map and the FitzHugh-Nagumo model [16, 17] have been considered as non-equilibrium open systems [18, 19].

In this paper, travelling wave solutions to reaction-diffusion systems are discussed from the standpoint based on chaos functions. Firstly, the F-KPP equation, which describes a model for the propagation of gene, is treated in Sect. 2 as a simplest reaction-diffusion system, and is transformed into a 2-D system of nonlinear differential equations. Then, 2-D solvable chaos maps for the 2-D system are derived from chaos functions in Sect. 3, and the bifurcation diagrams are numerically obtained to find the system parameter for generating limit cycles with discrete and chaotic properties. Finally, the discrete properties are presented by showing the so-called entrainment and synchronization in Sect. 4, and by illustrating the propagation of stable limit cycles as travelling waves on a new phase plane corresponding to the original plane. The last section is devoted to conclusions.

2 Reaction-Diffusion Systems

As is known, reaction-diffusion systems are mathematical models, and appear in the sciences of physical, chemical, biological and ecological phenomena [8]. The simplest one-component reaction-diffusion equation has the following form;

$$\frac{\partial u}{\partial t} = D\frac{\partial^2 u}{\partial x^2} + R(u), \quad u = u(x, t), \tag{1}$$

where D is a diffusion coefficient, and $R(u)$ is a reaction term. Generally, the solutions to (1) are known to include the formation of travelling waves and self-organized patterns [20, 21]. As a reaction term, $R(u) = ru(1-u)$ gives the Fisher-Kolmogorov, Petrovsky and Piskounov (F-KPP) equation for population dynamics and genetics, given by

$$\frac{\partial u}{\partial t} = D\frac{\partial^2 u}{\partial x^2} + ru(1-u), \tag{2}$$

where r is the reaction coefficient, and has been proposed originally as a model for propagation of gene [7]. In addition, the Eq. (2) describes flame propagation, spreading of some reaction fronts and stable travelling waves with respect to small perturbations [8, 22]. Here, if $R(u) = 0$, (1) represents a pure diffusion process, and if $D = 0$, (2) gives the logistic equation and the logistic map [12, 13, 19].

For the travelling waves of (2), we introduce a solution

$$u(x,t) \equiv f(z) \equiv f(x - vt) \tag{3}$$

with a constant velocity v for $z = x - vt, \ v > 0$. Then, (2) at $D = 1$ for simplicity reduces to a second-order nonlinear differential equation;

$$f''(z) + vf'(z) + rf(z)(1 - f(z)) = 0, \tag{4}$$

and by using a transformation $g(z) \equiv -f'(z)$, (4) is rewritten into a 2-D system of the first-order differential equations as

$$f'(z) = -g(z), \tag{5}$$

$$g'(z) = rf(z)(1 - f(z)) - vg(z), \tag{6}$$

which has two fixed points $(f^*, g^*) = (0,0), (1,0)$ on the $f(z) - g(z)$ phase plane with $z = x - vt$. Therefore, the F-KPP Eq. (2) can be transformed to the $f(z) - g(z)$ phase plane consisting of the independent variable z and dependent variables $f(z)$ and $g(z)$, from the $x - u(x,t)$ plane of (2). By introducing the difference method to find a 2-D solvable chaos map corresponding to the 2-D model (5) and (6);

$$\frac{df}{dz} \approx \frac{f_{n+1} - f_n}{\Delta z}, \frac{dg}{dz} \approx \frac{g_{n+1} - g_n}{\Delta z} \tag{7}$$

with $f_n \equiv f(z), g_n \equiv g(z)$ and a step $\Delta z > 0$, we have the following 2-D map from (5)–(7) as

$$f_{n+1}(z_{i+1}) = f_n(z_i) - (\Delta z)g_n(z_i), \tag{8}$$

$$g_{n+1}(z_{i+1}) = r(\Delta z)f_n(z_i)(1 - f_n(z_i)) + (1 - v(\Delta z))g_n(z_i), \tag{9}$$

where z_i is the discrete variable, and the passage from a point $(f_n(z_i),\ g_n(z_i))$ to the next one $(f_{n+1}(z_{i+1}),\ g_{n+1}(z_{i+1}))$ with the step $\Delta z \equiv z_{i+1} - z_i$ on the $f_n - g_n$ plane can be considered as a mapping.

Furthermore, for the reaction term $R(u)$ of (1), it is important to note that the Newell-Whitehead-Segel (NWS) equation has $R(u) = u(1 - u^2)$ describing the Rayleigh-Benard convection [9, 10], and the FitzHugh-Nagumo (FHN) model has $R(u) = u(1 - u)(u - a)$ with a parameter a stating the propagation of neural cells [16, 17], which are widely discussed as reaction-diffusion systems.

3 2-D Solvable Chaos Maps

We introduce the following chaos solutions consisting of chaos functions as

$$f_n(z) = a\cos(2^n z), \tag{10}$$

$$g_n(z) = b\sin^2(2^n z), \tag{11}$$

and conditions

$$\left(\frac{1}{a^2}\right) f_n^2(z) + \left(\frac{1}{b}\right) g_n(z) = 1, \tag{12}$$

$$z \neq \pm m\pi/2^l \tag{13}$$

with coefficients $\{a \neq 0,\ b \neq 0\}$ and finite positive integers $\{l,\ m\}$, and derive the 2-D solvable chaos map;

$$f_{n+1}(z_{i+1}) = -2\Big(\frac{a}{b}\Big) g_n(z_i) + a, \tag{14}$$

$$g_{n+1}(z_{i+1}) = \left(\frac{4}{a^2}\right) f_n^2(z_i) g_n(z_i), \tag{15}$$

$$\left(\frac{1}{a^2}\right) f_n^2(z_i) + \left(\frac{1}{b}\right) g_n(z_i) = 1, \tag{16}$$

where the discrete variable z_i satisfies the condition (13), by choosing the step $\Delta z \equiv z_{i+1} - z_i = \pi/431 \approx 0.0073$ for numerical iterations [18]. Thus, the passage from a point $(f_n(z_i),\ g_n(z_i))$ to the next one $(f_{n+1}(z_{i+1}),\ g_{n+1}(z_{i+1}))$ is considered as a mapping, and here it should be noticed that the nonlinear condition (12) is applied for deriving the linear Eq. (14). We set $a = 1$ and $b = 1$ for simplicity, and generalize the 2-D chaos map (14) and (15) with (16) by introducing a system parameter ε, $0 \leq \varepsilon \leq 4.0$. Then, we find the following 2-D chaotic maps;

$$\text{Case1}: \; f_{n+1}(z_{i+1}) = -2g_n(z_i) + 1, \tag{17}$$

$$g_{n+1}(z_{i+1}) = \varepsilon f_n^2(z_i) g_n(z_i), \tag{18}$$

$$\text{Case 2}: \; f_{n+1}(z_{i+1}) = -2g_n(z_i) + 1, \tag{19}$$

$$g_{n+1}(z_{i+1}) = \varepsilon f_n^2(z_i)(1 - f_n^2(z_i)), \tag{20}$$

$$\text{Case 3}: \; f_{n+1}(z_{i+1}) = -2g_n(z_i) + 1, \tag{21}$$

$$g_{n+1}(z_{i+1}) = \varepsilon g_n(z_i)(1 - g_n(z_i)), \tag{22}$$

$$\text{Case 4}: \; f_{n+1}(z_{i+1}) = -2g_n(z_i) + 1, \tag{23}$$

$$g_{n+1}(z_{i+1}) = \varepsilon(1 - f_n^2(z_i))(1 - g_n(z_i)). \tag{24}$$

Therefore, it is found that the four 2-D maps (Cases 1–4) have chaos solutions (10) and (11) at $\varepsilon = 4.0$, and the solutions $f_n(z_i)$ and $g_n(z_i)$ are numerically unstable for $\varepsilon > 4.0$. Equations (20), (22) and (24) are derived and generalized by using the condition (16) and setting ε instead of the coefficient $(4/a^2)$ in (15), respectively. It is found that Cases 2-4 correspond to the 2-D map (8) and (9) of the F-KPP equation [7], the NWS equation [9, 10] and the FHN model [16, 17] with respect to the form of reaction term, respectively.

Here, for finding limit cycles of reaction-diffusion equations, it is interesting to note that the Van der Pol oscillator is given by

$$\ddot{x} - \varepsilon(1 - x^2)\dot{x} + x = 0, \tag{25}$$

which represents a model for a simple vacuum tube oscillator circuit with a nonlinear damping term [23], and is rewritten by a transformation $y \equiv -\dot{x}$ to the 2-D model;

$$\dot{x} = -y, \tag{26}$$

$$\dot{y} = \varepsilon(1 - x^2)y + x. \tag{27}$$

Then, by the difference method, we have the following 2-D map from (26) and (27) as

$$x_{n+1}(t_{i+1}) = x_n(t_i) - (\Delta t)y_n(t_i), \tag{28}$$

$$y_{n+1}(t_{i+1}) = \varepsilon(\Delta t)(1 - x_n^2(t_i))y_n(t_i) + (\Delta t)x_n(t_i) + y_n(t_i) \tag{29}$$

with $\Delta t \equiv t_{i+1} - t_i$, which gives a mapping, and corresponds to the nonlinear term of Case 1. Similarly, the Duffing oscillator has the form;

$$\ddot{x} + \delta\dot{x} + \alpha x + \beta x^3 = 0, \tag{30}$$

where coefficients $\{\alpha,\ \beta,\ \delta\}$ are real constants [24, 25], and we have the following 2-D map by the difference method as

$$x_{n+1}(t_{i+1}) = x_n(t_i) - (\Delta t)y_n(t_i), \tag{31}$$

$$y_{n+1}(t_{i+1}) = (\Delta t)(\alpha x_n(t_i) + \beta x_n^3(t_i)) + (1 - \delta(\Delta t))y_n(t_i), \tag{32}$$

which corresponds to the nonlinear term of Case 2 given by (19) and (20). It is important to note that the nonlinear term in (18) of Case 1 corresponds to that in (29) derived from the Van der Pol oscillator, which has been discussed on limit cycle [26, 27].

4 Bifurcation Diagrams and Discrete Limit Cycles

The 2-D solvable chaos map (14) and (15) with (16) is generalized by introducing the system parameter ε, $0 \le \varepsilon \le 4.0$, and we have derived the 2-D chaotic maps given by (17)-(24), which have chaos solutions (10) and (11) at $\varepsilon = 4.0$. Therefore, bifurcation diagrams of $f_n(z_i)$ and $g_n(z_i)$ for Cases 1–4 are calculated on the $\varepsilon - f_n$ plane and the $\varepsilon - g_n$ plane, respectively, and are shown in Fig. 1a–d. In the numerical calculation, we carry out 200 iterations for the 2-D maps of Cases 1–4, and drop the first 150 iterations to show the remaining 50 subsequent values of $f_n(z_i)$ and $g_n(z_i)$ at each value of ε.

The bifurcation diagrams have chaotic regions, and it is found from the numerical calculation that Case 1 of Fig. 1a given by (17) and (18) generates discrete limit cycles depending on the system parameter ε as presented in Fig. 2a–d, and however Cases 2–4 of Fig. 1b–d have no limit cycles. As illustrated in Fig. 2a–d, we have discrete

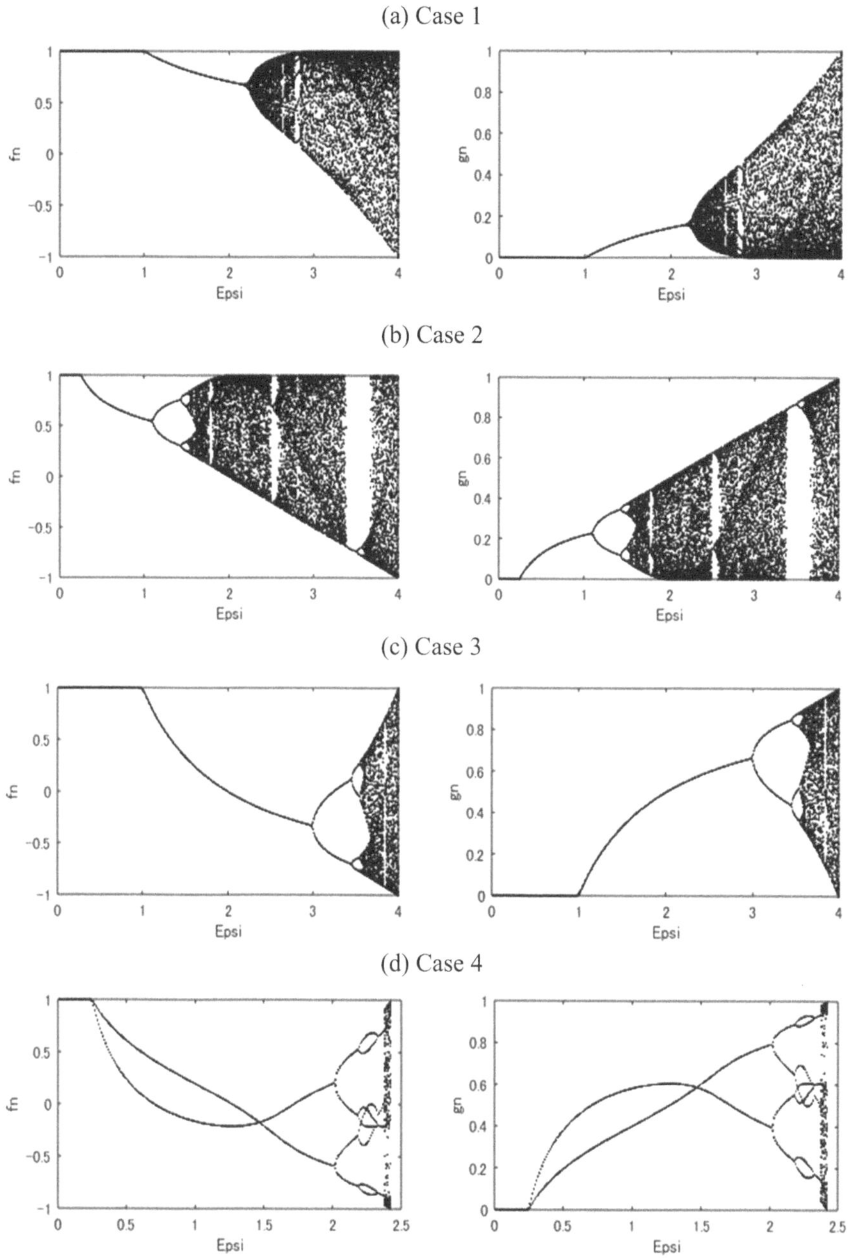

Fig. 1 Bifurcation diagrams of the 2-D maps (Cases 1–4) with a system parameter ε; **a** $f_{n+1} = -2g_n + 1$ and $g_{n+1} = \varepsilon f_n^2 g_n$, **b** $f_{n+1} = -2g_n + 1$ and $g_{n+1} = \varepsilon f_n^2(1 - f_n^2)$, **c** $f_{n+1} = -2g_n + 1$ and $g_{n+1}(z_{i+1}) = \varepsilon g_n(1 - g_n)$, and **d** $f_{n+1} = -2g_n + 1$ and $g_{n+1} = \varepsilon(1 - f_n^2)(1 - g_n)$

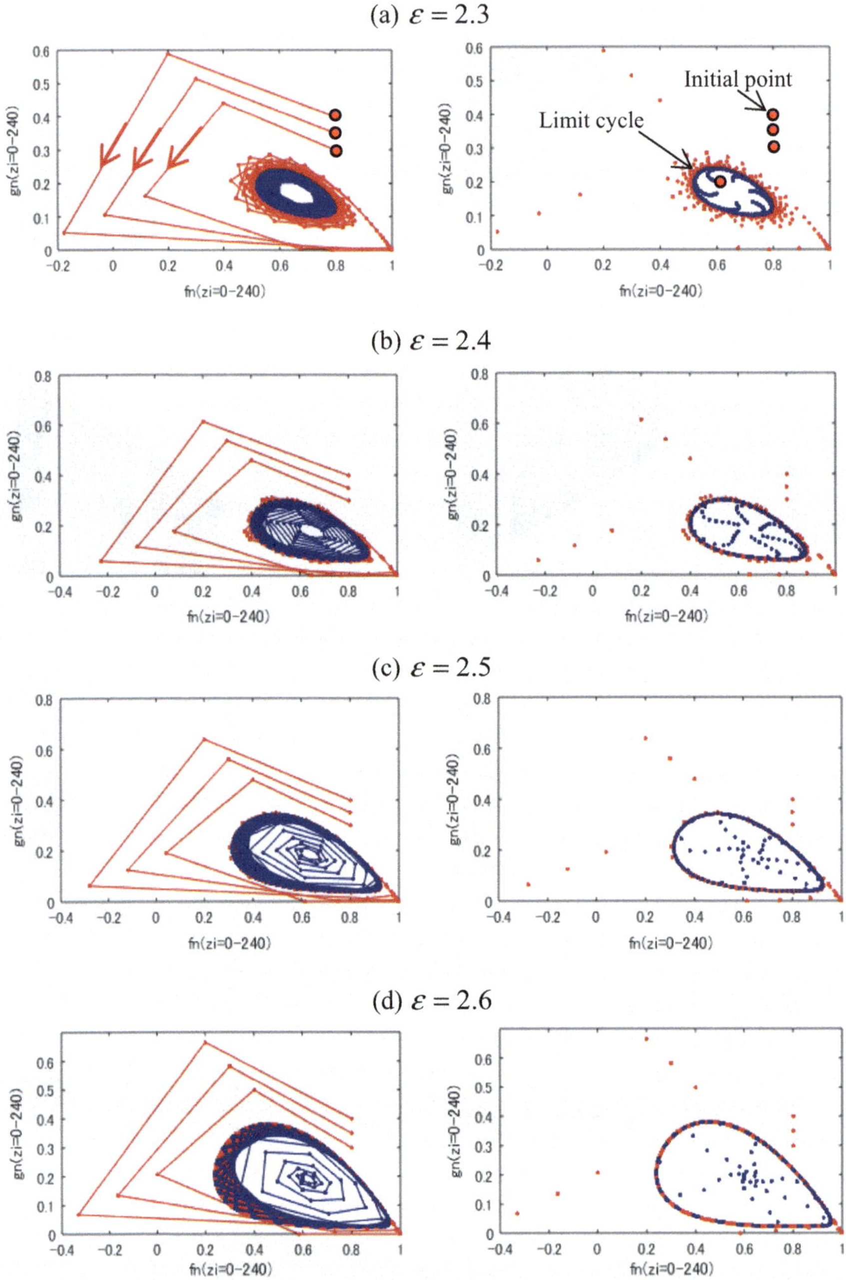

Fig. 2 Discrete and stable limit cycles obtained numerically for the 2-D map (Case 1); $f_{n+1} = -2g_n + 1$ and $g_{n+1} = \varepsilon f_n^2 g_n$ at **a** $\varepsilon = 2.3$, **b** $\varepsilon = 2.4$, **c** $\varepsilon = 2.5$ and **d** $\varepsilon = 2.6$

and stable limit cycles for the 2-D map of Case 1 given by (17) and (18); (a) $\varepsilon = 2.3$, (b) $\varepsilon = 2.4$, (c) $\varepsilon = 2.5$ and (d) $\varepsilon = 2.6$ for the system parameter ε. Especially, the orbit of one chemical cell and the sequence of collision points are shown for four initial points in Fig. 2, where one point $(f_0, g_0) = (0.6, 0.2)$ is in the inside of limit cycle, and three points $(f_0, g_0) = (0.8, 0.3)$, $(0.8, 0.35)$, $(0.8, 0.4)$ are in the outside of limit cycle, respectively. Here, it should be emphasized that orbits of the outside initial points are drawn into the limit cycle, and the orbit of the inside initial point is synchronized with the limit cycle in time, which are the so-called entrainment and synchronization [28–30]. The MATLAB program for Fig. 2a–d is shown in Appendix.

Then, time series of $f_n(z_i)$ and $g_n(z_i)$ for the 2-D map of Case 1 given by (17) and (18) are presented at $\varepsilon = 2.3$ and $\varepsilon = 2.5$ in Figs. 3 and 4, with two initial points $(f_0, g_0) = (0.6, 0.2)$ and $(f_0, g_0) = (0.8, 0.3)$ by blue and red lines, respectively.

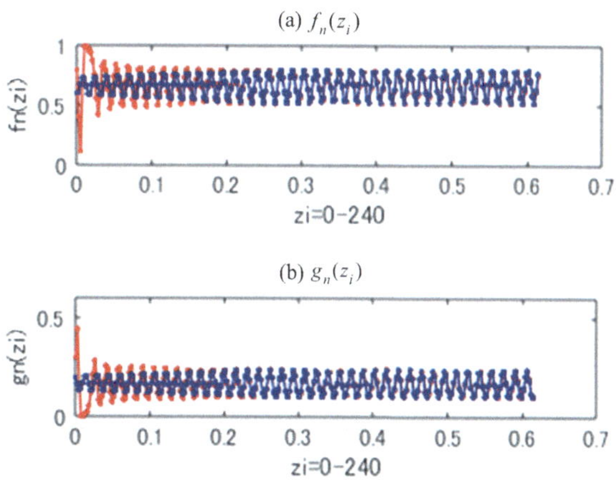

Fig. 3 Time series of $f_n(z_i)$ and $g_n(z_i)$ for Fig. 2 **a** Case 1 at $\varepsilon = 2.3$ with initial points; $(f_0, g_0) = (0.6, 0.2)$ in blue, and $(f_0, g_0) = (0.8, 0.3)$ in red

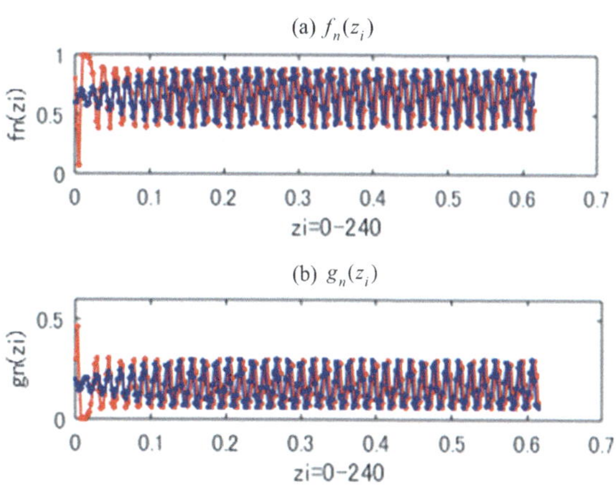

Fig. 4 Time series of $f_n(z_i)$ and $g_n(z_i)$ for Fig. 2 **a** Case 1 at $\varepsilon = 2.5$ with initial points; $(f_0, g_0) = (0.6, 0.2)$ in blue, and $(f_0, g_0) = (0.8, 0.3)$ in red

It is found that the time series converge to the orbit with stable amplitude and phase in Fig. 3a, b, and however the time series shown in Fig. 4a, b converge to the amplitude and have a phase shift as ε increases, where it should be noticed that the system parameter ε is the coefficient of nonlinear term in Case 1 of (17) and (18), and the time series become unstable for $\varepsilon > 4.0$.

From the numerical result, the discrete limit cycles are stable and have chaotic properties. Then, for travelling wave solutions to the F-KPP Eq. (2), we consider the propagation of limit cycles in the following, by introducing the solution (3) treated for the travelling waves of (2). That is, the F-KPP Eq. (2) has the $x - u(x,\ t)$ plane, and can be transformed into the 2-D map (8) and (9) of $f_n(z_i)$ and $g_n(z_i)$ by the solution (3); $u(x,\ t) = f(z) = f(x - vt)$ with a constant velocity v. On the other hand, we have derived the 2-D chaotic map given by (17) and (18) of $f_n(z_i)$ and $g_n(z_i)$, and have calculated stable limit cycles with chaotic dynamics on the $f_n - g_n$ phase plane of the map (17) and (18) as shown in Fig. 2. Therefore, if we assume the following discrete functions from the solution (3) as

$$u(x_n,\ t_i) \equiv f_n(z_i) \equiv f(x_n(t_i) - v(\Delta t)),\ \ \Delta t \equiv t_{i+1} - t_i, \tag{33}$$

then we can introduce a term $\beta(\Delta f)$ as the term $v(\Delta t)$ of (33) for equivalent travelling distance on the $f_n - g_n$ phase plane of the map (17) and (18);

$$v(\Delta t);\ \beta(\Delta f),\ \Delta f \equiv f_{n+1}(z_{i+1}) - f_n(z_i), \tag{34}$$

where β is a parameter of the travelling step Δf corresponding to the travelling velocity v of (3) and (33).

The propagation of discrete limit cycles with $\varepsilon = 2.3$ is presented numerically in Fig. 5a, b; the orbit of one chemical cell in (a) and the sequence of collision points in (b) with the initial point $(f_0,\ g_0) = (0.6,\ 0.2)$ on the $f_n - g_n$ plane, respectively. It is interesting to note here that the limit cycle with $\beta = 0.0$ (black) has the initial point in the inside, and however, for $\beta = 0.4$ (blue) and $\beta = 0.8$ (red), the initial point is in the outside of limit cycle, where the parameter β on the $f_n - g_n$ phase plane corresponds to the velocity v on the $x - u(x,\ t)$ plane of the F-KPP Eq. (2). Thus, the limit cycle is found to propagate forward or backward, depending on $\beta > 0$ or $\beta < 0$, and to be stable numerically for $\beta < 1.0$, under iterations with the step $\Delta f \equiv f_{n+1}(t_{i+1}) - f_n(t_i) = \pi/431 \approx 0.0073$ and by adding the value of $\beta(\Delta f)$ to $f_n(z_i)$ in each numerical iteration.

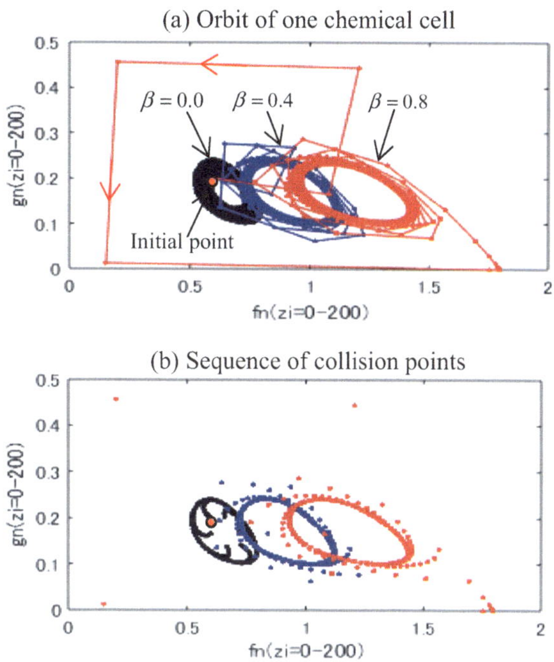

Fig. 5 Propagation of limit cycles as travelling waves on the $f_n - g_n$ phase plane corresponding to the original $x - u(x,\ t)$ plane; **a** Orbit of one chemical cell, **b** Sequence of collision points of limit cycles with a parameter β and an initial point $(f_0,\ g_0) = (0.6,\ 0.2)$

5 Conclusions

We have considered travelling wave solutions to reaction-diffusion systems from the standpoint based on chaos functions. The F-KPP equation, which describes a model for the propagation of gene, is firstly treated, and bifurcation diagrams of the 2-D chaotic maps with a system parameter ε for the reaction-diffusion systems are obtained in order to find limit cycles. Then, the discrete limit cycles with chaotic dynamics are presented in a wide interval of ε, and finally the propagation of limit cycles is discussed under the assumption of a new $f_n - g_n$ phase plane corresponding to the original $x - u(x,\ t)$ plane of the F-KPP equation. From the result calculated numerically in this paper, the stable limit cycles may correspond to the travelling waves observed widely in chemical reactions of reaction-diffusion systems.

Appendix

```
% MATLAB program for Fig. 2 (a)-(d) by S. Kawamoto
% initial conditions
EPSI=zeros(1,4);
X1=zeros(239,239);
Y1=zeros(239,239);
XX1=zeros(1,240);
YY1=zeros(1,240);
X2=zeros(239,239);
Y2=zeros(239,239);
XX2=zeros(1,240);
YY2=zeros(1,240);
X3=zeros(239,239);
Y3=zeros(239,239);
XX3=zeros(1,240);
YY3=zeros(1,240);
X4=zeros(239,239);
Y4=zeros(239,239);
XX4=zeros(1,240);
YY4=zeros(1,240);
X01=0.6;
Y01=0.2;
X02=0.8;
Y02=0.3;
X03=0.8;
Y03=0.35;
X04=0.8;
Y04=0.4;
% system parameter EPSI(J), 2-D chaos map and figures
for J=1:4, EPSI(J)=2.2+0.1.*J;
    for I=1:239
        for N=1
            X1(I,N)=-2*Y01+1;
            Y1(I,N)=EPSI(J).*(X01)^2*Y01;
            X2(I,N)=-2*Y02+1;
            Y2(I,N)=EPSI(J).*(X02)^2*Y02;
            X3(I,N)=-2*Y03+1;
            Y3(I,N)=EPSI(J).*(X03)^2*Y03;
            X4(I,N)=-2*Y04+1;
            Y4(I,N)=EPSI(J).*(X04)^2*Y04;
        end
```

```
            for N=2:I
                X1(I,N)=-2*Y1(I,N-1)+1;
                Y1(I,N)=EPSI(J).*(X1(I,N-I))^2*Y1(I,N-1);
                X2(I,N)=-2*Y2(I,N-1)+1;
                Y2(I,N)=EPSI(J).*(X2(I,N-I))^2*Y2(I,N-1);
                X3(I,N)=-2*Y3(I,N-1)+1;
                Y3(I,N)=EPSI(J).*(X3(I,N-I))^2*Y3(I,N-1);
                X4(I,N)=-2*Y4(I,N-1)+1;
                Y4(I,N)=EPSI(J).*(X4(I,N-I))^2*Y4(I,N-1);
            end
        end
        for I=1
            XX1(I)=X01;
            YY1(I)=Y01;
            XX2(I)=X02;
            YY2(I)=Y02;
            XX3(I)=X03;
            YY3(I)=Y03;
            XX4(I)=X04;
            YY4(I)=Y04;
        end
        for  I=2:240
            XX1(I)=X1(I-1,I-1);
            YY1(I)=Y1(I-1,I-1);
            XX2(I)=X2(I-1,I-1);
            YY2(I)=Y2(I-1,I-1);
            XX3(I)=X3(I-1,I-1);
            YY3(I)=Y3(I-1,I-1);
            XX4(I)=X4(I-1,I-1);
            YY4(I)=Y4(I-1,I-1);
        end
        figure('Position', [100 100 350 200])
        plot(XX2,YY2,'-r.','MarkerFaceColor','r','MakerSize',7); hold on
        plot(XX3,YY3,'-r.','MarkerFaceColor','r','MakerSize',7); hold on
        plot(XX4,YY4,'-r.','MarkerFaceColor','r','MakerSize',7); hold on
        plot(XX1,YY1,'-b.','MarkerFaceColor','b','MakerSize',7); hold off
        xlabel('fn(zi=0-240)'); ylabel('gn(zi=0-240)')
        figure('Position', [100 100 350 200])
        plot(XX2,YY2,'r.','MarkerFaceColor','r','MakerSize',7); hold on
        plot(XX3,YY3,'r.','MarkerFaceColor','r','MakerSize',7); hold on
        plot(XX4,YY4,'r.','MarkerFaceColor','r','MakerSize',7); hold on
        plot(XX1,YY1,'b.','MarkerFaceColor','b','MakerSize',7); hold off
        xlabel('fn(zi=0-240)'); ylabel('gn(zi=0-240)')
    end
```

References

1. M.J. Ablowitz, H. Segur, *Solitons and the Inverse Scattering Transform* (SIAM, Philadelphia, 1981)
2. F.C. Moon, *Chaotic and Fractal Dynamics* (Wiley, New York, 1992)
3. H. Peitgen, H. Jurgens, D. Saupe, *Chaos and Fractals – New Frontiers of Science* (Springer, New York, 1992)
4. A. Scott, *Nonlinear Science* (Routledge, London, 2005)
5. E. Mosekilde, Y. Maistrenko and D. Postnov. *Chaotic Synchronization: Applications to Living Systems*. Nonlinear Science Vol. 42, World Scientific, Singapore, 2002
6. Ed. By C. H. Skiadas and C. Skiadas. *Handbook of Application of Chaos Theory*. Chapman and Hall/CRC Press, 2016
7. R.A. Fisher, The wave of advance of advantageous genes. Ann. Eugenics **7**, 355–369 (1937)
8. G. Nicolis, A.D. Wit, Reaction-diffusion systems. Scholarpedia **2**(9), 1475 (2007)
9. A.C. Newell, J.A. Whitehead, Finite bandwidth, finite amplitude convection. J. Fluid Mech. **38**, 279–303 (1969)
10. L.A. Segel, Distant side-walls cause slow amplitude modulation of cellular convection. J. Fluid Mech. **38**, 203–224 (1969)
11. J.-S. Guo, Y. Morita, Entire solutions of reaction-diffusion equations and an application to discrete diffusive equations. Discrete and Continuous Dynamical Systems **12**, 193–212 (2005)
12. P.F. Verhulst, Mathematical researches into the law of population growth increase. Nouveaux Mémoires de l'Académie Royale des Sciences et Belles-Lettres de Bruxelles **18**, 1–42 (1845)
13. R.M. May, Simple mathematical models with very complicated dynamics. Nature **261**, 459–467 (1976)
14. S. Kawamoto, 2-D and 3-D solvable chaos maps. Chaotic Modeling and Simulation (CMSIM) **1**, 107–118 (2017)
15. S. Kawamoto, Chaotic time series by time-discretization of periodic functions and its application to engineering. Chaotic Modeling and Simulation (CMSIM) **2**, 193–204 (2017)
16. R. FitzHugh, Impulses and physiological states in theoretical models of nerve membrane. Biophys. J. **1**, 445–466 (1961)
17. J. Nagumo, S. Arimoto, S. Yoshizawa, An active pulse transmission line simulating nerve axon. Proc. IRE. **50**, 2061–2070 (1962)
18. S. Kawamoto, The FitzHugh-Nagumo model and 2-D solvable chaos maps. Chaotic Modeling and Simulation (CMSIM) **3**, 269–283 (2018)
19. S. Kawamoto. The FitzHugh-Nagumo model and spatiotemporal fractal sets based on time-dependent chaos functions. *11th Chaotic Modeling and Simulation International Conference*, Springer Proceedings in Complexity, ed. by C. H. Skiadas and I. Lubashevsky: 159–173, Springer Nature Switzerland AG, 2019
20. A.N. Zaikin, A.M. Zhabotinsky, Concentration wave propagation in two-dimensional liquid-phase self-oscillating system. Nature **225**, 535–537 (1970)
21. A.T. Winfree, Spiral waves of chemical activity. Science **175**, 634–636 (1972)
22. H.P. McKean, Application of Brawnian motion to the equation of Kolmogorov-Petrovskii-Piskunov. Comm. Pure Appl. Math. **28**, 323–331 (1975)
23. B. Van der Pol, J. Van der Mark, Frequency demultiplication. Nature **120**(3019), 363–364 (1927)
24. Y. Ueda. Random phenomena resulting from nonlinearity: In the system described by Duffing's equation. *Int. J. Non-linear Mechanics* 20: 481–491, 1985: Translated from *Trans. IEEJ* 98-A: 167-173, 1978
25. Y. Ueda, Randomly transitional phenomena in the system governed by Duffing's equation. J. Statistical Physics **20**, 181–196 (1979)
26. S.H. Strogatz, *Nonlinear Dynamics and Chaos* (Westview Press, Boulder, 1994)
27. T. Kanamaru, Van der Pol oscillator. Scholarpedia **2**(1), 2202 (2007)
28. Y. Kuramoto, *Chemical Oscillations* (Waves and Turbulence. Springer, Berlin, 1984)

29. A. Pikovsky, M. Rosenblum, Synchronization. Scholarpedia **2**(12), 1459 (2007)
30. E. Mosekilde, J. L. Laugesen and Z. T. Zhusubaliyev. On the transition to phase synchronized chaos. *Handbook of Application of Chaos Theory*, ed. by C. H. Skiadas and C. Skiadas: 39–61. Chapman and Hall/CRC Press, 2016

Decision-Making in a Context of Uncertainty

Abdelbaki Laidoune

Abstract The decision process is considered as the key element for the smooth behavior of systems. For non-complex or linear systems, this process usually follows clear rules or requirements, so that the decision will be easy to make without any difficulty. In the case of complex (non-linear) systems, the decision-making process is repeatedly challenged problems because of the interactions of several factors at the same time and because of the hazards characterizing these factors. This situation will worsen further if the system goes out of its normal operating range to fall into uncertainty. In this paper, we will try to make our contribution, to arrive at simplifying the decision-making process, while playing on human factors (ability, anticipation, risk-taking, etc.). As this process can be improved by the development of intelligent decision support tools.

Keywords Decision process · Complex systems · Intelligent decision support tools · Rules and prescriptions · Decision making

1 Introduction

The decision-making process is the cornerstone in the operation of any system. This process can be very simple for linear systems that obey an uncomplicated logic, where the effects of a crisis can be elucidated in the manner of falling dominoes [1]. In complex systems, this process could become more complicated, especially when the system goes out of its normal operating range and into a state of uncertainty or even a crisis whose consequences can not be predicted or controlled. Currently, the majority of the systems dominating the domains of the daily life, are complex systems, characterized by the interaction of several factors, the distribution of the information on several decision centers, which makes the coordination of the actions and the decision making difficult [2]. In this communication, we will offer some

A. Laidoune (✉)
Health & Safety Institute, Industrial Prevention Research Laboratory (LRPI), University of Batna 2 (Mostafa Ben Boulaid), Fesdis, Batna, Algeria
e-mail: a.laidoune@univ-batna2.dz

C. H. Skiadas and Y. Dimotikalis (eds.), *12th Chaotic Modeling and Simulation International Conference*, Springer Proceedings in Complexity,
https://doi.org/10.1007/978-3-030-39515-5_13

suggestions, to simplify the decision process. This can be done by improving both aspects: human and technical. As far as the human side is concerned, the operators must be chosen according to criteria corresponding to the job profile, namely: the capacity for synthesis and analysis, good reasoning, intuition capacity, anticipatory capacity, recovery capacity, etc. From a technical point of view, systems must use experience feedback to improve their performance by correcting defects encountered during operation. As they must be designed in a way that allows for multiple decision alternatives. The systems must be error tolerant and allow the recovery of anomalies (feedback loops).

(1) **Complex systems**:

Complex systems currently represent a large part of the vital installations of our daily life: industrial processes, transport networks, urban management, etc. Unlike linear systems where the evolution of events, obeys a logical and simple sequence, complex systems consist of several interacting subsystems, forming an often complex architecture. This makes their evolution unpredictable and does not follow a classical linear relationship cause-effect, but it is circular characterized by feedbacks of collective behaviors and emergent (macroscopic) properties on the behavior of (microscopic) elements (Fig. 1). These systems are characterized by dynamic environments, in opposition to the static environments that characterize linear systems, the physical or structural properties of these dynamic environments, are likely to change with or without human intervention [3]. The components of these systems will collectively modify their environment, which in turn will force them to modify their states or possible behaviors. In a complex system, knowing the properties and behavior of isolated elements is not enough to predict the overall behavior of the system [4]. Despite the degree of their organization, these systems are still vulnerable

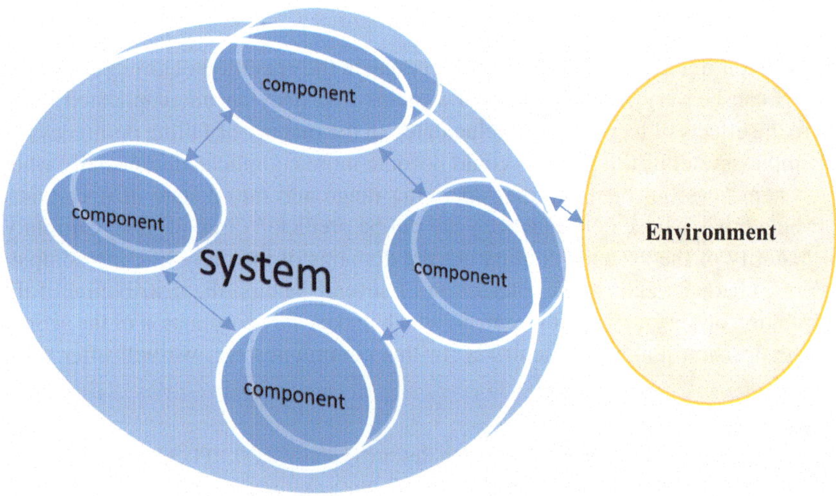

Fig. 1 Complex system

to the hazards, constraints and challenges imposed by the environment. In complex systems, the operation is always governed by formal requirements defined by the designers, leaving only small margins for the initiatives of human operators. But, when the system goes out of its normal operating range to switch into abnormal situations that can evolve into uncertain situations, the problem of decision making is acute. Faced with these uncertainties, maintaining the stability, performance and availability of the system in a changing and unstable environment is a challenge that forces the proponents of these systems to constantly reconfigure their means and methods of work, to adapt to changing situations. This constant reconfiguration of resources will require a great deal of dynamic capacity to adapt with these uncertain environments [5].

(2) **uncertainty in complex systems**:

Generally, the uncertainty is due to the presence of certain hazards that disrupt the functioning of the systems, to which is added the complexity of the architecture of these systems combined with the disturbances of the environment. This uncertainty will worsen further with the unpredictable behavior of human operators [6]. Uncertainty can be exacerbated by pressures on the system as a result of requirements from different parties: regulatory requirements, safety requirements, user requirements, etc. [7]. Lack of resources could also affect some system functions, such as maintenance, renovations and upgrades, which complicates its operation (outages, disruptions, incidents, etc.) [8]. The poor control of the techniques by the human operators will have a negative impact on the functioning of the system by errors which can evolve towards uncertain or even catastrophic situations [9]. The traffic and information processing component, if not adequately addressed, can complicate communication between different stakeholders in the various spheres of the system and create confusion during unpredictable events. Rigidity in the requirements for operators to perform certain tasks requiring decision-making, will further affect their freedom of action, which drives the system into uncertainty [10]. Finally, the technical means such as: man-machine interfaces, communication media, decision-support software, if they are not adapted with the technological evolutions, can compromise the functioning of the system and make the realization very delicate tasks. This can lead operators to misinterpretations that sometimes lead to uncertainty [11].

(3) **improvement of the decision in the contexts of uncertainty**:

To create better enabling conditions that help decision-makers in an optimal and efficient way, we need to act on several factors: human, technical, organizational, and so on.

(3.1) **On the human level**:

Human operators who will be assigned to positions requiring a level of decision-making responsibility, must be selected according to profiles compatible with the required qualifications (ability to analyze and summarize, ability to intuit and deduce, ability anticipation, recovery capacity, etc.) [12], evoked the controlled attention skills in complex cognitive tasks, through which the operator tries to resist interference from the instantaneous surge of a large amount of information, by blocking access to unnecessary or irrelevant information. In addition, these operators must undergo periodic training focused on intervention in uncertain situations and must practice simulation exercises to test their ability to respond to unpredictable situations [13]. Operators must train on co-operation and mutual co-ordination and avoid hasty and risky behavior. To successfully coordinate actions in order to arrive at a timely decision, good communication between interveners on the one hand and between them and the hierarchy is necessary.

(3.2) **At the technical level**:

The decision process could be improved by the design of:

- Error-tolerant systems with multiple choices for intervention;
- Systems characterized by flexibility in procedures and allowing freedom of action for interveners;
- Systems with feedback loops to recover errors and anomalies;
- Systems with interfaces facilitating communication with the human operator.

Finally, we must point out the importance of modeling and simulation techniques in the calculation and verification of many low cost assumptions, compared to the actual development of equipment. Thus, many security defects are identified and corrected as early as this design phase [14].

(3.3) **use feedback to improve decision support systems**:

Feedback can be used from situations of incidents or accidents experienced in the past and creating situations of uncertainty. Information about these situations must be stored in a database. After processing this information selectively, it will be possible to configure devices offering several actions to facilitate decision-making [15]. These actions may concern the aspects inherent to the correction of certain system design defects (corrective actions), or may concern the development and implementation of intelligent tools in the form of decision support software (Fig. 2).

2 Conclusion

In conclusion, it can be said that the decision-making process in uncertain situations, especially when dealing with complex systems, requires the consideration of several factors. The human factor is the cornerstone of this process because human is the main actor in the design, operation and maintenance of the systems. Thus human operators must be selected according to criteria compatible with the functions to be assumed. Continuous training and simulation exercises are needed to test the capabilities of the responders to the disturbances that can evolve to uncertain situations. On the technical side, several solutions can be proposed, in the form of flexible decision support devices and allowing the recovery of errors due to bad decisions. Improved interfaces

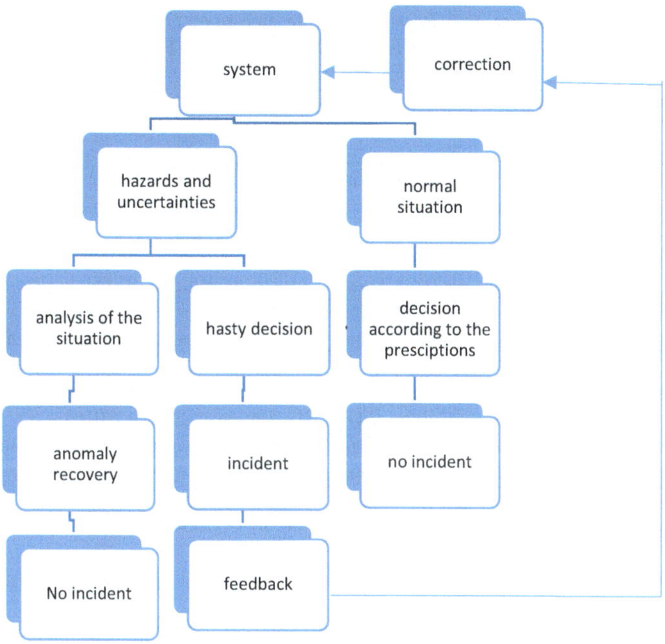

Fig. 2 Decision process

and modes of communication can facilitate a good interpretation of information which facilitates the decision process.

References

1. J.C. Ramo, *The Age of the Unthinkable—Why the New World Disorder Constantly Surprises Us and What We Can Do About It* (Little, Brown and Company, New York, 2009), p. 219
2. P. Lagadec, Terra incognita, INHESJ, file on decision making in uncertainty. LIREC **58**, 6–12 (2019)
3. R. Samurçay, J.-M. Hoc, From work analysis to decision support specification in dynamic environments. French Psychol. **33**(3), 187–196 (1988). (in French)
4. A. Lesne, Robustness: confronting lessons from physics and biology. Biol. Rev. **83**, 509–532 (2008)
5. C. Wang, P. Ahmed, Dynamic capabilities, a review and research agenda. Int. Manage. Rev. **9**, 31–51 (2007)
6. E. Marsden, Justifying safety interventions based on uncertain foresight: empirical evidence. In *Conference: Enhancing Safety: the Challenge of Foresight* (European Commission Joint Research Centre, Ispra, Italy, 2017), https://www.researchgate.net/
7. A. Bekkouche, Uncertainty and decision-making in an uncertain world. in Conference Paper presented in 2nd Bechar International Civil Engineering Seminar 2015, Bechar University (2015), https://www.researchgate.net/

8. E. Martinais. *Uncertainty as a Resource for the Decision. The Case of the Prevention of Industrial Risks. Practices of the Decision Under Uncertainty* (FONCSI, Paris, France, 2016). (in French)
9. V. Steyer, H. Laroche, The virus of doubt: decision and sensemaking in a crisis cell. French review of management, Dossier "the factory of the decision". **38**(225), 167–186 (2012). (in French)
10. A. Laidoune, H. El Rahal Gharbi Med, Analysis testing of sociocultural factors influence on human reliability within sociotechnical systems: the algerian oil companies. Saf. Health Work **7**(3), 194–200 (2016)
11. J. Reason, Managing the management risk: new approaches to organisational safety. In *Reliability and Safety in Hazardous Work Systems* eds. by B. Wilpert, T. Quale (Hove Lawrence Erlbaum, 1993), pp. 7–22
12. A.R. Conway, R.W. Engle, Working memory and retrieval: a resource-dependent inhibition model. J. Exp. Psychol. Gen. **123**(4), 354 (1994)
13. T. Reverdy, A. Rohrich, *Uncertainty and Resilience in Technological Projects* (Foundation Papers for an Industrial Safety Culture (FONCSI), Toulouse, France, 2016)
14. M. Mazeau, *Consideration of Security When Designing Complex Systems* (Summer School Scientific Risk Management, ALBI, France, 1999), https://www.previnfo.net/sections.php?op=viewarticle&artid=18. Avalaible 6–10 Sept 1999
15. A. Bisseret, Representation and expert decision, ed. by D.L. Octares (Toulouse, France, 1995)

Microwave Oven Plasma Reactor Moding and Its Detection

V. J. Law and D. P. Dowling

Abstract Over the last 20 years microwave power supplies in domestic microwave ovens have increasingly found applications in plasma reactors for processing of functional carbon-based nanostructures for engineering materials, electronics and biomedical applications. However, the packaged magnetrons used in the microwave ovens are known to suffer from moding due to frequency pushing and pulling, both of which may limit the efficiency of the plasma treatment process. This paper explores patent records for high voltage doubler circuits, coupled to the cathode filament heater circuit as the magnetron pushing source and the multimode resonant cavity plasma load as the pulling source. These circuits are compared with microwave oven plasma reactor circuits published in peer reviewed journals. This study highlights that a number of academic research groups have investigated power supply design parameters such as the input transformer voltage, as well as the power level using different capacitors values, as a result the transformer output is poorly described. Identification of moding within the plasma reactors due to magnetron warm-up time and changing cavity load conditions is also poorly reported. This work attempts to address this information gap on microwave oven plasma processes, through the extraction of reports on a packaged magnetron warm-up times and near-field *E*-probe mode measurement within the Cambridge Fluid Systems MRC 200 plasma reactor using a argon and hydrogen plasma.

Keywords Microwave oven · Drive circuit pushing · Load pulling · Plasma

1 Introduction

Since the domestic microwave oven became widely available, engineers have converted the oven's multi-mode resonant cavity (MRC) space, that was originally intended for cooking of food stuffs [1, 2], to one that can rapidly synthesise a wide

V. J. Law (✉) · D. P. Dowling
School of Mechanical and Materials Engineering, University College Dublin, Belfield, D04 V1W8 Dublin 4, Ireland
e-mail: viclaw66@gmail.com

C. H. Skiadas and Y. Dimotikalis (eds.), *12th Chaotic Modeling and Simulation International Conference*, Springer Proceedings in Complexity,
https://doi.org/10.1007/978-3-030-39515-5_14

range of organic compounds [3]. In 1989, the domestic microwave oven underwent another conversion to one of plasma etching and cleaning of substrates [4]. The availability of a low-cost microwave source in the form of an oven (typically 50–60 Euros for a basic unit in 2018), minimal oven space, and ease of use are arguably the driving force for these conversions. The wide spread availability of the microwave oven has also enabled the reuse of the oven's magnetron and its drive circuit in more advanced plasma processing. These processes include: microwave chemical deposition of diamond-like carbon films [5], plasma cleaning of ion implant ceramic insulators [6, 7], in the semiconductor industry [8]. As well as the synthesis of ZnO and Zn nanoparticles [9] and Graphene using suspector initiated plasmoid chemistry [10]. Microwave plasma dissociation of atomic hydrogen from water [11], plasma processing of solid waste [12], and the production of synthetic gas form hydrocarbon fuels [13–15]. In addition, preionization of Ohmic plasma within a TOKAMAK [16], and the construction of atmospheric pressure plasma sources [17, 18]. A monograph in '*The Bell Jar*' journal also highlighted the use of oven microwave components for excitation of plasma within electronic values as well as its use to produce down-stream plasmas [19].

In 2018 the authors of this current paper published a series of review papers on the conversion of the domestic microwave oven and its subsequent calibration [20], microwave oven plasma reactor processing of organic and biomaterials [21] and the use of the microwave oven plasma reactor as a prototyping tool [22]. These three papers citied 38 papers relating to microwave oven plasma processing of materials, however only 20 stated the type of microwave oven used and within this group the methodology of describing the power conditions used varied from one paper to another, therefore it is difficult to make a meaningful comparison of the power condition used. A feature of many of these papers however was the under reporting of experimental conditions, as most of the papers focused on the plasma process rather than the how the magnetron drive circuitry influences the plasma process itself. This inability to compare the microwave oven plasma processing conditions applied arises from the fact that each oven manufactures (e.g. Daewoo, LG Electronics, Panasonic Electronics and Samsung etc.,) uses a different magnetron type and drive circuit. Moreover the drive circuits used are designed to defrost and cook food; not for plasma processing of thermally sensitive materials. The decision on which oven to choose can be viewed as: what is available rather than how the oven is driven? In contrast to plasma processing papers, publications on the subject physics-based modelling of microwave oven heating of food detailed the oven's applied power and duty cycle [23–26] and references therein due to the fundamental need to understand how the applied electromagnetic energy is converted to heat.

It has been widely demonstrated that the applied averaged power to the MRC is readily controlled by using a half-wave or full-wave pulsed direct current drive circuits. The control is achieved by altering the value a capacitor within the high voltage (HV) section of the microwave oven transformer (MOT), or by changing the duty-cycle envelope of the pulsed direct current waveform at the low voltage side of the (MOT). In the latter case however the peak applied power becomes significantly different to the applied averaged power at short processing times of <1 min. This short processing time is typical for microwave oven plasma reactors used for prototype batch processing of ZnO and Zn using susceptor initiated plasmoid chemistry [9], irreversible sealing of microfluidic devices at the centimetre scale [27] and the selective synthesis of allotropic carbon-based nanostructures [single and multi-walled carbon nanotubes (CNT and MWCNT) and onion-like nanostructure's (OLNs)] [28–31]. To progress from the prototype stage to full-scale manufacturing by using the same microwave oven design therefore potentially could lock-in the disadvantage of the drive circuit: namely, moding within the magnetron warm-up period, abrupt pulse width modulation (repeatedly turned on and off) times and changes in the cavity mode as the process progresses. Here the mode of a magnetron and cavity is characterised by having bounded smooth changes in its electric field patterns or distributions, which are distinct from a neighbouring mode, and where changes from one mode to another is termed moding [32]. In today's chaos theory the term moding can also be equated to a chimera state [33]. Understanding these mode changes under short plasma processing conditions, may allow superior microwave oven plasma processes to be engineered in terms of process stability, product yield and product selectivity at the nano-scale level. Thereby produce a scaled-out production-line, rather than a scaled-up production-line [34].

The aim of this paper is twofold. First to compare a number patents on microwave oven magnetron drive circuits which are designed to prevent magnetron moding (Sect. 2) with microwave oven plasma reactor drive circuits designed and constructed within academic institutions (Sect. 3). Section 4 examines mode production [35], with the free running magnetron (Sect. 4.1) and plasma mode detection, while operating under empty cavity conditions and with monatomic argon (Ar) and diatomic hydrogen (H_2) plasma conditions (Sect. 4). Section 5 provides an overall summary of this paper.

2 US Patents: Slot-Hole Cavity Magnetron and Packaged Magnetron Drive Circuits

The US patent record office is a rich resource of historical information about the magnetron. For example a US patent search using the words of magnetron moding found 260,000 hits; and M. Leconte 2010 paper used 887 US patents for a heuristic review of the cavity magnetron [36]. In this work 20 US patents and 1 Russian patent published between 1946 and 2013 are used to describe the development of the slot-hole cavity magnetron and its successor (the packaged magnetron) for use in a microwave oven. The patent selection is far from complete and is only used as an example. The patents are selected on the basis of how practical problems were encountered, identified and overcome (Table 1).

An early example of a magnetron was the slot-hole cavity magnetron that was developed mid 1930s and then on into the mid 1950s. These magnetrons were capable of producing pulse powers of 100 kW with primary uses in the radio broadcasting industry and 10-cm RADAR [37–39]. But when employed in early commercial

Table 1 Magnetron and microwave oven US, in addition to 1 Russian patent

Title	Inventor	Number	File date	Issued date	Moding reference
8 Slot-hole cavity magnetron					
High frequency electrical oscillator	J. T. Randall H. A. H. Boot	2,542,966	Aug 20, 1941	Feb 20, 1951	Yes
High Frequency magnetron	P. L. Spencer	2,408,35	Dec 31, 1941	Sep 24, 1946	Yes
High frequency electrical oscillator	J. Sayers	2,546,870	Feb 9, 1945	Mar 27, 1951	Yes
Microwave oven					
Method of treating foodstuffs	P. L. Spencer	2,495,429	Oct 8, 1945	Jan 24, 1950	
Coffee brewing	P. L. Spencer	2,601,067	Mar 24, 1948	Jun 17, 1952	
Microwave oven vending control					
Microwave oven control	J. T. Lamb	3,168,637	June 14, 1962	Feb 2, 1965	
Voltage doubler					
Power supply circuit for continuous wave magnetron operated by pulse direct current	A. E. Feinberg	3,396,342	Aug 16, 1965	Aug 6, 1968	
Packaged magnetron					
Control systems for regulating the current in a magnetron tube	L. Blok et al.	3,302,060	Jul 6, 1964	Jan 31, 1967	
Microwave magnetron	J. R. Mins	3,739,225	Apr 24, 1972	Jun 12,1973	
Magnetron	T. Koinuma	3,809,950	Feb 14, 1973	May 7, 1974	
Defrost					

(continued)

Table 1 (continued)

Title	Inventor	Number	File date	Issued date	Moding reference
Microwave oven defrost circuit	J. T. Lamb	3,842,233	May 31, 1973	Oct 15, 1974	
Microwave oven supply circuit	S. Nagamoto	3,943,317	Jul 15, 1974	Mar 9, 1976	
Filament holding circuit					
Magnetron power supply and cathode heater circuit	W. C. Hickman	3,392,309	May 24, 1965	Jul 9, 1968	
Interlock circuitry for microwave oven	H. F. Chapell	3,624,334	Apr 1 1970	Nov 30, 1971	
Microwave oven having controllable frequency microwave power source	T. Nobue S. Kusunoki	4,415,789	Dec 2, 1981	Nov 5, 1983	Yes
Magnetron mode detection	S. A. Levie W. E. Taylor	4,501,767	Sep 7, 1982	Mar 12, 1985	Yes
Power control circuitry for magnetron	P. H. Smith	4,620,078	Oct 14, 1984	Oct 28, 1986	Yes
Filament power compensation for magnetron	P. H. Smith et al.	4,835,353	Dec 28, 1987	May 30, 1989	Yes
Microwave oven plasma reactor					
Microwave plasma etching machine and method of etching	A. Ribner	4,804,431	Nov 3, 1987	Feb 14, 1989	
Three phase supply					
Magnetron variable power supply with moding prevention	Chuck Daley et al.	5,571,439	Apr 27, 1995	Nov 5, 1996	Yes
Magnetron generator	V. N. Tikhonov et al.	RU 2,480, 890	Dec 12, 2011	Apr 27, 2013	

microwave ovens [1, 2] their normal warm uptime produced delays when reheating/cooking of food and beverages at peak vending times. To overcome these delays a cathode filament holding circuit was needed to hold the heater circuit 'ON' after the reheating/cooking had finished. In this way at peak vending times the waiting between consecutive food sales become shorten [40]. Typically, if the reheating or cooking time was 1 min then the holding time was around 5 min.

The packaged magnetron used in today's microwave oven owes its success to the development of the vane anode design which greatly reduced machining costs, and the integration of top and bottom permanent magnetic (thus removing the need for a electromagnetic control circuit) and the replacement of water cooling systems by an air cooling design [41–43]: all of which brought down the weight and cost of the magnetron, along with the complexity of its control circuitry.

Arguably the most common design choice for the modern package magnetron, for use in the standard domestic microwave oven is the half-wave voltage doubler circuit. The circuit provide a HV output in the 5–6 kV range. In the following circuits the mains supply voltage to the MOT is 110/240 Vac, 50/60 Hz, sine-wave and the HV windings (T_{1a}) produces 2.7 kV and the voltage doubler circuit provides 5.4 kV prior to connection to the magnetron. This circuit is based upon A. F. Feinberg's 1968 US patent 3,396,342 [44], and further exploited for plasma processing by Salvador et al. in 1995 [5], Choe et al. 2000 [16], Chaichumporn et al. 2011 [17] and Law and Dowling 2018 [20]. A monograph '*plasma experiments with commercial gas tubes and some ideas for microwave oven conversions*' published in '*The Bell Jar*' 4(2) 1997 [19] is further examples of the circuit in plasma production. Within this body of work a number of circuit diagrams with descriptions of how the circuits operate can be found.

The operation of the voltage doubler circuit is described in this section with the aid of Figs. 1, 2, 3, 4 and 5. During the first positive half-cycle the voltage from the MOT increases as capacitor C_1 is charged and accordingly forward biasing the diode D_1. During the capacitor charging time there is no voltage to the magnetron because the current takes the course of least resistance. In effect the magnetron becomes short circuited. In the second half-cycle, the voltage swings into the negative at the transformer windings, at this point the transformer winding (T_{1a}) and the charged capacitor are now essentially two energy sources: that is the −2.7 kV across the MOT winding adds to the −2.7 kV stored in the capacitor to produce a sum voltage of −5.4 kV and which is applied to the magnetron. It is this voltage which is now rectified into a pulsed square-wave DC voltage with a duty-cycle (D) = 50%, where the on-period (t_{on}) is equal to the off-period (t_{off}). This is because the circuit generates an output only during the negative half-cycle of the transformer's output voltage. It should be noted however that once this voltage is connected to the magnetron cathode, the internal capacitance of the magnetron reduces the voltage maximum.

The patent by Albert E. Feinberg [44] is one of a number, which proposed developments on the full-wave rectification circuit which uses a second HV capacitor-diode pair (C_2 *and* D_2). Choe et al. [16] used this circuit for preionization of an Ohmic plasma within a TOKAMAK. How the additional capacitor-diode pair accomplish full-wave rectification can be understood by referring to Fig. 1b and the following

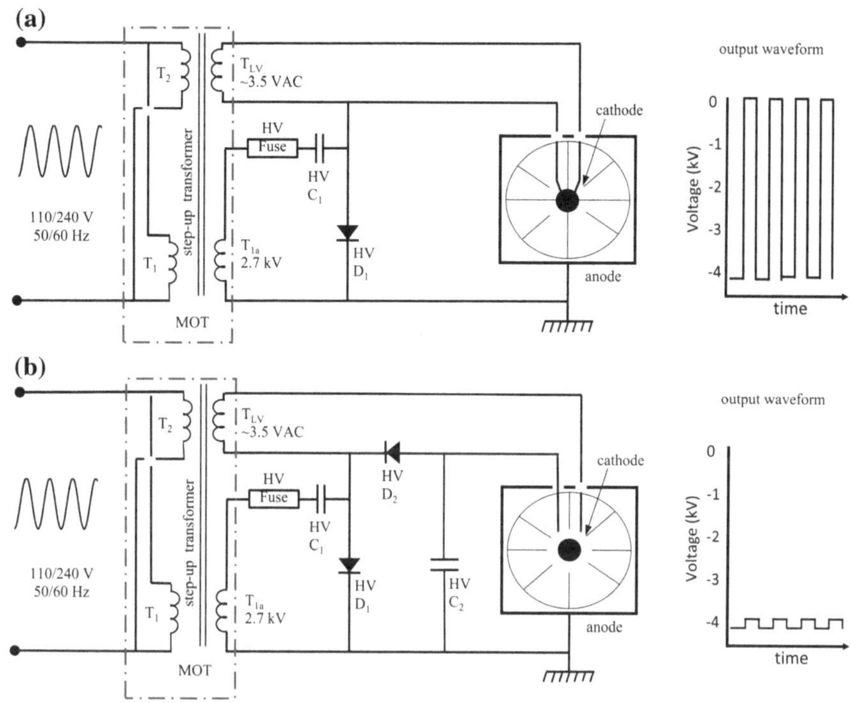

Fig. 1 Albert E. Feinberg's half-wave rectification voltage doubler (**a**), full-wave rectification voltage doubler (**b**). Fan and magnetron temperature senor removed for clarity [44]

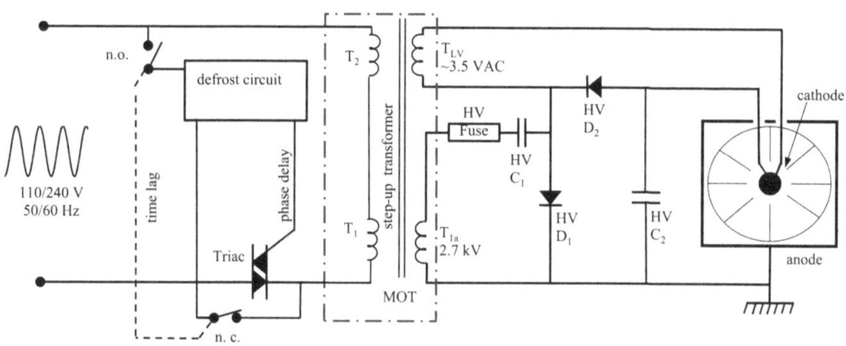

Fig. 2 S. Nagamoto's defrost circuit. Fan and magnetron temperature sensor removed for clarity [46]

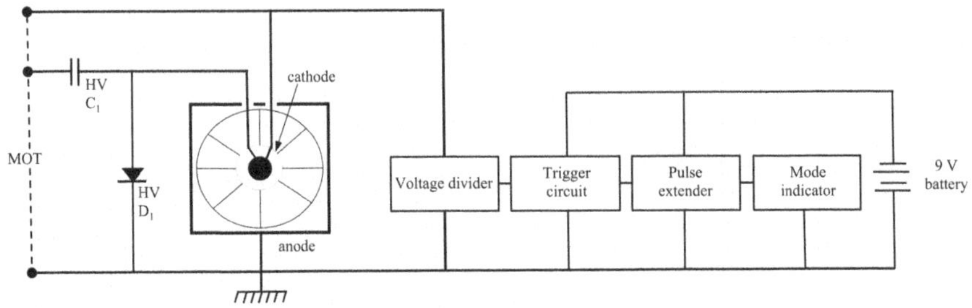

Fig. 3 The Levie and Taylor magnetron mode detection probe (US patent 4,501,767) connected across the magnetron HV terminals. MOT removed for clarity

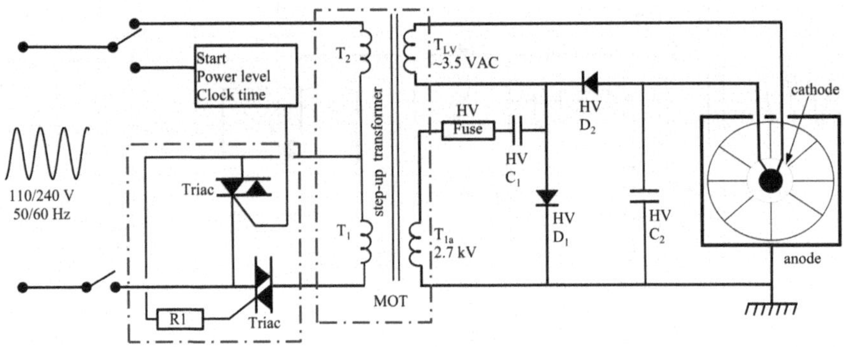

Fig. 4 P. H. Smith's two triac circuit (US patent 4,620,078). Fan and magnetron temperature sensor removed for clarity

text. During the first positive half-cycle the voltage from the MOT increases as C_1 is charged and accordingly forward biasing the D_1. In the second half-cycle, the voltage swings into the negative at T_{1a}, but now the second diode (D_2) charges C_2 so now the voltage across C_2 is doubled to 5.4 kV. In this case, however, the HV diodes D_1 and D_2 are active during the first and second half-cycle periods to produce a non-smoothed DC level of −5.4 kV, and when applied to the magnetron, it receives two pulses per cycle though at a lower voltage than 5.4 kV due to the capacitance of the magnetron.

Both half-wave and full-wave pulse sequence are intended for cooking purposes, but when defrosting of frozen food without cooking is required, it is necessary to reduce the input power. This is achieved by turning the magnetron on and off, by interrupting the pulsed waveform. Thereby reducing the average applied power to the magnetron and hence the ovens MRC in which the food is placed. Cycling on and off the pulsed waveform varies the cathode filament voltage, therefore a change in the cathode-anode space-charge formation. The process of cycling tends to damage circuit components leading to a reduction in the magnetron's working lifetime and the production of moding. In 1974, John Lamb was granted the US patent 3,842,233

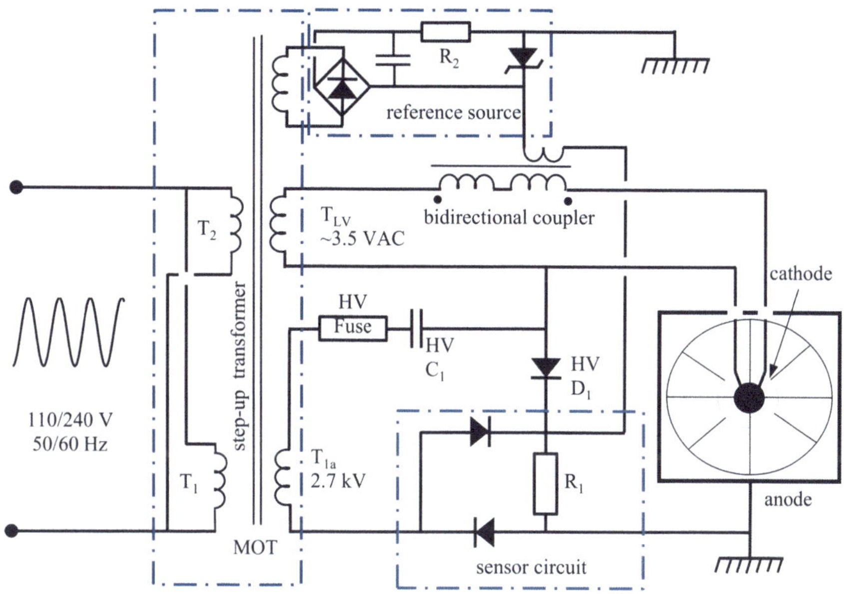

Fig. 5 P. H. Smith et al. filament compensation circuit. Fan and magnetron temperature sensor removed for clarity

that goes some way to reduce this problem [45]. In his patent a two-way switch is used to select between the cook and defrost mode of operation. In the defrost mode a motor with a plurality of cam operation is used to interrupt the mains supply voltage along with a current limiting resister to prevent high current surges.

Advancement over the electromechanical design came in the Schun-ichi. Nagamoto 1976 patent [46] in this a bidirectional triode thyristor (triac), along with defrost timer circuit (comprising an astable multivibrator) is used to control the on-off state at the gate of the bidirectional triode thyristor. Figure 2 shows a simplified schematic of this circuit. The combined electronics allows the mains voltage to be pulsed at a duty-cycle = 50% with T_{on} and T_{off} time periods much greater (typically 10 s of seconds) than t_{on} and t_{off}. With a shaped envelope imposed on to the voltage double dc pulsed waveform, the time varying power level to the magnetron and on to the MRC becomes suitable for defrosting of frozen food without cooking the food. Two additional advantages of this circuit are: first, a time-lag can be introduced when switching between the cooking state and defrost state so preventing logic circuit lock-up and second, phase coupling of the power source under constant HV transformer remanence conditions can suppress current surges to zero, or below that of normal current.

In Figs. 1a, b and 2, the cathode filament heater circuit is integral to the voltage doubler circuit. This simple integration has a number of disadvantages. First, to avoid magnetron moding the precise voltage and current (typically 3.3–3.5 VAC, and 10 Amps) needs to be matched to magnetron model used. Secondly, this approach means that the duty-cycle that controls the voltage doubler waveform also provides the same duty cycle to the magnetron filament causing it to undergo considerable variation in operating temperature as the cathode surface cools down in the T_{off} period. It therefore became necessary to use a filament holding circuit to control or delay the anode operation during the magnetron warm-up period. In 1965, Wallace C. Hickman's US patent described the effective use of how to characteristic the magnetrons anode voltage cut-off point (which is rated of 95% of most magnetrons) to prevent stress and damage to the cathode surface before fully heated [47]. By 1970, Harry F. Chapel used cooperating interlocks that are connected a digital circuit to control the cathode heater circuit [48].

In 1983, Nobue and Kusunoki outlined a microwave oven comprising: a heating cavity, a controllable sold-state variable frequency (915 ± 13 MHz) power source with a detector (antenna called here as a *E*-probe) for detecting the intensity of the electric field in the cavity, and control means for setting the microwave source at the frequency as determined by the intensity of the electric field [49]. In the context of microwave oven plasma reactors that employ a packaged magnetron as the power source, the significance of this pattern is the use of a non-contact *E*-probe for detecting the modulus of the electric strength at the cavity wall.

By 1985, Levie and Taylor's US patent magnetron moding detection probe was issued [50]. The probe was designed as a temporary HV probe that counted the frequency of mode changes in the magnetron and being temporary was aimed at the microwave oven service industry, rather than being incorporated into standard commercial ovens. Figure 3 provides a schematic of the probe circuit.

The probe works by stepping-down the magnetron anode voltage to a voltage level for the remaining circuit. The remaining circuit comprises a trigger circuit that detects voltage transient as the mode changes, an extender circuit that lengthens the transient pulse, a counter circuit that displays the accumulated number transients within a given test time, and a 9 V battery supply power to the trigger, extender and counter circuits. A preset number of counts per test time is then used to gauge whether the magnetron is defective (aged) and therefore need to be changed out.

Peter H. Smith's 1986 patent further describes a two triac arrangement which overcomes filament cooling and moding within the packaged magnetron [51], see Fig. 4 where for illustrative purposes a Toshiba 2M170 magnetron is used. The patent also described how the magnetron transformer windings T_1 and T_2 being connected in series and 2 bidirectional triode thyristors are used to switch-out part of the T_1 winding. The switching process provides two power states to the MOT. The first state (standby state) supplies voltage to the magnetron that is below the minimum anode voltage level for microwave energy generation. The second state (power on state) switches in the full turns windings of T_1 that allows full voltage to be supplied to the anode. In either instance, the filament voltage is sufficient to maintain the filament temperature high enough to avoid moding in the magnetron.

A further patent by Smith in 1989 described a filament compensation circuit that includes a sensor for voltage doubler circuit [52]. A constant voltage source and a bidirectional coupler circuit that stabilizes the filament voltage against changes in the pulse width applied to the transformer winding T_1 is varied. Figure 5 shows a simplified schematic of this circuit.

In 1989, Aaron Ribiner was awarded the US patent 4,804,431 '*Microwave plasma etching machine and method of etching*' [4]. In this patent a Whirlpool microwave oven, model 12OOXS is used as the example. The patent embodiments take the form of a potentiometer placed in series with the MOT main winding (T_1) and which is used to adjust the high voltage to the magnetron and thereby change the intensity of the microwave energy. Other modifications such as the use of a variac, SCR, or transistor where also suggested. No other form of modification to the filament heater circuit was given. However, it is reasonable to suggest that some form of filament compensation circuitry was incorporated in the original whirlpool drive circuit design.

By 1996, Chuck Daley et al. published his three-phase power voltage doubler circuit for the prevention of moding within UV plasma source that are employed in UV curing applications [53]. As stated within the patent the three-phase circuit is based on Smiths filament compensation circuit [52]. Daley's patent is most relevant to microwave over plasma reactors as it is one of the first patents to consider plasma quenching in the pulse chain off-periods (for 50 Hz, t_{off} ~ 10 ms; and 60 Hz t_{off} ~ 8.0 ms). More recently (2013), Tikhonov et al. further enhanced the patent from Daley et al's three-phase circuit within a Russian patent [54].

3 Packaged Magnetron and Their Power Supplies Reported by Academic Institutes

This section provides an alternative view to Sect. 2 in that it reports on magnetron power supplies used by academics for the generation of microwave plasmas. Purpose built microwave ovens plasma reactors are also reported.

Salvadori et al. [5] reused an un-named packaged magnetron (rated at 2.5 GHz and 700 W) as a microwave energy source for plasma assisted chemical vapour deposition. They also reused a 0.85 μF HV capacitor and HV diode to recreate a half-wave drive voltage doubler circuit that included voltage and current sensing circuitry. A stated normal 520 W for 20 h was used in the syntheses of the diamond films.

In 1997 the editor of '*The Bell Jar*' Journal wrote on the subject of reusing microwave oven components for the production of plasmas [19]. These simple experiments use the 2M172J (2.45 GHz, 600 W) package magnetron to excite plasma within the O84-G argon filled vacuum tube. In addition, the production of low-pressure downstream plasma is described where the low pressure is generated by using a modest rotary refrigeration compressor. The magnetron drive circuit descried is the half-wave voltage doubler design were the MOT input voltage is '*throttled back*' using a variac.

Law and Tait reported on the use of the AWI AM717 packaged magnetron (2.45 GHz, 850 W with a 3.3 V filament circuit) driven by a half-wave voltage doubler circuit within the Cambridge Fluid systems MRC 200 plasma reactor for cleaning of ion implant ceramic insulators [6, 7]. These two papers are further reviewed using

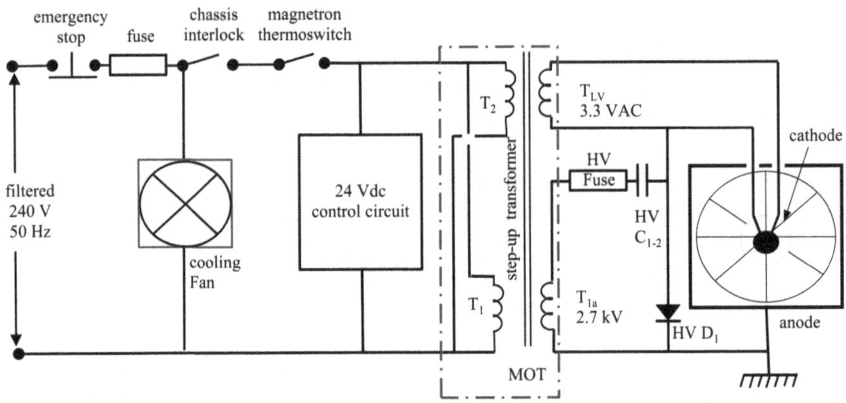

Fig. 6 Law and Tait's [6, 7] half-wave voltage doubler circuit. The magnetron power level is controlled by varying the value $C_{1\text{–}2}$

previously un-reported commercial information in 2018 [20]. In this purpose built plasma reactor, the power level is controlled by switching the capacitance value of $C_{1\text{–}2}$ from to 0.6 to 1.5 μF. A schematic of the circuit used is shown in Fig. 6.

In the same year as [6, 7] Law reported on the mode characterization of the MRC 200 reactor under both empty cavity and plasma conditions using a near-field E-probe [35]. The reactor has a cavity metal wall linear edge dimension of $a = 300$ mm, $b = 250$ mm and $L = 350$ mm with a coaxial plasma cylindrical volume (V_p, 9 L) aligned perpendicular to the waveguide iris. In terms of microwave free space wavelength the plasma volume length is 3 times λ_o and a diameter of 1.5 times λ_o. Figure 7 shows the E-probe and its mode detection circuit. The E-probe is formed as a monopole antenna (length $< \lambda_o/2\pi$) and is constructed from 1 mm diameter stainless steel wire that is coupled at one end to a neon lamp and the other end set to protrude through a 2 mm diameter aperture in the top of the cavity and mid way along the longitudinal axis of the cylindrical plasma volumes. A photodiode is then used to convert the lamp emission into a digital mV signal. The insertion of the E-probe close to the inner surface of cavity wall provides a spatially fixed voltage modulus at the wall that indicates the changing cavity Q-factor as the plasma is turned-on and -off.

The theoretical maximum unloaded Q-factor (Q_u) of the cavity is defined in Eq. 1.

$$Q_u = \frac{2V_c}{\delta A_c} \tag{1}$$

where V_c is the cavity volume, δ is the electrical skin depth at the cavity wall and by A_c is the cavity wall area. Thus, Q_u is defined as the ratio of stored energy in cavity to the energy loss to the cavity walls. Under plasma conditions the cavity Q-factor falls and so is redefined as Q_p (Eq. 2) where V_p is the plasma volume.

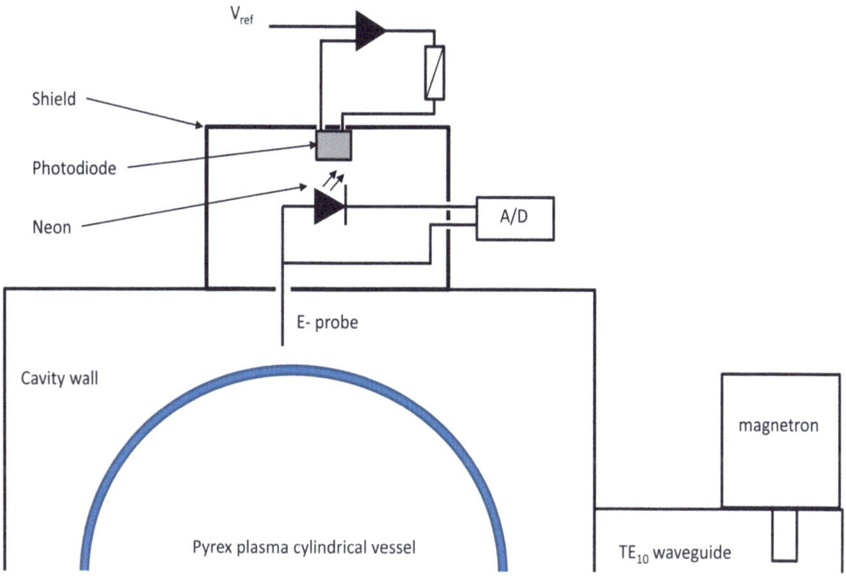

Fig. 7 Schematic of *E*-probe position within the MRC 200 cavity as looking along the longitude axis of the cylindrical plasma vessel. For clarity, only the upper half of the cavity, plasma vessel, and TE_{10} waveguide is shown

$$Q_p = \frac{2(V_c - V_p)}{(\delta A_c) + V_p} \tag{2}$$

In 2000, Choe et al. reused a LG Electronics packaged magnetron (2.45 GHz, 500 W with a 3.5 V, 11 Amp filament circuit) for preionization of Ohmic plasma within a TOKAMAK [16]. They employed both a half-wave and full-wave voltage doubler circuits. In the full-wave circuit they used capacitor values of $C_1 = 30\ \mu F$ and $C_2 = 14.5\ \mu F$, however they did not state the value of C_1 in the half-wave circuit. The cathode filament heater circuit proved insufficient for their needs therefore they resorted to a high current voltage supply.

Law and Macgearailt reported on the use of a Hitachi UHF ECR dual frequency (2 MHz and 2.45 GHz) plasma tool for the plasma etching of silicon devices at Intel Ireland [8]. In this system a Hitachi packaged magnetron model 2M121A (2.45 GHz, 1.450 W and 4.6 V, 14 Amp filament circuit) is used. In this commercial plasma system, the magnetron filament heater is on a separate circuit for preheating of the cathode thereby minimizing moding when the voltage doubler circuit is turned-on.

In 2008, James Smith at West Virginia University reported on the reuse of microwave oven components (MOT, package magnetron and doubler circuit) in conjunction with a variac to vary the applied voltage. This voltage was then used to operate a stack of metal on oxide field effect transistor (MOSFET) that controlled the firing of a magnetron with the aim to preionization of fuel mixture prior to

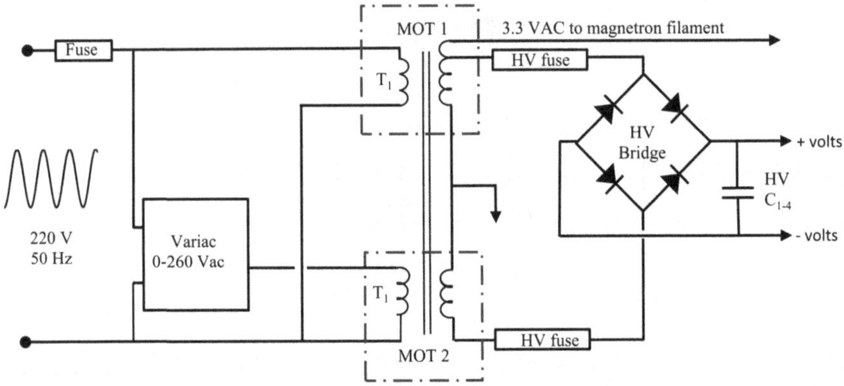

Fig. 8 Chaichumporn et al. [17] full-wave voltage doubler drive circuit. The magnetron power level is controlled by varying the Variac windings and the value of C_{1-4}

engine ignition within a multi-fuel (gasoline and jet fuel) engine [55]. Smith's circuits (described in both text and photographs) are aimed at the automotive sector however; the aim is comparable with the work of Choe et al. [16] thus, the use of a MOSFET stack provides a further example of controlling the PWM envelope that may be useful microwave oven plasma reactors.

By 2011, Chaichumporn et al. reused a Samsung Electronics model 2M218 packaged magnetron (2.45 GHz, 850 W with a 3.3 V filament circuit) to create an atmospheric plasma [17]. They used a full-wave voltage doubler circuit design using a HV bridge rectifier and HV capacitors network (~1.5 μF) all supplied by two MOT (one fixed and one variable to produce a combined HV of 3.3–6.6 kV. Their circuit is shown in Fig. 8. Using a calorimetric method for power calibration their system delivered a continuous microwave power between 0 and 250 W.

In 2013 Jung et al. reused an un-named packaged magnetron (2.45 GHz, 2 kW) for plasma dissociation of atomic hydrogen from water [11]. Unfortunately, no other details of the applied power or filament circuit were provided.

In 2017, Tikhonov et al. published on how to reuse the Samsung Electronic OM 75P packaged magnetron (2.45 GHz, 1 kW) for the production of an atmospheric plasma torch [18]. In their work they used a three phase voltage doubler circuit (one voltage doubler circuit per phase, which is the basis of their Russian patent [54]) along with the use of water cooling conversion kit to cool the magnetron. Although this work is outside the scope of this review (due to the three phase circuitry) the work is included for comparative purposes.

4 Mode Production

4.1 Mode Production Within a Free Running Magnetron

Typical package magnetrons have a free running frequency of $f_o = 2.45$ GHz ($\lambda_o = 12.24$ Hz) with a bandwidth of 200 MHz which can be pushed by changes in the drive circuit or is frequency pulled by changes in the standing wave ratio (SWR) at the magnetron's output antenna. The latter is located in the TE_{10} waveguide that connects the magnetron to the MRC of the microwave oven. The SWR mathematical expression of the non-uniformity of the electromagnetic field within the TE_{10} waveguide is given in Eq. 3. In this expression, the 1+ corresponds to the addition of the incident and reflected wave and 1− corresponds to subtraction incident and reflected wave and where $|\Gamma|$ is the modulus of the voltage [current] ratio within the waveguide.

$$SWR = \left(\frac{1 + |\Gamma|}{1 - |\Gamma|}\right) \tag{3}$$

This attribute makes the magnetron the idea power oscillator for auto-impedance matching to varying modes within the MRC. For more advance plasma reactor designs frequency pulling is used as a means of process monitoring and control [8, 56, 57].

From the very start of the slot-hole cavity magnetron development [37, 38] it was found that the magnetron's output signal changed from pulse to pulse, both in frequency and phase. The term moding arises from the magnetron's ability to jump between modes and therefore different frequencies and phase. In 1941 James Sayers took an interest in these mode jumps. He found that due to the multiplicity of resonators used a number of modes could be generated either separately or simultaneously. Moreover, these may be characterized by the mutual phase displacements between two successive cavities [39]. Fortunately the number of possible modes is limited as the magnetron's anodes resonator cavities and the interaction space between the anode and cathode is a bounded system. See Fig. 9a. For an even number of N-cavities the mode is characterized either by an integer n, defined as the total electrical phase change around the anode block measured in revolutions (N/2); or by the phase difference ($\Delta\phi$) between successive cavities. For an anode block with N cavities these two quantities ($\Delta\phi$ and n) are related by the Eq. (4)

$$\Delta\phi_n = \frac{2\pi n}{N} \tag{4}$$

In the special case of mode $n = 4$, successive segments are in anti-phase (that is they have equal and opposite potential, where $\Delta\phi = \pi$) and is found to be the most stable mode. This is because a stable 4 spoke space-charge cloud is formed in the cathode-anode interaction space and which rotates around anode. To enhance the

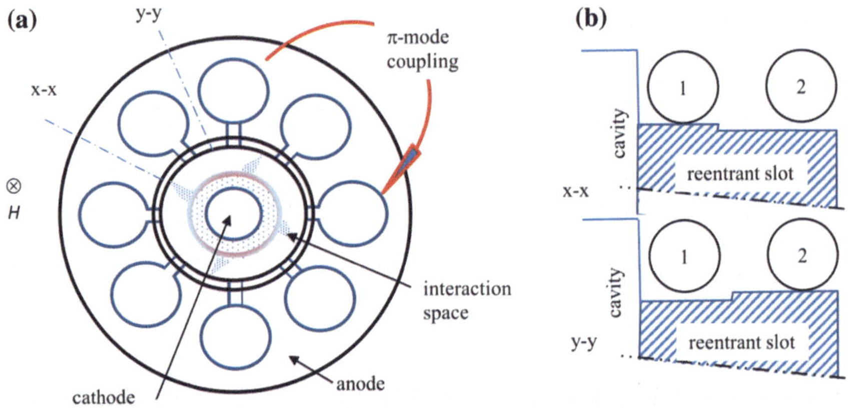

Fig. 9 **a**, **b** Cross-sectional view of a cavity magnetron cathode-anode interaction space, plus π-mode space-charge cloud (**a**), double strapping arrangement at x-x and y-y (**b**)

stability of this mode by sufficient voltage separation from all other modes Sayers developed the anode double ring strapping system (π-strapping) where the first ring connects the even numbered segments and the second ring connects the odd numbered segments (Fig. 9b) so fixing the phase difference between the segments to a value of π (180°): in effect making two coupled oscillator systems that are connected through the phase delay of a slow-wave structure within the interaction space.

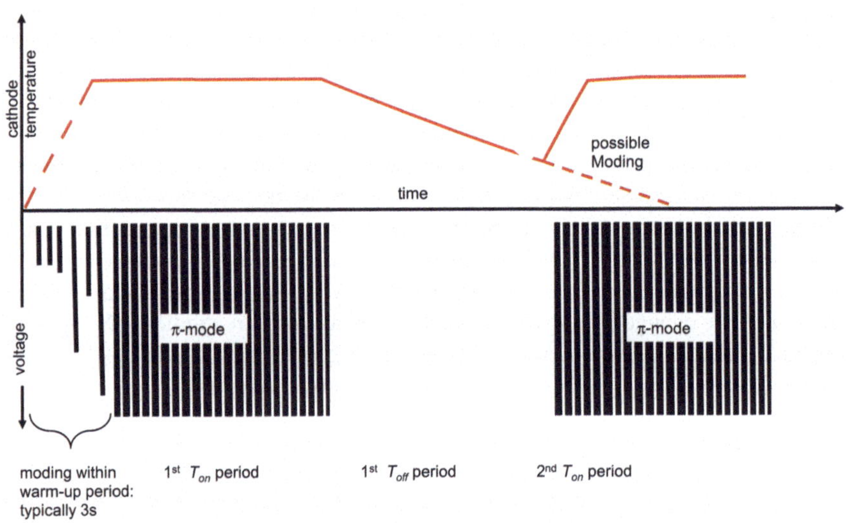

Fig. 10 Schematic of the initial 1st and 2nd magnetron voltage pulse envelopes and filament temperature. The moding regions are also shown

Now consider the production of non π-modes when using a coupled voltage doubler and cathode filament heater circuit operating with a pulse width modulation duty cycle of 50% ($T_{on} = T_{off}$). Figure 10 shows a schematic of the anode voltage and filament temperature as a function time for the warm-up period and the first two T_{on} periods. In initial stages of this period, the drive circuit begins to heat the cathode surface to release electrons into the anode-cathode interaction space to form the space-charge cloud. The warm-up process takes a few seconds to stabilizes the π-mode after which the full rated filament and anode voltage is applied allowing first T_{on} period to generate magnetron full output power for a given applied anode voltage.

For comparison purposes Geedipalli et al. has measured the G. E., Inc Lousivilel KY: model JE635WW03 warm-up time as 5 s [58]. Whist modelling the heating of a rotating object within a microwave oven (Panasonic, model NN-SD767W) Raj et al. has measured the magnetron warm-up time to be in the order of 2.5 s [59, 60].

This period is followed by the first T_{off} period where the drive circuit is automatically turned-off. In this period, the cathode surface cools resulting in a reduction in electron emission and weakening of space-charge formation. As previously commented if the T_{off} period becomes longer than 15–30 s the cathode cools down to a situation where non π-modes are supported.

4.2 Plasma Mode Production

The use of microwave ovens for the generation of Ar plasma have been reported for the use in-liquid plasma syntheses of carbon-based nanomaterials [28, 29 and 30], the use of susceptors to initiate nanomaterials plasmoid chemistry [9, 10] as well as the decomposition of methane hydrate [31]. In these reactions Ar gas was used as the carrier gas and H_2 and carbon monoxide being the main reaction products. Singh and Jarvis have also reported on the reaction of H_2 plasma with a mixture of hydrocarbons for the plasma syntheses of carbon nanostructures [61]. In this section, *E*-probe measurements of an empty-cavity mode (Sect. 4.2.1) and the plasma (Ar-H_2) cavity mode (Sect. 4.2.2) are compared and discussed [35].

4.2.1 Calibration of Empty Cavity Mode

Figure 11 provides a rendition of the empty-cavity mode calibration at atmospheric pressure and applied power of 250 W. The time period are as follows: magnetron turn-on at $t = 0$ s; *E*-probe first sensing the magnetron electric field within the cavity at t ~ 1 s; Maximum electric field at t ~ 1.5 s; and the 50 Hz modulated between E_1 and E_2 levels is the cavity-mode at t ~ 1.5–6 s. The upper level (E_1) represents the steady-state power per cycle 250 W and equates to a maximum of 4 J per pulse. The lower limit (E_2) represents the power stored in the cavity volume per cycle when

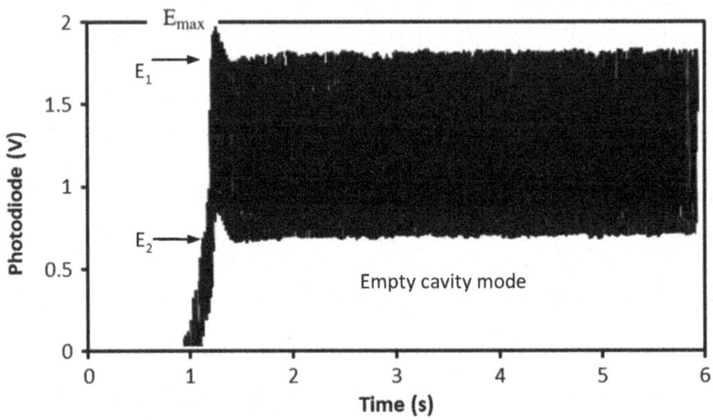

Fig. 11 Calibration of empty-cavity mode: The time components are: packaged magnetron warm-up time, plasma ignition, and quasi steady-state mode

the magnetron is turned-off. Finally, the difference between E_1 and E_2 represents the power lost to the cavity wall per cycle.

4.2.2 Ar-H_2 Plasma Mode

Figure 12a shows the Ar plasma mode at a pressure of 10 mbar (1 k Pascal) with an applied power of 350 W. Here it is seen that the probe only detects an electric field in the time period tp. In this period, the electric field strength builds up to a maximum of 49 mV at which point the plasma is struck and the electric field collapses. The absence of the electric field at the cavity wall after the plasma has struck implies that plasma is absorbing the applied power.

Figure 12b shows the Ar-H_2 plasma mode at the same pressure and applied power with a gas flow ratio is 50%. For this mode, the time trace closely resembles the modulated empty cavity mode, however the plasma is visibly reduced in size to form a sphere of approximately 120 mm diameter. This sphere is reduces further in diameter as the gas pressure is increased until it is extinguished at to 20 mbar. These results suggest that the Ar-H_2 plasma is absorbing little of the applied power.

4.2.3 Mode Discussion

The contrasting Ar and Ar-H_2 plasma modes may be understood in terms of the competition between positive ion generation and loss that is driven by diffusion of electrons to the wall where they are quenched under microwave breakdown conditions, [62, 63 and 64]. See Eq. 5.

$$1 = \frac{(V_i - V_a)}{D_o/\Lambda^2} \tag{5}$$

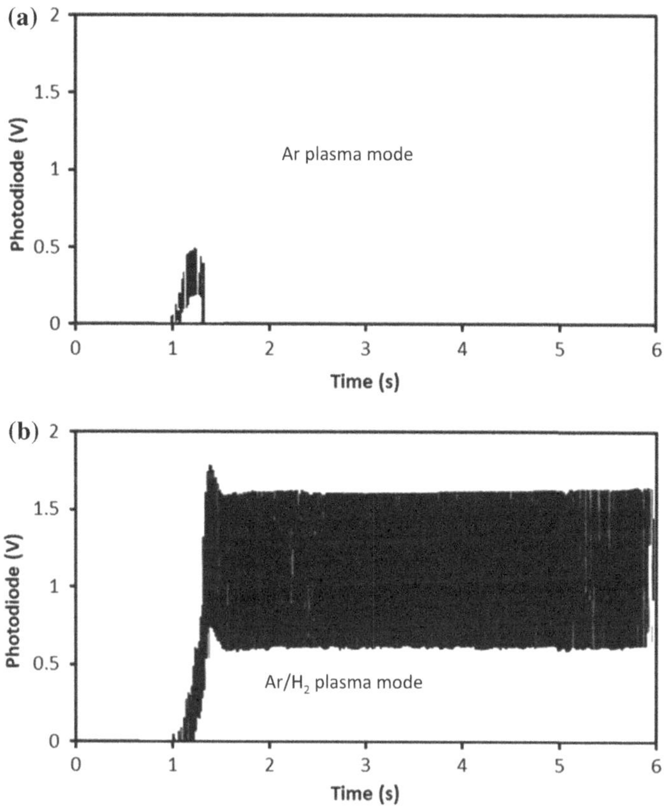

Fig. 12 **a** *E*-probe measurement of 10 mbar Ar plasma. **b** *E*-probe measurement of 10 mbar Ar-H_2 plasma

In this equation: V_i is the ionization frequency, V_a is the electron loss frequency, hence the term $(V_i - V_a)$ is the net electron production frequency. The term D_o/Λ^2 is the diffusion frequency to the vessel wall; and Λ is the characteristic diffusion length (cube root of the reaction vessel dimensions) and is equal to 3.7 cm in the MRC 200 plasma reactor.

The mathematical construct of Eq. 5 implies the following: a low average ratio is maintained by the presence of electron attachment or by fast ambipolar diffusion rate, whereas a high average ratio is maintained by the absence of electron attachment or by slow ambipolar diffusion rates.

Now consider Ar and H_2 that are both ready ionized ($Ar + e > Ar^+ + 2e$ at 15.8 eV); and ($H_2 + e \rightarrow H_2^+ + 2e$ at 13.6 eV) but due to their differing mass (u = 39.8 and 1.8) they have very contrasting ambipolar diffusion coefficient (Ar ~ 80 cm s^{-1}-Torr and H_2 ~ 700 cm s^{-1}-Torr [62]). Thus, the fast diffusion rate of H_2 would yield low average and therefore a weak plasma density with respect to the Ar plasma: and therefore transparent to the electromagnetic radiation. The contrast between the uniformity of Ar plasma and localized Ar-H_2 plasma may also be attributed to the difference the ambipolar diffusion coefficients. In addition to the presence of long lived Ar excited and long lived metastable species ($Ar^+ + e \rightarrow Ar^* + h\nu$) that can carry the energy out of the reactor before quenching occurs.

5 Summary

The first aim of this paper is to review patents (particularly those from the USA) of microwave oven power supply circuits and mode detection circuits and compare them with equivalent microwave oven plasma reactor circuits published by academic research groups. It is found that based on a review of US Patent Office patents, that they provide a comprehensive historical development of the magnetron power supply designed for extended time periods (where the filament warm-up time is not an operation issue). It also demonstrates how the issue of short repeated operation time periods required for the operation of microwave ovens for the cooking of food stuffs was addressed, particularly where warm-up times becomes a significant fraction of the cooking time. A survey of academic reuse of microwave components for microwave oven plasma reactors reveals that the power supplies are modified using either; a variac transformer at the MOT input or by varying the voltage doubler capacitor value to control the magnetron cathode voltage. These two approaches however do not address inadequate filament warm-up periods that lead to moding and enhanced magnetron aging. This survey reveals the information gap between the domestic microwave oven patent records and the academic research community that use microwave oven components for plasma processing. In the case of the paten records, it is acknowledgment that moding within the packaged magnetron warm-up period and at prolonged off periods may lead to inadequate heating of foodstuff. To overcome these processing problems, the voltage doubler circuit and coupled filament heating circuit have evolved overtime to mitigate moding. These improved circuits however do help to prevent moding within a loaded MRC. It is also the author's view that knowledge on the impact of these circuit modifications needs to be strengthened within the academic community, working on microwave oven plasma generation.

In the case of carbon-based nanostructure in-liquid plasma processing [28–30] and susceptor initiated plasmoid chemistry [9, 10] where the ionization time is of the order of 15–120 s, magnetron moding due to its warm-up period changes the MRC loading. It is also reasonable to assume that plasma induced diamond-like films for example using naphthalene and graphite [65] and C_2H_4 [66] as the precursor at very high powers (1200 W) and low processing times (5 s) will undergo moding. The phenomena of moding at the start of the plasma process needs to be controlled for process repeatability and increased yield are to be achieved on an economic scale. Combining both US patent office records and academic published papers provides a resource which allows both industrial and academic plasma researchers to enhance the performance of microwave oven plasma reactors for use in prototyping and production line scale-out, in advance of upgrading to more advanced and costly microwave systems.

The paper has also highlighted the potential/benefits of three mode detector circuits. These are; an antenna (*E*-probe) placed in a microwave oven cavity for frequency tuning a sold-state variable (915 ± 13 MHz) power source [49]. A contact probe for counting the rate of mode changes within an aged cavity-magnetron operating in the 2.45 GHz range [50]. In addition, an *E*-probe for characterizing a packaged-magnetron (operating in the 2.45 GHz range) warm-up time and the plasma working gas composition [35]. In this case plasma containing: Ar, and H_2. For plasma containing O_2, and CF_4 also see [35].

Following on from this overview, the authors propose that a strategy of scale-out of the academic microwave over plasma reactor rather than scale-up to costly industrial microwave plasma reactor provides an economic processing route to the plasma synthesis of advanced functional nanostructures.

Acknowledgements This publication has emanated from research supported of Enterprise Ireland through the Irish Composites Centre (IComp). The Authors declare that there is no conflict of interest regarding the publication of this paper.

References

1. P.L. Spencer, Method of treating foodstuffs. US Patent 2,495,429 (issued Jun 24, 1950)
2. P.L. Spencer, Coffee brewing. US patent 2,601,067 (issued Jun 17, 1952)
3. R.N. Gedye, W. Rank, K.C. Westaway, The rapid synthesis of organic compounds in microwave ovens. Can. J. Chem. **6**(1), 17–26 (1988)
4. A. Ribner, Microwave plasma etching machine and method of etching. US Patent 4,804,431, (issued Feb 14, 1989)
5. M.C. Salvadori, V.P. Mammana, O.G. Martins, F.T. Degasperi, Plasma-assisted chemical vapour deposition in a tunable microwave cavity. PSST **4**(3), 489–494 (1995)
6. V.J. Law, D. Tait, Contaminated ceramic plasma cleaning. Eur Semicond. **19**(9), S38–S41 (1997)
7. V.J. Law, D. Tait, Microwave plasma cleaning of ion implant ceramic insulators. Vacuum **49**(4), 273–278 (1998)
8. V.J. Law, N. Macgearailt, Visualization of a dual frequency plasma etch process. Meas. Sci. Technol. **18**(3), 645–649 (2007)
9. A. Irzh, I. Genish, L. Klein, L.A. Solovyov, A. Gedanken, Synthesis of ZnO and Zn nanoparticles in microwave plasma and their deposition on glass slides. Langmuir **26**(8), 5976–5984 (2010)
10. P.H. Talemi, G.P. Simon, Preparation of graphene nanowalls by a simple microwave-based method. Carbon **48**, 3993–4000 (2010)
11. Y.H. Jung, S.O. Jang, H.J. You, Hydrogen generation from the dissociation of water using microwave plasmas. Chin. Phys. Lett. **3**(6), 065204 (2013)
12. N. Ismail, F.N. Ani, A review on plasma treatment for the processing of solid waste. Jurnal Teknologi **72**(5), 41–49 (2015)
13. E. Chaiya, P. Khongkrapan, N. Tippayawong, Use of non-thermal microwave plasma for syngas production from dry reforming of compressed biomethane. Int. J. App. Eng. Res. **9**(20), 6835–6842 (2014)

14. P. Khongkrapan, P. Thanompongchart, N. Tippayawong, T. Kiatsiriroat, Microwave plasma assisted pyrolysis of refuse derived fuels. Cent. Eur. J. Eng. **4**(1), 72–79 (2014)
15. N. Tippayawonga, E. Chaiya, P. Thanompongchart, P. Khongkrapan, Sustainable energy from biogas reforming in a microwave discharge reactor. Proc. Eng. **118**, 120–127 (2015)
16. W. Choe, G.-C. Kwon, J. Kim, J. Kim, S.-J. Jeon, S. Huh, Simple microwave preionization source for ohmic plasmas. Rev. Sci. Instrum. **71**(7), 2728–2732 (2000)
17. C. Chaichumporn, P. Ngamsirijit, N. Brkoonklin, K. Eaiprasetsak, M. Fuangfoong, Design and construction of 2.45 GHz microwave plasma source at atmospheric pressure. Proc. Eng. **8**, 94–100 (2011)
18. V.N. Tikhonov, S.N. Aleshin, I.A. Ivanov, A.V. Tikhonov, The low-cost microwave plasma sources for science and industry applications. J. Phys. Conf. Ser. **927**, 012067 (2017)
19. Some resources and ideas for plasma experiments: plasma experiments with commercial gas tubes and some ideas for microwave oven conversions. The Bell Jar. **4**(2) (1997). http://www.belljar.net/plasma.htm (Accessed June 2019)
20. V.J. Law, D.P. Dowling, Converting a microwave oven into a plasma reactor: a review. Int. J. Chem. Eng. **2018**, 12 (2018)
21. V.J. Law, D.P. Dowling, Domestic microwave oven and fix geometry waveguide applicator processing of organic compounds and biomaterials: a review. Glob. J. Res. Eng. (A). **18**(2), Version 1.0, 19 (2018)
22. V.J. Law, D.P. Dowling, The domestic microwave oven as a prototype tool. Anal. Chem. Ind. J. (2018)
23. S. Birla, Effect of magnetron frequency on heating pattern in domestic oven, in *Conference Presentations and White Papers: Biological Systems Engineering*, vol. 54 (2010)
24. S.P. Yeong, M.C. Law, C.C.V. Lee, Y.S. Chan, Modelling batch microwave heating of water. IOP Conf. Ser. Mater. Sci. Eng. **217**, 012035 (2107)
25. Z. Zhang, T. Su, S. Zhang, Shape effect on the temperature field during microwave heating process. J. Food Qual. **2018**, 24 (2018)
26. S. Dąbrowska, T. Chudoba, J. Wojnarowicz, W. Łojkowski, Current trends in the development of microwave reactors for the synthesis of nanomaterials in laboratories and industries: a review. Crystals **8**(10), 379 (2018)
27. A.Y.N. Hui, G. Wang, B. Lin, W.-T. Chan, Microwave plasma treatment of polymer surface for irreversible sealing of microfluidic devices. Lab Chip **5**, 1173–1177 (2005)
28. S. Nomura, H. Toyota, S. Mukasa et al., Production of hydrogen in a conventional microwave oven. J. Appl. Phys. **106**, 073306 (2009)
29. S. Nomura, H. Yamashita, H. Toyota et al., Simultaneous production of hydrogen and carbon nanotubes in a conventional microwave oven, in *International Symposium on Plasma Chemistry (ISPC19)*, Bochum, Germany, vol. 65 (2009)
30. H. Toyota, S. Nomura, S. Mukasa, A practical electrode for microwave plasma processes. Int. J. Mater. Sci. Appl. **2**(3), 83–88 (2013)
31. I. Rahim, S. Nomura, S. Mukasa, H. Toyota, Decomposition of methane hydrate for hydrogen production using microwave and radio frequency in-liquid plasma methods. Appl. Therm. Eng. **90**, 120–126 (2015)
32. V.J. Law, D.P. Dowling, Electronic valve instabilities and mode jumps. CMSIM J. **1**, 3–35 (2017)
33. V.J. Law, D.P. Dowling, Magnetron modes and the chimera states, in *Fractional Dynamics and Anomalous Transport in Plasma Science*, Chap 2, ed. by C. H. Skiadas (Springer, 2018), pp. 35–65. ISBN 978-3-030-04482-4
34. R.C. Lo, Application of microfluidics in chemical engineering. Chem. Eng. Process. Tech. **l**, 1002 (2013)
35. V.J. Law, Microwave near-field plasma probe. Vacuum **51**(3), 463–468 (1998)
36. M. Leconte, Statistical study of magnetron patents in the early years of electronics between 1920 and 1945; heuristic focusing around the discovery of the cavity magnetron, in *Proceedings of IEEE CAVMAG 2010*, pp. 11–16 (2010). https://doi.org/10.1109/cavmag.2010.5565570

37. N.F. Alekseev, D.D. Malairov, I.B. Bensen, Generation of high-power oscillations with a magnetron in the centimeter band. Proc. IRE **32**, 136–139 (1944)
38. J.T. Randall, H.A.H Boot, High frequency electrical oscillator. US 2,542,966 (Issued Feb 20, 1951)
39. J. Sayers, High frequency electrical oscillator. US 2,546,870 (Issued Mar 27, 1951)
40. J.T. Lamb, Microwave oven control. US 3,168,637 (Issued Feb 2, 1965)
41. L. Blok et al., Control systems for regulating the current in a magnetron tube. US 2,302,060 (Issued Jan 31, 1967)
42. J.R. Mins, Microwave magnetron. US 3,739,225 (Issued Jun 12, 1973)
43. T. Koinuma, Magnetron. US 3,809,950 (Issued May 7, 1974)
44. A.E. Feinberg, Power supply circuit for continuous wave magnetron operated by pulse direct current US 3,396,342 (Issued Aug 6, 1968)
45. J.T. Lamb, Microwave oven defrost circuit. US 3,842,233 (Issued Oct 15, 1974)
46. S. Nagamoto. Microwave oven supply circuit. US 3,943,317 (Issued Mar 9, 1976)
47. W.C. Hickman, Magnetron power supply and cathode heater circuit. US 3,392,309 (Issued Jul 9, 1968)
48. H.F. Chapell, Interlock circuitry for microwave oven. US 3,624,334 (Issued Nov, 1971)
49. T. Nobue, S. Kusunoki, Microwave oven having controllable frequency microwave power source. US. Patent 4,415,789 (Issued Nov 5, 1983)
50. S.A. Levie, W.E. Taylor, Magnetron mode detection. US 4,501,767 (Issued Mar 12, 1985)
51. P.H. Smith, Power control circuitry for magnetron. US 4,620,078 (Issued Oct 18, 1986)
52. P.H. Smith, T.R. Payne, F. Reising Jr., Filament power compensation for magnetron. US 4,835,353 (Issued May 30, 1989)
53. C. Daley, R.J. Sweetman, C. Lenny, Magnetron variable power supply with moding prevention. US. 5,571,439 (Issued Nov 5, 1996)
54. V.N. Tikhonov, D.V. Pugashkin, J.A. Chetokin, Magnetron generator. RU 2,480, 890 (Issued Apr 27, 2013)
55. J. Smith, Integration of microwave plasma ignition into a multi-fuel engine. Contract No. FA8651-07-1-00082008. http://www.dtic.mil/dtic/index.html (Accessed June 2019)
56. A. Pagliarani, A.J. Kenyon, N.F. Thornhill, E. Sirisena, K. Lee, V.J. Law, Process harmonic pulling in a RIE plasma-tool. Electr. Lett. **42**(2), 120–121 (2006)
57. V.J. Law, Process induced oscillator frequency pulling and phase noise within plasma systems. Vacuum **82**(6), 630–638 (2008)
58. S.S.R. Geedipalli, V. Rakesh, A.K. Datta, Modeling the heating uniformity contributed by a rotating turntable in microwave ovens. J. Food Eng. **82**, 359–368 (2007)
59. J.D. Raj, S. Birla, K. Pitchai, J. Subbiah, D. Jones, Modeling of microwave heating of a rotating object in domestic oven, in *Proceedings of COMSOL conference*, Boston, USA (2011)
60. J.D. Raj, S. Birla, J. Subbiah, S. Shanmugasundaram, S. Dhanasekaran, Modelling coupled rotation and microwave heating of an object in a domestic microwave oven. Inst. J. Eng. Res. Appl. **6**(3), 15–18 (2016)
61. R. Singh, A.L.L. Jarvis, Microwave plasma enhanced chemical vapour deposition growth of carbon nanostructures. S. Afr. J. Sci. **106**(5-6), 4 (2010)
62. A.D. MacDonald, S.C. Brown, High frequency gas discharge breakdown in hydrogen. Phys. Rev. **76**, 1634–1639 (1949)
63. M.A. Biondi, Diffusion cooling of electrons in ionized gases. Phys. Rev. **93**, 1136–1140 (1954)
64. L. Thomas, J.L. Jauberteau, I. Jauberteau, J. Aubreton, J. Catherinot, Characterization of argon-hydrogen microwave plasma discharge using a atomic hydrogen source. Effect of hydrogen dilution on atomic hydrogen production. Plasma Chem. Plasma Process. **17**(2), 193–205 (1997)
65. R. Bajpai, L. Rapoport, K. Amsalem, H.D. Wagner, Rapid growth of onion-like carbon nanospheres in a microwave oven. Crst. Eng. Comm. **18**, 230–239 (2016)
66. C. Pakpum, k Kanchiang, Fabrication of DLC nanoparticle clusters by μ-wave oven based plasma reactor with acetylene diluted in air precursor. Appl. Surf. Sci. **458**, 100–110 (2018)

Unexpected Properties of Open Quantum Graphs and Microwave Networks

Michał Ławniczak, Jiří Lipovský, Szymon Bauch and Leszek Sirko

Abstract We will discuss some peculiar properties of open quantum graphs and microwave networks. In particular, we will consider graphs which do not obey the Weyl's law $N(R) = \frac{\mathcal{L}}{\pi}R + O(1)$, where $O(1)$ is a function which for R going to infinity is bounded by a constant. The Weyl's law directly links the counting function $N(R)$ of the number of resonances with the square root of energy k, $0 < \text{Re}\,(k) < R$, and the total length $\mathcal{L}$ of the internal edges of a graph. Graphs which do not obey the Weyl's law are called non-Weyl graphs. For standard boundary conditions a Weyl graph becomes a non-Weyl graph if a balanced vertex is introduced in which the numbers of infinite leads and internal edges are the same. We show that our experimental results obtained in the frequency range $\nu = 0.3{-}2.2$ GHz are in excellent agreement with the numerical calculations yielding the resonances of the networks simulating Weyl and non-Weyl graphs. We also demonstrate that the numerical calculations performed in the frequency range $\nu = 0.3{-}2.8$ GHz for the Weyl and non-Weyl graphs are in the agreement with the Weyl's and non-Weyl's laws, respectively.

1 Introduction

It is well known that a model of a quantum graph can be successfully used to describe the behavior of a quantum particle on a physical network [1–3]. In this model the Hamiltonian is acting as the negative second derivative with appropriate coupling conditions at the vertices. We consider the case of boundary conditions, which is the most physical one—the standard conditions, also called the Neumann boundary conditions. These conditions impose the continuity of the function's value and vanishing

M. Ławniczak · S. Bauch · L. Sirko (✉)
Institute of Physics, Polish Academy of Sciences, Aleja Lotników 32/46, 02-668 Warszawa, Poland
e-mail: sirko@ifpan.edu.pl

J. Lipovský
Faculty of Science, Department of Physics, University of Hradec Králové, Rokitanského 62, 50003 Hradec Králové, Czechia

C. H. Skiadas and Y. Dimotikalis (eds.), *12th Chaotic Modeling and Simulation International Conference*, Springer Proceedings in Complexity,
https://doi.org/10.1007/978-3-030-39515-5_15

of the sum of its outgoing derivatives at each vertex. For more details on quantum graphs, we refer the reader to the book [4] and the references therein. It turned out that quantum graphs can be modeled by microwave networks. It is possible because both systems are described by the same equation, namely, the telegrapher's equation for microwave networks is formally equivalent to the one-dimensional Schrödinger equation describing quantum graphs [5, 6]. These peculiar properties of microwave networks (graphs) were successfully used, e.g., in [7] where a famous question asked by Mark Kac "Can one hear the shape of a drum?" was investigated. It was originally posed in the context of dissipationless isospectral systems. The paper [7] extends it to scattering systems such as isoscattering ones. Moreover, in the recent paper [8] the missing level statistics were analyzed in various applications. It is important to point out that there are also some other useful model systems which are often used to simplify analyzes of complicated quantum systems. Such examples are microwave flat cavities simulating two-dimensional quantum billiards [9–14] and experiments with Rydberg atoms strongly driven by microwave fields [15–21].

There are two main approaches to resonances in mathematical physics. One possibility is to study the poles of the meromorphic continuation of the resolvent to the second Riemann sheet, called *resolvent resonances*. The second notion are so-called *scattering resonances*, obtained as poles of the continuation of the determinant of the scattering matrix. These two approaches are for quantum graphs closely related. One can prove (see e.g., [22, 23]) that the set of resolvent resonances is equal to the union of the set of scattering resonances and the set of eigenvalues with eigenfunctions which have support on the compact (internal) part of the graph. Since we do not have these eigenvalues for the quantum graphs with external leads and rationally unrelated lengths of the edges, we can interchange both notions.

If we consider the compact graph (which has no infinite leads), the resolvent resonances coincide with the eigenvalues. It is well known that the counting function of the number of eigenvalues with the square root of energy k in the interval $(0, R)$ satisfies the Weyl's law

$$N(R) = \frac{\mathcal{L}}{\pi} R + \mathcal{O}(1), \tag{1}$$

where $\mathcal{L}$ is the sum of the lengths of the graph's edges and $\mathcal{O}(1)$ is a function which in the limit $R \to +\infty$ is bounded by a constant of order of 1. If one attaches infinite leads and obtains a non-compact graph, the same asymptotics would be expected to hold also for the number of resolvent resonances encircled in the semi-circle of radius R and center at the origin in the k-plane with $\operatorname{Re} k > 0$, with $\mathcal{L}$ denoting only the sum of the lengths of the internal edges of the graph. A surprising observation by Davies and Pushnitski [24] yields that this is not the case for all the graphs. There are graphs for which the coefficient of the leading term of the asymptotics is smaller than expected. Such graphs are called non-Weyl graphs [25], while the graphs satisfying the asymptotics (1) are called Weyl graphs.

Let us briefly review the theoretical results on non-Weyl quantum graphs. In the seminal paper [24] the authors found a simple geometric condition, which distinguishes Weyl and non-Weyl graphs, for graphs with standard boundary conditions.

The graph is non-Weyl if and only if there exists a vertex for which the number of internal and external edges (infinite leads) is the same. This result was generalized in [26] and the condition which distinguishes non-Weyl graphs was found for general boundary conditions. The properties of magnetic quantum graphs were researched in the paper [27], its main result is that the presence of the magnetic field cannot change a non-Weyl graph into the Weyl one but it can change the coefficient of the leading term of a non-Weyl graph. Non-Weyl graphs were later studied in more detail in [28] and the bounds on the coefficient of the leading term of the asymptotics were found.

In this paper, we show the experimental results reported in the recent publication [25] which verified the geometric condition introduced by Davies and Pushnitski obtained for two similar microwave networks, one Weyl and the other one non-Weyl, and measured their scattering matrices in the frequency range $\nu = 0.3-2.2$ GHz. However, the numerical calculations were extended to 2.8 GHz in this paper.

2 Simulations of Quantum Graphs Using Microwave Networks

Quantum graphs can be considered as idealizations of physical networks in the limit where the lengths of their elements are much bigger than their widths. There are many successful applications which model a broad range of physical problems, see, e.g., [29], as well as experimental realizations of quantum graphs. Using contemporary epitaxial techniques it is possible to design and fabricate quantum nanowire networks [30, 31].

A seminal paper by Hul et al. [5] shows that quantum graphs can be successfully simulated by microwave networks composed of microwave junctions and coaxial cables. In the present investigations the SMA-RG402 coaxial cables were used. The SMA-RG402 cable consists of an inner conductor of radius $r_1 = 0.05$ cm, which is surrounded by a concentric conductor of inner radius $r_2 = 0.15$ cm. The space between them is filled with Teflon with a dielectric constant $\varepsilon \simeq 2.06$. Below the cut-off frequency of the TE_{11} mode $\nu_c \simeq \frac{c}{\pi(r_1+r_2)\sqrt{\varepsilon}} \simeq 33$ GHz [32], where c denotes the speed of light in vacuum, only the fundamental TEM mode can propagate inside a coaxial cable. It should be stressed, that the lengths of the bonds of the corresponding quantum graph are given not by the geometric lengths ℓ_i^g of the coaxial cables, but by the optical lengths $\ell_i = \ell_i^g\sqrt{\varepsilon}$.

Therefore, one can experimentally investigate the properties of a quantum graph by constructing microwave networks which have the same topology and boundary conditions at the vertices. This system is extremely fruitful in both experimental and theoretical studies of chaotic quantum systems. There is a vast literature on the topic including many spectral and scattering problems of microwave networks, see e.g., [5, 6, 8, 33–37]. Microwave networks can also simulate various quantum chaotic systems with the spectral properties described in the Random Matrix Theory by the three main symmetry classes: Gaussian orthogonal ensemble (GOE) [5, 6],

Gaussian unitary ensemble (GUE) [5, 8, 35, 38] and Gaussian symplectic ensemble (GSE) [36].

Let us first consider a compact metric graph which consists of the set of V vertices and B bonds (edges). For a vertex with index i we define its valency b_i. This means that this vertex is connected to the other vertices by b_i bonds. The propagation of the wave on each bond is governed by the one-dimensional Schrödinger equation. The spectral properties of the graph are given by the set of its edge lengths and the boundary conditions at the vertices which connect the amplitudes and the first derivatives of the waves meeting at the vertex. We will only consider standard boundary conditions, which prescribe the continuity of the function's values and the fact that the sum of the outgoing derivatives at each vertex vanishes. Since there clearly is no balanced vertex, these graphs are standard examples of Weyl graphs.

To obtain non-Weyl graphs from Weyl graphs infinite external leads have to be connected to a subset of vertices of this compact graph. As already stated in the text above, a Weyl graph becomes a non-Weyl graph if we introduce a balanced vertex. A vertex j is denoted as a balanced vertex if the number of internal edges attached to j is equal to the number of external edges (infinite leads) attached to j. In the opposite case, when the number of external leads attached to this vertex is greater or smaller than the number of internal edges attached to j, we call this vertex unbalanced. Such vertices are used, e.g., in the recent photovoltaic smartwire connection technology [39].

If there is not an additional symmetry of the graph, the counting function of the number of resonances of a non-Weyl graph can be expressed as

$$N_{nW}(R) = \frac{\mathcal{L}'}{\pi} R + \mathcal{O}(1)\,, \tag{2}$$

where $\mathcal{L}' = \mathcal{L} - \ell_s$ is the effective size of a non-Weyl graph and ℓ_s is the length of the shortest bond outgoing from the balanced vertex.

The properties of the Weyl and non-Weyl graphs were experimentally tested using two graphs shown in Figs. 1a and 2a. Both graphs have the same length $\mathcal{L} = \sum_{i=1}^{7} \ell_i$. In the graph in Fig. 1a there are two unbalanced vertices attaching infinite leads L_1^∞ and L_2^∞. Therefore, it is a Weyl graph. The non-Weyl graph in Fig. 2a contains one balanced vertex to which two leads L_1^∞ and L_2^∞ are attached. The black broken lines denote the leads. According to the formula (2) we expect that the non-Weyl graph in Fig. 2a has smaller number of resonances. The microwave networks were connected to the VNA with the two elastic microwave cables which is equivalent to attaching of two infinite leads L_1^∞ and L_2^∞ to quantum graphs.

Such configurations of the experimental systems allow for measuring the scattering matrix $\hat{S}(\nu)$:

$$\hat{S}(\nu) = \begin{pmatrix} S_{11}(\nu) & S_{12}(\nu) \\ S_{21}(\nu) & S_{22}(\nu) \end{pmatrix}, \tag{3}$$

which relates the amplitudes of the incoming and outgoing waves of frequency ν in both leads.

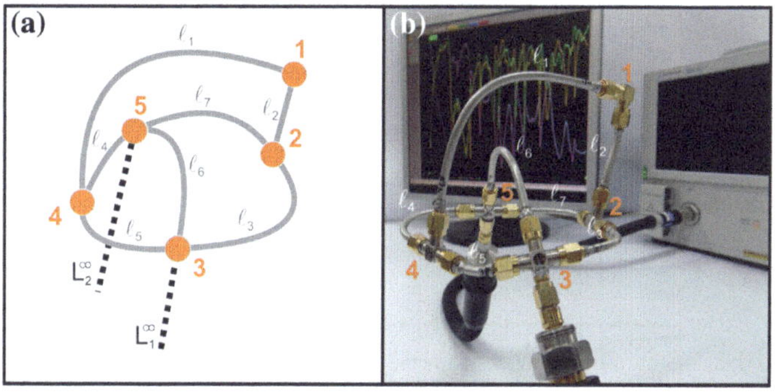

Fig. 1 Panel **a** shows the scheme of the Weyl graph. The Weyl graph contains two unbalanced vertices with infinite leads L_1^∞ and L_2^∞. Panel **b** shows the corresponding Weyl microwave network (graph) constructed from microwave coaxial cables and joints. The optical lengths of both graphs are the same and are equal to $\mathcal{L} = \sum_{i=1}^{7} \ell_i$. The microwave network is connected to the VNA with two microwave cables which is equivalent to attaching of two infinite leads L_1^∞ and L_2^∞ to the quantum graph

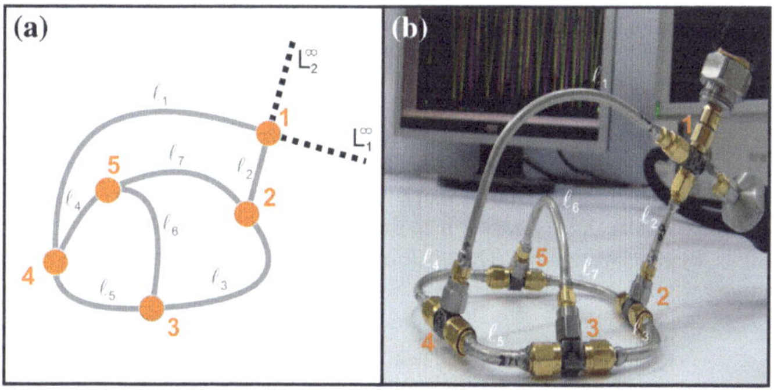

Fig. 2 Panel **a** shows the scheme of the non-Weyl graph. The graph is also characterized by the same length as the Weyl graph $\mathcal{L} = \sum_{i=1}^{7} \ell_i$. The non-Weyl graph contains one balanced vertex with two attached leads L_1^∞ and L_2^∞, which are marked by black broken lines. Panel **b** shows the corresponding non-Weyl microwave network (graph). The optical lengths of the graphs in panels (**a**) and (**b**) are the same and are equal to $\mathcal{L}$. Also here, the microwave network is connected to the VNA with two microwave cables which are equivalent to the two infinite leads L_1^∞ and L_2^∞ in the non-Weyl graph in the panel (**a**)

Two corresponding microwave networks constructed from microwave coaxial cables are shown in Figs. 1b and 2b.

The proof that the effective size of the non-Weyl graph is $\mathcal{L}' = \mathcal{L} - \ell_s$ where $\ell_s = \ell_2$ is the length of the shortest edge outgoing from the balanced vertex 1 in Fig. 2a, b can be found in [25]. In order to carry out the measurements of the two-port scattering matrix $\hat{S}(\nu)$ we connected the vector network analyzer (VNA) Agilent E8364B to the microwave networks shown in Figs. 1b and 2b and measurements were done in the frequency range $\nu = 0.3$–2.2 GHz (see Figs. 3 and 4).

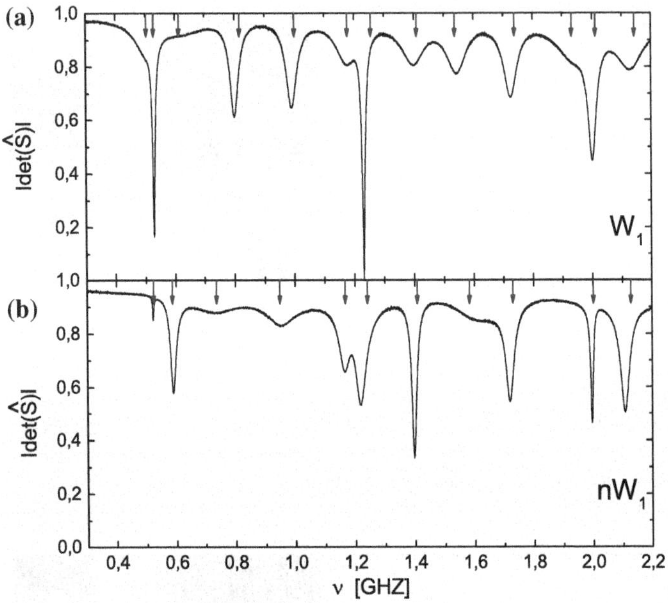

Fig. 3 Panel **a** shows the amplitude of the determinant of the scattering matrix $|\det(\hat{S}(\nu))|$ of the experimentally studied Weyl network W_1 with the length $\mathcal{L}_1 = 0.999$ m, which contains two unbalanced vertices in the frequency range 0.3–2.2 GHz. Panel **b** shows the amplitude of the determinant of the scattering matrix $|\det(\hat{S}(\nu))|$ of the experimentally studied non-Weyl network nW_1 with the effective size $\mathcal{L}'_1 = 0.896$ m in the same frequency range. There is one balanced vertex in the non-Weyl network nW_1. The positions of the expected resonances obtained from the theory are marked with blue arrows

Two pairs of the Weyl and non-Weyl networks were considered, W_1 and nW_1, and W_2 and nW_2, with the lengths $\mathcal{L}_1 = 0.999$ m and $\mathcal{L}_2 = 1.151$ m, respectively (see Table 1).

The length of the shortest bond ℓ_s ongoing from a balanced vertex of the two non-Weyl networks nW_1 and nW_2 is ℓ_2, hence we can easily compute the effective sizes of these networks and obtain $\mathcal{L}'_1 = 0.896$ m and $\mathcal{L}'_2 = 0.972$ m, respectively. The uncertainties of the bonds' lengths of the networks are due to the preparation of coaxial microwave cables.

We show the amplitudes $|\det(\hat{S}(\nu))|$ of the determinants of the scattering matrices of the experimentally studied Weyl and non-Weyl networks in the frequency range 0.3–2.2 GHz in Figs. 3 and 4. Figure 3a, b show the comparison of the experimental results obtained for the Weyl and non-Weyl microwave networks W_1 and nW_1 containing $\ell_s = \ell_2 = 0.103$ m bond. Figure 3a, b prove that the presence of the balanced vertex dramatically changes the spectrum of the non-Weyl network and also lowers the number of observed resonances. For the Weyl network W_1 from the Weyl's formula (1) in the frequency range 0.3–2.2 GHz we should expect $N \simeq 13$ resonances while for the non-Weyl network nW_1 the formula (2) gives $N \simeq 11$ resonances. Indeed, the theoretical predictions are in good agreement with the experiment. In

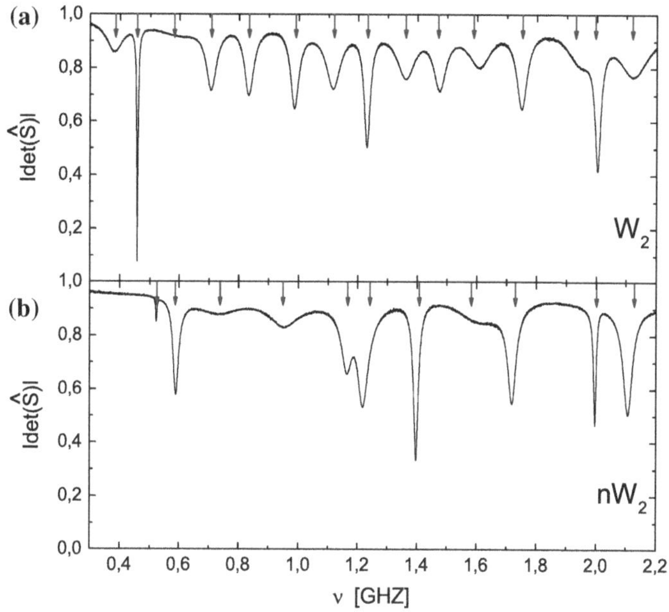

Fig. 4 Panel **a** shows the amplitude of the determinant of the scattering matrix $|\det(\hat{S}(\nu))|$ of the experimentally studied Weyl network with the length $\mathcal{L}_2 = 1.151$ m, which contains two unbalanced vertices in the frequency range 0.3–2.2 GHz. Panel **b** shows the amplitude of the determinant of the scattering matrix $|\det(\hat{S}(\nu))|$ of the experimentally studied non-Weyl network nW_2 with the effective size $\mathcal{L}'_2 = 0.972$ m in the same frequency range. There is one balanced vertex in the non-Weyl network nW_2. The positions of the expected resonances obtained from the theory are marked with blue arrows

Table 1 The optical lengths of the bonds of the microwave Weyl and non-Weyl networks, W_1 and nW_1, and W_2 and nW_2, with the lengths $\mathcal{L}_1 = 0.999$ m and $\mathcal{L}_2 = 1.151$ m, respectively

Networks W_1 and nW_1	Networks W_2 and nW_2
$\ell_1 = 0.127 \pm 0.001$ m,	$\ell_1 = 0.203 \pm 0.001$ m,
$\ell_2 = 0.103 \pm 0.001$ m,	$\ell_2 = 0.179 \pm 0.001$ m,
$\ell_3 = 0.130 \pm 0.001$ m,	$\ell_3 = 0.130 \pm 0.001$ m,
$\ell_4 = 0.225 \pm 0.001$ m,	$\ell_4 = 0.225 \pm 0.001$ m,
$\ell_5 = 0.116 \pm 0.001$ m,	$\ell_5 = 0.116 \pm 0.001$ m,
$\ell_6 = 0.171 \pm 0.001$ m,	$\ell_6 = 0.171 \pm 0.001$ m,
$\ell_7 = 0.127 \pm 0.001$ m,	$\ell_7 = 0.127 \pm 0.001$ m.

Fig. 3a we observe 13 resonances (as the Weyl's formula predicts) while in Fig. 3b we see only 11 resonances (also with correspondence to the non-Weyl formula). To test more versatilely the applicability of the modified Weyl's formula (2) we compared our experimental results for the positions of resonances with the theoretical ones (blue arrows) obtained using the method of pseudo-orbits expansion. The result of this comparison is very satisfying.

3 The Resonance Conditions

Using the method of pseudo-orbit expansion one can find the resonance conditions for the Weyl and non-Weyl graphs, for more details see [28, 40, 41]. We obtain the formula

$$\det[\hat{I}_{2N} - \exp(\mathrm{ik}\hat{\mathrm{L}})\hat{\Sigma}] = 0\,, \tag{4}$$

where $\hat{I}_{2N}$ is $2N \times 2N$ identity matrix, N is the number of the internal edges of the graph (in our case $N = 7$) and the matrix $\hat{L} = \mathrm{diag}(\ell_1, \ell_2, \ldots, \ell_7, \ell_1, \ldots, \ell_7)$, where we denote the lengths of the edges $\ell_1 = \ell(1,4)$, $\ell_2 = \ell(1,2)$, $\ell_3 = \ell(2,3)$, $\ell_4 = \ell(4,5)$, $\ell_5 = \ell(3,4)$, $\ell_6 = \ell(3,5)$, $\ell_7 = \ell(2,5)$. Such k which give zeros of the left hand side of the above equation are the resonances.

We should stress that for open systems resonances are poles of the determinant of the scattering matrix [42] which occur at complex wave numbers $k_r = \frac{2\pi}{c}(\nu_r - i\Delta\nu_r)$. Here, ν_r and $2\Delta\nu_r$ can be associated with the positions and the widths of resonances, respectively. The scattering matrix $\hat{\Sigma}$ for the non-Weyl and Weyl graphs and the resonance conditions imposed by (4) are discussed in the details in [25].

The results of the calculations presented in Fig. 3a, b (blue arrows) show that there should be 13 and 11 resonances, respectively, in the frequency range 0.3–2.2 GHz, which is in agreement with both the experiment and the number of resonances predicted by the formulas (1) and (2). Therefore, we show that if the balance vertex is present in the non-Weyl network nW_1, the number of resonances is lowered in comparison with the network W_1 by 2. We stress that also the position of the resonances obtained both theoretically and experimentally are in very good agreement.

In Fig. 4a, b we present the comparison of the experimental results obtained for the Weyl and non-Weyl microwave networks W_2 and nW_2 of the lengths $\mathcal{L}_2 = 1.151$ m, which contain the bond with the length $\ell_s = \ell_2 = 0.179$ m. They are longer than the microwave networks W_1 and nW_1. In this case from the Weyl's formula (1) we expect to have $N \simeq 15$ resonances while for the non-Weyl network nW_2 the expected number of the resonances is $N \simeq 12$. As we can see from the experimental figures, we can identify 15 resonances for the Weyl network W_2 and 12 resonances the non-Weyl network nW_2 in agreement with the formula (2). The positions of the theoretically computed resonances are marked with blue arrows. The correspondence between the theoretical and experimental results is for both types of networks also very good. We can see that due to the presence of the balanced vertex the number of the resonances in the non-Weyl network is lowered in comparison to the network W_2 by 3.

In this paper we also report the results of the numerical calculations performed for the Weyl and non-Weyl graphs W_1 and nW_1, and W_2 and nW_2, with the lengths $\mathcal{L}_1 = 0.999$ m and $\mathcal{L}_2 = 1.151$ m, respectively, in a broader frequency range $\nu = 0.3-2.8$ GHz. Table 2 shows the positions of the calculated resonances. For the Weyl and non-Weyl graphs W_1 and nW_1 in the frequency range 0.3–2.8 GHz we found 17 and 15 resonances while for the graphs W_2 and nW_2 19 and 16 resonances, respectively. These results are in the agreement with the Weyl's and non-Weyl's laws, respectively.

Table 2 The positions of resonances ν_r of the microwave Weyl and non-Weyl networks, W_1 and nW_1, and W_2 and nW_2, with the lengths $\mathcal{L}_1 = 0.999$ m and $\mathcal{L}_2 = 1.151$ m, respectively, calculated in the frequency range 0.3–2.8 GHz

Resonance no.	ν_r (GHz) of W_1	ν_r (GHz) of nW_1	ν_r (GHz) of W_2	ν_r (GHz) of nW_2
1	0.504	0.524	0.384	0.454
2	0.528	0.587	0.456	0.579
3	0.613	0.736	0.582	0.715
4	0.816	0.948	0.707	0.830
5	1.001	1.166	0.834	0.996
6	1.177	1.242	0.989	1.170
7	1.256	1.408	1.118	1.346
8	1.409	1.583	1.232	1.470
9	1.537	1.730	1.361	1.582
10	1.738	2.001	1.470	1.752
11	1.932	2.127	1.588	1.989
12	2.011	2.321	1.751	2.116
13	2.143	2.460	1.928	2.237
14	2.384	2.557	1.994	2.372
15	2.503	2.743	2.120	2.508
16	2.567		2.314	2.695
17	2.782		2.380	
18			2.508	
19			2.641	

4 Conclusions

To sum up, the microwave networks simulating quantum graphs were used to show that some graphs do not obey the Weyl's law. Such graphs are called non-Weyl graphs. We recall the experimental data from [25] obtained in the frequency range $\nu = 0.3-2.2$ GHz which are in excellent agreement with the numerical calculations coming from the theoretical predictions of pseudo-orbit expansion theory. We show that the number of resonances of the measured system may significantly depend on the way how it is connected to the external world. Moreover, in the numerical calculations we extended the frequency range up to $\nu = 2.8$ GHz. We show that these numerical results are also in the agreement with the Weyl's and non-Weyl's laws.

Acknowledgements This work was supported in part by the National Science Centre, Poland, Grant No. 2016/23/B/ST2/03979 and the National Centre for Research and Development, Grant POIR.04.01.04-00-0144/17. J. Lipovský was supported by the research programme "Mathematical Physics and Differential Geometry" of the Faculty of Science of the University of Hradec Králové.

References

1. T. Kottos, U. Smilansky, Phys. Rev. Lett. **79**, 4794 (1997)
2. T. Kottos, U. Smilansky, Ann. Phys. **274**, 76 (1999)
3. Z. Pluhař, H.A. Weidenmüller, Phys. Rev. Lett. **112**, 144102 (2014)
4. G. Berkolaiko, P. Kuchment, *Introduction to Quantum Graphs*. Mathematical Surveys and Monographs, vol. 186 (2013), p. 270
5. O. Hul, S. Bauch, P. Pakoński, N. Savytskyy, K. Życzkowski, L. Sirko, Phys. Rev. E **69**, 056205 (2004)
6. M. Ławniczak, S. Bauch, L. Sirko, *Handbook of Applications of Chaos Theory*, eds. by C. Skiadas, C. Skiadas (CRC Press, Boca Raton, USA, 2016), p. 559
7. O. Hul, M. Ławniczak, S. Bauch, A. Sawicki, M. Kuś, L. Sirko, Phys. Rev. Lett **109**, 040402 (2012)
8. M. Białous, V. Yunko, S. Bauch, M. Ławniczak, B. Dietz, L. Sirko, Phys. Rev. Lett. **117**, 144101 (2016)
9. L. Sirko, P.M. Koch, R. Blümel, Phys. Rev. Lett. **78**, 2940 (1997)
10. H.-J. Stöckmann, *Quantum Chaos: An Introduction* (Cambridge University Press, Cambridge, 2000)
11. Y. Hlushchuk, A. Błędowski, N. Savytskyy, L. Sirko, Phys. Scr. **64**, 192 (2001)
12. R. Blümel, P.M. Koch, L. Sirko, Found. Phys. **31**, 269 (2001)
13. S. Hemmady, X. Zheng, E. Ott, T.M. Antonsen, S.M. Anlage, Phys. Rev. Lett. **94**, 014102 (2005)
14. B. Dietz, A. Richter, CHAOS **25**, 097601 (2015)
15. R. Blümel, A. Buchleitner, R. Graham, L. Sirko, U. Smilansky, H. Walther, Phys. Rev. A **44**, 4521 (1991)
16. M. Bellermann, T. Bergemann, A. Haffmann, P.M. Koch, L. Sirko, Phys. Rev. A **46**, 5836 (1992)
17. L. Sirko, S. Yoakum, A. Haffmans, P.M. Koch, Phys. Rev. A **47**, R782 (1993)
18. L. Sirko, P.M. Koch, Appl. Phys. B **60**, S195 (1995)
19. L. Sirko, A. Haffmans, M.R.W. Bellermann, P.M. Koch, Eur. Lett. **33**, 181 (1996)
20. L. Sirko, S.A. Zelazny, P.M. Koch, Phys. Rev. Lett. **87**, 043002 (2001)
21. L. Sirko, P.M. Koch, Phys. Rev. Lett. **89**, 274101 (2002)
22. P. Exner, J. Lipovský, *Equivalence of Resolvent and Scattering Resonances on Quantum Graphs*. Adventures in Mathematical Physics (Proceedings, Cergy-Pontoise 2006, vol. 447, R.I, Providence, 2007), p. 73
23. J. Lipovský, Acta Phys. Slovaca **66**, 265 (2016)
24. E.B. Davies, A. Pushnitski, Analysis and PDE **4**, 729 (2011)
25. M. Ławniczak, J. Lipovský, L. Sirko, Phys. Rev. Lett. **122**, 140503 (2019)
26. E.B. Davies, P. Exner, J. Lipovský, J. Phys. A: Math. Theor. **43**, 474013 (2010)
27. P. Exner, J. Lipovský, Phys. Lett. A **375**, 805 (2011)
28. J. Lipovský, J. Phys. A: Math. Theor. **49**, 375202 (2016)
29. S. Gnutzmann, U. Smilansky, Adv. Phys. **55**, 527 (2006)
30. K.A. Dick, K. Deppert, M.W. Larsson, T. Märtensson, W. Seifert, L.R. Wallenberg, L. Samuelson, Nat. Mater. **3**, 380 (2004)
31. K. Heo, E. Cho, J.E. Yang, M.H. Kim, M. Lee, B.Y. Lee, S.G. Kwon, M.S. Lee, M.H. Jo, H.J. Choi, T. Hyeon, S. Hong, Nano Lett. **8**, 4523 (2008)
32. N. Savytskyy, A. Kohler, S. Bauch, R. Blümel, L. Sirko, Phys. Rev. E **64**, 036211 (2001)
33. O. Hul, O. Tymoshchuk, S. Bauch, P.M. Koch, L. Sirko, J. Phys. A **38**, 10489 (2005)
34. M. Ławniczak, O. Hul, S. Bauch, P. Šeba, L. Sirko, Phys. Rev. E **77**, 056210 (2008)
35. M. Ławniczak, S. Bauch, O. Hul, L. Sirko, Phys. Rev. E **81**, 046204 (2010)
36. A. Rehemanjiang, M. Allgaier, C.H. Joyner, S. Müller, M. Sieber, U. Kuhl, H.-J. Stöckmann, Phys. Rev. Lett. **117**, 064101 (2016)
37. B. Dietz, V. Yunko, M. Białous, S. Bauch, M. Ławniczak, L. Sirko, Phys. Rev. E **95**, 052202 (2017)

38. M. Ławniczak, L. Sirko, Sci. Rep. **9**, 5630 (2019)
39. T. Söderströma, P. Papetb, Y. Yao, J. Ufheilc, *Smartwire Connection Technology* (White paper) (2014), http://www.meyerburger.com
40. R. Band, J.M. Harrison, C.H. Joyner, J. Phys. A Math. Theor. **45**, 325204 (2012)
41. J. Lipovský, Acta Phys. Polonica A **128**, 968 (2015)
42. T. Kottos, U. Smilansky, J. Phys. A: Math. Gen. **36**, 350–1 (2003)

Dynamics of a Cournot Duopoly Game with Differentiated Goods Between Public and Private Firms

Georges Sarafopoulos and Kosmas Papadopoulos

Abstract This paper investigates the dynamics of a nonlinear Cournot-type duopoly game with differentiated goods, linear demand and cost functions for two bounded rational players that have different objective functions. Specifically, the first player is a public company and cares about the social welfare and the second player is a private company which cares only about its own profit maximization. The game is modeled with a system of two difference equations. The stability analysis of the fixed points are analyzed and complex dynamic features including period doubling bifurcations of the unique Nash equilibrium is also investigated. Numerical simulations are carried out to show the complex behavior of the models' parameters. We show that a higher (lower) degree of the speed of adjustment and a lower (higher) degree of the parameter of product differentiation destabilize (stabilize) the economy. The chaotic features are justified numerically via computing Lyapunov numbers, sensitive dependence on initial conditions, bifurcation diagrams and strange attractors.

Keywords Cournot Duopoly game · Social welfare · Discrete dynamical system · Nash equilibrium · Stability · Bifurcation diagrams · Lyapunov numbers · Strange attractors · Chaotic behavior

1 Introduction

The Oligopoly market is a structure between monopoly (one firm) and perfect competition (many firms), where there are only a few numbers of sellers in this market producing homogeneous products. The dynamics of an oligopoly game are more

---

G. Sarafopoulos (✉) · K. Papadopoulos
Department of Economics, Democritus University of Thrace,
University Campus, 69100 Komotini, Greece
e-mail: gsarafop@econ.duth.gr

K. Papadopoulos
e-mail: kpapa@econ.duth.gr

C. H. Skiadas and Y. Dimotikalis (eds.), *12th Chaotic Modeling and Simulation International Conference*, Springer Proceedings in Complexity,
https://doi.org/10.1007/978-3-030-39515-5_16

complex because the firms must make a hypothesis not only about the consumers' behavior, but also about the competitors' reactions i.e. they form their expectations concerning in the way that their rivals will act. In 1838, the first formal theory of oligopoly has introduced by Augustin Cournot considering the case of naive expectations, where in every step (decision) each player (firm) assumes the last values that were chosen by the competitors without an estimation of their future reactions.

The players' expectations is an important factor in modeling economic phenomena. A producer can set his own rules (expectations) choosing one of many available techniques his production outputs' adjustment. In this study we find the dynamics of a duopoly game model where each player behaves with heterogeneous expectations (heterogeneous strategies). In this duopoly model each firm forms a strategy in order to compute his expected output. Each player, using his expectations rule adjusts his outputs towards the profit maximizing amount as target. Sarafopoulos and Papadopoulos [1–5]. Some authors as Agiza [6–8], Agliari et al. [9, 10], Bischi and Kopel [11], Kopel [12], Puu [13], Sarafopoulos [14], considered duopolies with homogeneous expectations for two players and study the variety of complex dynamics in these games, such as the appearance of strange attractors. Also, by Agiza and Elsadany [15, 16], Agiza et al. [8], Den Haan [17], Fanti and Gori [18], Tramontana [19], Zhang [20], models with heterogeneous agents were studied.

In real markets producers do not have the knowledge of the entire demand function, though it is possible that they have a high knowledge of technology, revealed by the cost function. Hence, it is more likely that players employ some local estimate of the demand. Naimzada and Ricchiuti [21] and Askar [22] have analyzed this issue. Firms that characterized as bounded rational players update their production strategies based on discrete time periods and by using a local estimate of their marginal profit. With such a local adjustment mechanism, the players are not requested to have a complete knowledge of the demand and the cost functions [15, 23, 20, 22]. In recent years, maximizing relative profit instead of absolute profit has aroused the interest of researchers as Tanaka [24], Elsadany [25]. In this study, the concept of generalized relative profit in a Cournot—type duopoly game with differentiated goods, linear demand and asymmetric cost functions has introduced. The article is organized as follows: In Sect. 2, the dynamics of the Cournot duopoly game with differentiated goods and generalized relative profit maximization is analyzed. We suppose that the players follow homogeneous expectations and use linear demand and asymmetric cost functions. Also, the existence and local stability of the equilibrium points are analyzed. Section 3 contains the numerical simulations that are used to reveal the complex dynamics via computing bifurcations diagrams, Lyapunov numbers, strange attractors and sensitive dependence on initial conditions.

2 The Game

2.1 The Construction of the Game

In this Cournot duopoly game there are two firms that produce differentiated goods and offer them at discrete time periods ($t = 0, 1, 2, \ldots$) on their common market. These two firms take decisions about their production quantities also at discrete-time periods ($t = 0, 1, 2, \ldots$). In addition it is considered that two players are homogeneous and more specifically, that both companies choose their productions rationally, following the same adjustment mechanism (bounded rational players). At every discrete period t, each player must forms an expectation of the rival's output of the next time period in order to determine the corresponding profit-maximization quantities for the next period $t + 1$. The different consideration in this study is that the first player is a public company and the second one is a private. So, the public company cares about the maximization of social welfare and the private company about its own profit-maximization only. It is supposed that q_1, q_2 are the productions of each player, then the inverse demand function (as a function of production quantities) is given by:

$$p_i = \alpha - q_i - dq_j, \quad \text{where } i, j = 1,\ 2 \text{ and } i \neq j \tag{1}$$

Also, p_i is the price of i firm's product and α is the positive parameter which expresses the size of market. So, for the two players it means:

$$p_1 = \alpha - q_1 - dq_2 \quad \text{and} \quad p_2 = \alpha - q_2 - dq_1 \tag{2}$$

In these equations, $d \in (-1, 1)$ is the parameter which expresses the differentiation degree between two the products. For positive values of the parameter d the larger the value, the less diversification there is between the two products. If $d = 0$, then each company participates in a monopoly game. On the other hand, the negative values of the differentiation parameter describe that two products are complementary.

We assume that both players use the same linear cost function:

$$C_i(q_i) = c \cdot q_i \tag{3}$$

which means that for two players the cost functions are the following:

$$C_1(q_1) = c \cdot q_1 \quad \text{and} \quad C_2(q_2) = c \cdot q_2 \tag{4}$$

To make our calculations easier we use the same positive cost parameter $c > 0$ for two players which is equal to the marginal cost of two players.

With these assumptions, it's easy to form the profit function for each player as follows:

$$\Pi_1(q_1, q_2) = p_1 \cdot q_1 - C_1(q_1) = (\alpha - c - q_1 - dq_2) \cdot q_1 \tag{5}$$

and

$$\Pi_2(q_1, q_2) = p_2 \cdot q_2 - C_2(q_2) = (\alpha - c - q_2 - dq_1) \cdot q_2 \tag{6}$$

Then, the marginal profits for the players at the point of the strategy space are given by:

$$\frac{\partial \Pi_1}{\partial q_1} = \alpha - c - dq_2 - 2q_1 \quad \text{and} \quad \frac{\partial \Pi_2}{\partial q_2} = \alpha - c - dq_1 - 2q_2 \tag{7}$$

At first we calculated the consumer surplus (CS) because it is contained in the social welfare that the first player cares about it and it is given by the equation:

$$CS = U(q_1, q_2) - p_1q_1 - p_2q_2 = \frac{1}{2}(q_1^2 + q_2^2) + d \cdot q_1q_2 \tag{8}$$

As it is noticed the first player is a public company which cares about the maximization of the social welfare (W) which is given by:

$$W(q_1, q_2) = CS + \Pi_1 + \Pi_2 = -\frac{1}{2}(q_1^2 + q_2^2) + (\alpha - c) \cdot (q_1 + q_2) - d \cdot q_1q_2 \tag{9}$$

and the marginal welfare is calculated as:

$$\frac{\partial W}{\partial q_1} = \alpha - c - q_1 - d \cdot q_2 \tag{10}$$

Both players follow the same strategy to decide their production quantities (homogeneous players) and they are characterized as bounded rational players. According to the existing literature it means that the first public company decide its production following a mechanism that is described by the equation:

$$\frac{q_1(t+1) - q_1(t)}{q_1(t)} = k \cdot \frac{\partial W}{\partial q_1}, \quad \text{with } k > 0 \tag{11}$$

and the second player (private company) who is also a bounded rational player follows a similar mechanism that is given by the equation:

$$\frac{q_2(t+1) - q_2(t)}{q_2(t)} = k \cdot \frac{\partial \Pi_2}{\partial q_2} \tag{12}$$

Through this mechanism each player increases his level of adaptation when his marginal utility is positive or decreases his level when his marginal utility is negative, where k is considered as the speed of adjustment for two players. The parameter k is positive ($k > 0$), and gives the extend variation production of the player i following a given utility signal.

The duopoly's dynamical system is described by:

$$\begin{cases} q_1(t+1) = q_1(t) + k \cdot q_1(t) \cdot \frac{\partial W}{\partial q_1} \\ q_2(t+1) = q_2(t) + k \cdot q_2(t) \cdot \frac{\partial \Pi_2}{\partial q_2} \end{cases}$$

or

$$\begin{cases} q_1(t+1) = q_1(t) + k \cdot q_1(t) \cdot [\alpha - c - d \cdot q_1(t) - 2q_2(t)] \\ q_2(t+1) = q_2(t) + k \cdot q_2(t) \cdot [\alpha - c - q_1(t) - d \cdot q_2(t)] \end{cases} \quad (13)$$

We investigate the effect of the parameter k (speed of adjustment) and parameter d (products' differentiation degree) on the dynamics of this system.

2.2 *Dynamical Analysis*

2.2.1 The Equilibriums of the Game

The equilibriums of the dynamical system (13) are obtained as the nonnegative solutions of the algebraic system:

$$\begin{cases} q_1^* \cdot \frac{\partial W}{\partial q_1} = 0 \\ q_2^* \cdot \frac{\partial \Pi_2}{\partial q_2} = 0 \end{cases} \quad (14)$$

which obtained by setting: $q_1(t+1) = q_1(t) = q_1^*$ and $q_2(t+1) = q_2(t) = q_2^*$.

- If $q_1^* = q_2^* = 0$, then the equilibrium position is:

$$E_0 = (0, 0) \quad (15)$$

- If $q_1^* = 0$ and $\frac{\partial \Pi_2}{\partial q_2} = 0$, then the equilibrium position is:

$$E_1 = \left(0, \frac{\alpha - c}{2}\right) \quad (16)$$

- If $q_2^* = 0$ and $\frac{\partial W}{\partial q_1} = 0$, then the equilibrium position is:

$$E_2 = (\alpha - c, 0) \tag{17}$$

- If $\frac{\partial W}{\partial q_1} = \frac{\partial \Pi_2}{\partial q_2} = 0$, it gives the system:

$$\begin{cases} \alpha - c - d \cdot q_1^* - 2q_2^* = 0 \\ \alpha - c - q_1^* - d \cdot q_2^* = 0 \end{cases} \tag{18}$$

whose solution is the Nash equilibrium:

$$E_* = \left(\frac{(\alpha - c)(2 - d)}{2 - d^2}, \frac{(\alpha - c)(1 - d)}{2 - d^2} \right) \tag{19}$$

with

$$(\alpha - c)(2 - d) > 0 \tag{20}$$

and

$$(\alpha - c)(1 - d) > 0 \tag{21}$$

2.2.2 Stability of Equilibriums

The stability of game's equilibriums is studying using the Jacobian matrix. The Jacobian matrix $J(q_1, q_2)$ along the variable strategy (q_1, q_2) is:

$$J(q_1, q_2) = \begin{bmatrix} f_{q_1} & f_{q_2} \\ g_{q_1} & g_{q_2} \end{bmatrix} \tag{22}$$

where:

$$f(q_1, q_2) = q_1 + k \cdot q_1 \cdot \frac{\partial W}{\partial q_1} \tag{23}$$

and

$$g(q_1, q_2) = q_2 + k \cdot q_2 \cdot \frac{\partial \Pi_2}{\partial q_2} \tag{24}$$

The Jacobian matrix becomes as:

$$J(q_1^*, q_2^*) = \begin{bmatrix} 1 + k \cdot \left(\frac{\partial W}{\partial q_1} + q_1^* \cdot \frac{\partial^2 W}{\partial q_1^2} \right) & k \cdot q_1^* \cdot \frac{\partial^2 W}{\partial q_1 \partial q_2} \\ k \cdot q_2^* \cdot \frac{\partial^2 \Pi_2}{\partial q_2 \partial q_1} & 1 + k \cdot \left(\frac{\partial \Pi_2}{\partial q_2} + q_2^* \cdot \frac{\partial^2 \Pi_2}{\partial q_2^2} \right) \end{bmatrix} \quad (25)$$

For the E_1 the Jacobian matrix becomes as:

$$J(E_0) = \begin{bmatrix} 1 + k \cdot (\alpha - c) & 0 \\ 0 & 1 + k(\alpha - c) \end{bmatrix} \overset{A=1+k\cdot(\alpha-c)}{=} \begin{bmatrix} A & 0 \\ 0 & A \end{bmatrix} \quad (26)$$

with $Tr = 2A$ and $Det = A^2$.

Forming the characteristic equation of $J(E_0)$ which is: $r^2 - Tr \cdot r + det = 0$, it is taken the unique eigenvalue: $r_1 = A = 1 + k \cdot (\alpha - c)$. It's clear that $|r_1| > 1$, so E_0 is unstable.

For E_1 the Jacobian matrix becomes as:

$$J(E_1) = \begin{bmatrix} 1 + k \cdot \frac{(\alpha - c)(2-d)}{2} & 0 \\ k \cdot q_2^* \cdot \frac{\partial^2 \Pi_2}{\partial q_2 \partial q_1} & 1 - k(\alpha - c) \end{bmatrix} \underset{C=1-k(\alpha-c)}{\overset{B=1+k\cdot\frac{(\alpha-c)(2-d)}{2}}{=}} \begin{bmatrix} B & 0 \\ D & C \end{bmatrix} \quad (27)$$

with $Tr = B + C$ and $Det = B \cdot C$.

From the characteristic equation of $J(E_1)$, we find two eigenvalues:$r_1 = B$ and $r_2 = C$. Since $r_1 = 1 + k \cdot \frac{(\alpha-c)(2-d)}{2}$ and $(\alpha - c)(2 - d) > 0$ Eq. (20), it's clearly seems that $|r_1| > 1$ and the equilibrium E_1 is unstable.

For E_2 the Jacobian matrix becomes as:

$$J(E_2) = \begin{bmatrix} 1 - k \cdot (\alpha - c) & k \cdot q_1^* \cdot \frac{\partial^2 W}{\partial q_1 \partial q_2} \\ 0 & 1 + k \cdot (\alpha - c) \cdot (1 - d) \end{bmatrix} \underset{F=1+k\cdot(\alpha-c)\cdot(1-d)}{\overset{E=1-k\cdot(\alpha-c)}{=}} \begin{bmatrix} E & G \\ 0 & F \end{bmatrix} \quad (28)$$

with $Tr = E + F$ and $Det = E \cdot F$. From the characteristic equation of $J(E_2)$, we find two eigenvalues: $r_1 = E$ and $r_2 = F$. Since $r_2 = 1 + k \cdot (\alpha - c) \cdot (1 - d)$ and $(\alpha - c) \cdot (1 - d) > 0$ Eq. (21), it's clearly seems that $|r_2| > 1$ and the equilibrium E_2 is unstable. The Jacobian matrix at E_* becomes as:

$$J(E_*) = \begin{bmatrix} 1 + k \cdot q_1^* \cdot \frac{\partial^2 W}{\partial q_1^2} & k \cdot q_1^* \cdot \frac{\partial^2 W}{\partial q_1 \partial q_2} \\ k \cdot q_2^* \cdot \frac{\partial^2 \Pi_2}{\partial q_2 \partial q_1} & 1 + k \cdot q_2^* \cdot \frac{\partial^2 \Pi_2}{\partial q_2^2} \end{bmatrix} = \begin{bmatrix} 1 - k \cdot q_1^* & -k \cdot d \cdot q_1^* \\ -k \cdot d \cdot q_2^* & 1 - 2k \cdot q_2^* \end{bmatrix} \quad (29)$$

with $Tr = 2 - kq_1^* - 2kq_2^*$ and $Det = 1 - 2 - kq_1^* - 2kq_2^* + k(2 - d^2)q_1^*q_2^*$.

We study the stability of Nash equilibrium E_* using the three conditions that make the equilibrium position E_* locally asymptotically stable when they are satisfied simultaneously:

(i) $1 - Det > 0$

$$\text{(ii)} \quad 1 - \mathrm{Tr} + \mathrm{Det} > 0$$
$$\text{(iii)} \quad 1 + \mathrm{Tr} + \mathrm{Det} > 0 \tag{30}$$

The condition (ii) gives:

$$1 - \mathrm{Tr} + \mathrm{Det} > 0 \Leftrightarrow \left(2 - \mathrm{d}^2\right) \cdot \mathrm{k}^2 \cdot \mathrm{q}_1^* \mathrm{q}_2^* > 0 \tag{31}$$

and it's always satisfied because $\left(2 - \mathrm{d}^2\right) \cdot \mathrm{k}^2 > 0$.

From the first condition (i) is given:

$$1 - \mathrm{Det} > 0 \Leftrightarrow \mathrm{q}_1^* + 2\mathrm{q}_2^* - \left(2 - \mathrm{d}^2\right) \cdot \mathrm{k} \cdot \mathrm{q}_1^* \mathrm{q}_2^* > 0 \tag{32}$$

Finally, the condition (iii) becomes as:

$$1 + \mathrm{Tr} + \mathrm{Det} > 0 \Leftrightarrow \left(2 - \mathrm{d}^2\right) \cdot \mathrm{q}_1^* \mathrm{q}_2^* \cdot \mathrm{k}^2 - 2\left[\mathrm{q}_1^* + 2\mathrm{q}_2^*\right]\mathrm{k} + 4 > 0 \tag{33}$$

Proposition *The discrete dynamical system* Eq. (13) *has a locally asymptotically stable Nash Equilibrium if:*

$$q_1^* + 2q_2^* - \left(2 - d^2\right) \cdot k \cdot q_1^* q_2^* > 0$$

and

$$\left(2 - d^2\right) \cdot q_1^* q_2^* \cdot k^2 - 2\left[q_1^* + 2q_2^*\right]k + 4 > 0$$

2.2.3 Stability Conditions Focusing on the Parameter K (Speed of Adjustment)

From the condition (i) focusing on the parameter k we take the following inequality:

$$0 < \mathrm{k} < \frac{1}{2 - \mathrm{d}^2} \cdot \left(\frac{1}{\mathrm{q}_2^*} + \frac{2}{\mathrm{q}_1^*}\right) \tag{34}$$

The condition (iii) is the following:

$$\left(2 - \mathrm{d}^2\right) \cdot \mathrm{q}_1^* \mathrm{q}_2^* \cdot \mathrm{k}^2 - 2\left[\mathrm{q}_1^* + 2\mathrm{q}_2^*\right]\mathrm{k} + 4 > 0 \tag{35}$$

And its discriminant is positive:

$$\Delta = 4\left(\mathrm{q}_1^* - 2\mathrm{q}_2^*\right)^2 + 16\mathrm{d}^2 \cdot \mathrm{q}_1^* \mathrm{q}_2^* > 0 \tag{36}$$

so the condition (iii) is satisfied if:

$$k \in (0, k_1) \cup (k_2, +\infty) \tag{37}$$

where

$$k_{1,2} = \frac{2(q_1^* + q_2^*) \pm \sqrt{\Delta}}{2(2 - d^2) \cdot q_1^* q_2^*} \tag{38}$$

are its two positive roots.

2.2.4 Stability Conditions Focusing on the Parameter D (Differentiation Degree)

From the condition (i) focusing on the parameter d we take the following inequality:

$$H \cdot d^2 + 3(1 - H) \cdot d + 2(H - 2) < 0, \quad \text{where } H = k(\alpha - c) \tag{39}$$

Which is satisfied if:

$$d \in (d_1, d_2) \tag{40}$$

where

$$d_{1,2} = \frac{3(H - 1) \pm \sqrt{H^2 - 2H + 9}}{2H} \tag{41}$$

The condition (iii) becomes as:

$$(H - 2) \cdot \left[(H + 2)d^2 - 3H \cdot d + 2(H - 2)\right] > 0 \tag{42}$$

and it is clear that: $H \neq 2$, because if: $H = 2$, then the Eq. (39) cannot become true. So:

- If $H > 2$ it is satisfied when:

$$d \in (-1, d_3) \cup (d_4, 1) \tag{43}$$

- If $H < 2$ it is satisfied when:

$$d \in (d_3, d_4) \tag{44}$$

where:

$$d_{3,4} = \frac{3H \pm \sqrt{H + 32}}{2(H + 2)} \tag{45}$$

3 Numerical Simulations

3.1 Focusing on the Parameter K (Speed of Adjustment)

In this section various numerical results focusing on the parameter k, including bifurcation diagrams, Lyapunov numbers and sensitive dependence on initial conditions [12] are presented. For this reason some fixed values to other parameters as: $\alpha = 5$, $c = 1$ and $d = 0.5$ are chosen. So, as a result we find that $q_1^* \simeq 3.42$ and $q_2^* \simeq 1.14$ and the stability conditions become as:

$$0 < k < 0.49 \tag{46}$$

The stability space is made including the main two parameters we will focus on, the parameters k (speed of adjustment) and the parameter d (the differentiation degree between two products). This two-dimensional space (Fig. 1) is obtained by the common spaces that are appeared by the two stability conditions of game's Proposition, setting specific values for the other parameters $\alpha = 5$ and $c = 1$. As it

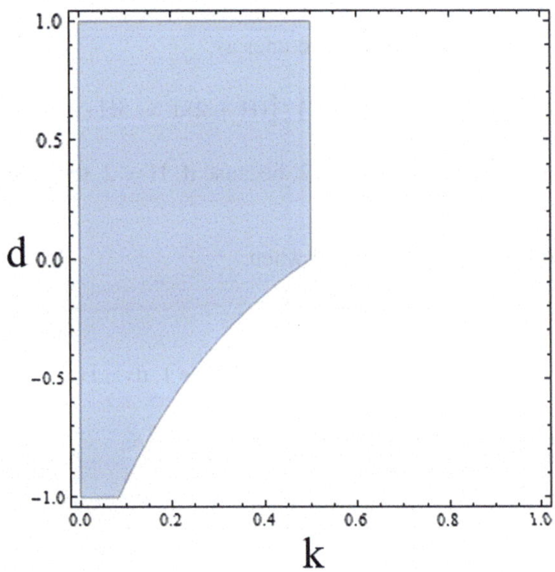

Fig. 1 Stability space between the parameter k (horizontal axis) and the parameter d (vertical axis) for $\alpha = 5$ and $c = 1$

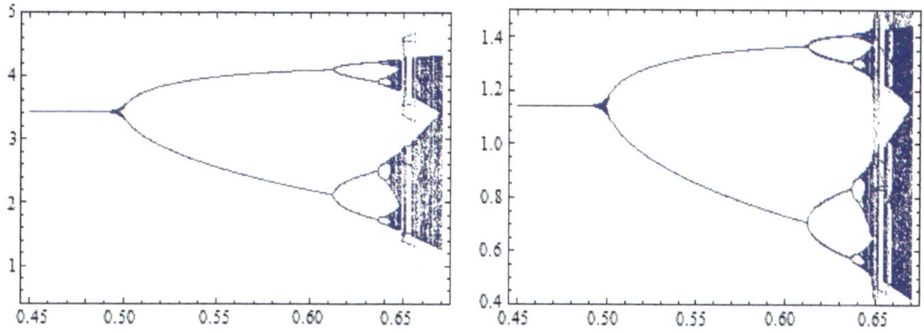

Fig. 2 Bifurcation diagrams with respect to the parameter k against the variables q_1 (left) and q_2 (right) taking 400 iterations of map Eq. (13) for $\alpha = 5$, c = 1 and d = 0.5

seems, for very small values of the parameter k (speed of adjustment) the equilibrium is stable for each value of the parameter d (differentiation degree).

By the bifurcation diagrams of the parameter k against the variables q_1^* and q_2^* that are shown in Fig. 2 it is clear that the equilibrium undergoes a flip bifurcation at k = 0.49. Increasing the speed of adjustment, a stable two-period cycle emerges for $0.49 < k < 0.61$ is implied. As long as the parameter k reduces a four-period cycle, cycles of highly periodicity and a cascade of flip bifurcations that ultimately lead to unpredictable (chaotic) motions are observed when k is larger than 0.64.

This unpredictable (chaotic) behavior of the system Eq. (13) is visualized with the useful tool of Lyapunov numbers (Fig. 3) (i.e. the natural logarithm of Lyapunov exponents) as a function of the parameter of interest. Figure 3 shows the Lyapunov numbers of the orbit of the system of Eq. (13) for $\alpha = 5$, c = 1, d = 0.5 and k = 0.67

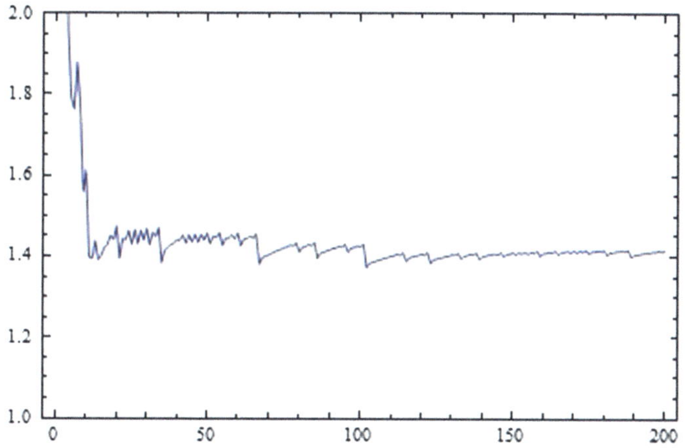

Fig. 3 Lyapunov numbers of the orbit of (0.1, 0.1) with 2000 iterations of the map Eq. (13) for $\alpha = 5$, c = 1, d = 0.5 and k = 0.67

(outside the stability space). It is known that if the Lyapunov number is greater than 1, one has evidence for chaos.

Another characteristic of deterministic chaos is the sensitivity dependence on initial conditions. To show the sensitivity dependence on initial conditions of the system Eq. (13), two orbits with initial points (0.1, 0.1) and (0.101, 0.1) respectively are computed. Figure 4 shows the sensitivity dependence on initial conditions for q_1-coordinate for the two orbits, of the system Eq. (13), are plotted against the time setting the specific parameter values and $\alpha = 5$, $c = 1$, $d = 0.5$ and $k = 0.67$. As it is shown at the beginning the two time series are indistinguishable; and after a number of iterations, differences between them are appeared and build up rapidly (Fig. 5).

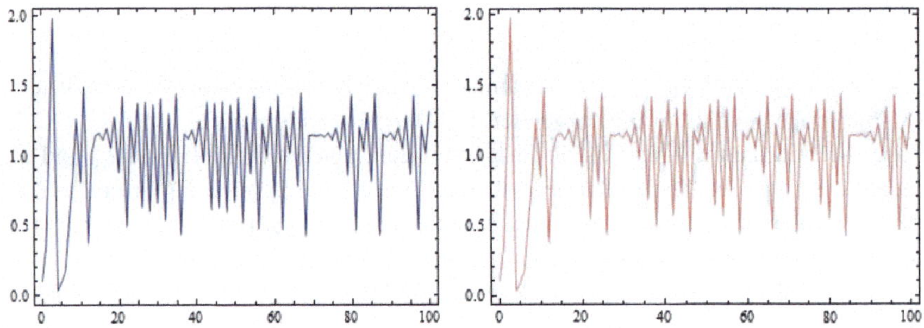

Fig. 4 Sensitive dependence on initial conditions for q_1-coordinate plotted against the time: the orbit of (0.1, 0.1) (left) and the orbit of (0.101, 0.1) (right) of the system Eq. (13) for $\alpha = 5$, $c = 1$, $d = 0.5$ and $k = 0.67$

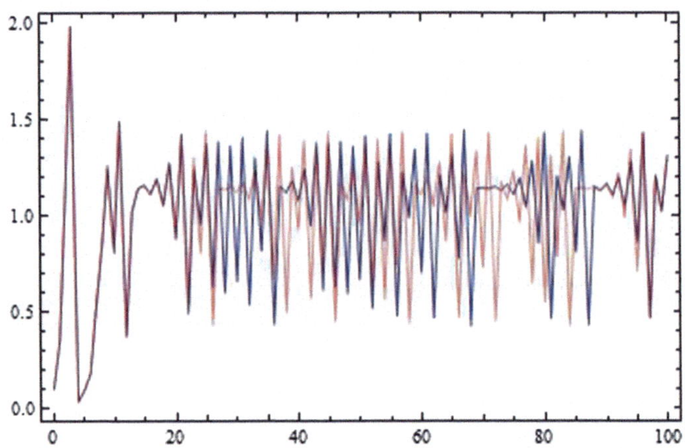

Fig. 5 Two previous graphs of Fig. 4 in one

3.2 Focusing on the Parameter D (Differentiation Degree)

Also, various numerical results here are presented showing the chaoticity, including its bifurcations diagrams, strange attractors, Lyapunov numbers and sensitive dependence on initial conditions providing some numerical evidence for the chaotic behavior of the system Eq. (13), as a consequence of change in the parameter differentiation degree d.

We study the local stability properties of the equilibrium points, taking specific values for the other parameters: $\alpha = 5$, $c = 1$ and $k = 0.45$ and with these assumptions it means that: $H = 1.8 < 2$.

So, as a result we find that $d_1 \simeq -0.14$ and $d_2 \simeq 1.48 > 1$ so the first stability condition become as:

$$d \in (-0.14, 1) \tag{47}$$

Also, for the second stability we find that: $d_3 \simeq -0.07$ and $d_4 \simeq 1.49 > 1$ and the stability interval for the parameter d is:

$$d \in (-0.07, 1) \tag{48}$$

So the total stability condition for the parameter d is:

$$-0.07 < d < 1 \tag{49}$$

This previous stability result is verified by the bifurcation diagrams of q_1 (left) and q_2 (right) (Fig. 6), focusing to the parameter d.

As it is shown, when all parameters take fixed values and only the parameter d varies, the game's structure becomes complicated through period doubling bifurcations. Also, more complex bounded attractors are appeared which are aperiodic cycles of higher order or chaotic attractors. In Fig. 7 the strange attractors of for $\alpha = 5$, $c = 1$, $k = 0.45$, $d = -0.3$ (left) and $d = -0.26$ (right) are plotted.

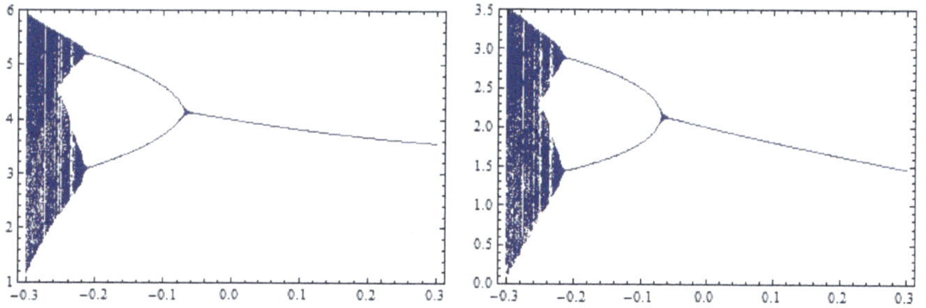

Fig. 6 Bifurcation diagrams with respect to the parameter d against the variables q_1 (left) and q_2 (right) with 400 iterations of the map Eq. (13) for $\alpha = 5$, $c = 1$ and $k = 0.45$

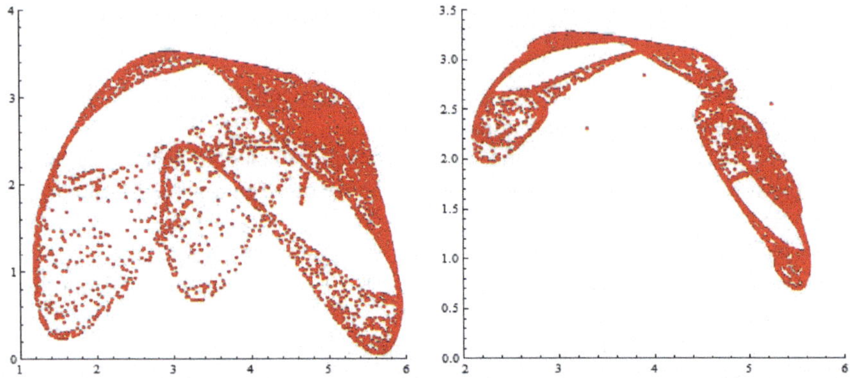

Fig. 7 Strange attractors (phase portrait) of the orbit of (0.1, 0.1) taking 2000 iterations for the map Eq. (13) for $\alpha = 5$, $c = 1$, $k = 0.45$ and $d = -0.3$(left) and $d = -0.26$(right)

Also, using the same values of the other fixed parameters, but now for $d = -0.26$, and initial conditions (0.1, 0.1), as it shown the graph of Lyapunov numbers as an evidence for chaotic behavior of the system when the parameter d takes values outside the stability space (Fig. 8).

Finally, we choose two different initial conditions (0.1, 0.1) (left) and (0.101, 0.1) (right) for the same parameters ($\alpha = 5$, $c = 1$, $k = 0.45$ and d—0.26), x-coordinate of the two orbits, for the system plotted against the time. From Fig. 9, as it is shown at the beginning the two time series are indistinguishable; and after a number of iterations, differences between them are appeared and build up rapidly (Fig. 10).

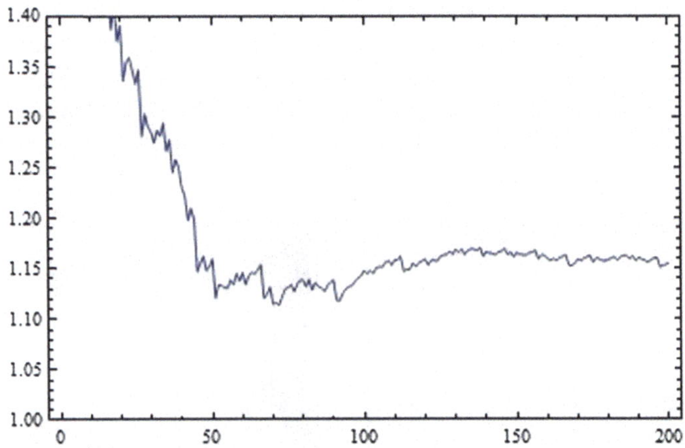

Fig. 8 Lyapunov numbers graph of the orbit of (0.1, 0.1) for 2000 iterations of the map Eq. (13) for $\alpha = 5$, $c = 1$, $k = 0.45$ and $d = -0.26$

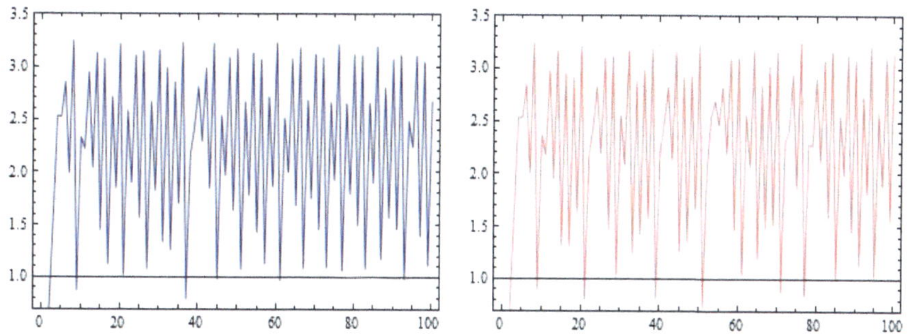

Fig. 9 Sensitive dependence on initial conditions for q_1-coordinate plotted against the time: the orbit of (0.1, 0.1) (left) and the orbit of (0.101, 0.1) (right) of the system Eq. (13) for $\alpha = 5$, $c = 1$, $k = 0.45$ and d—0.26

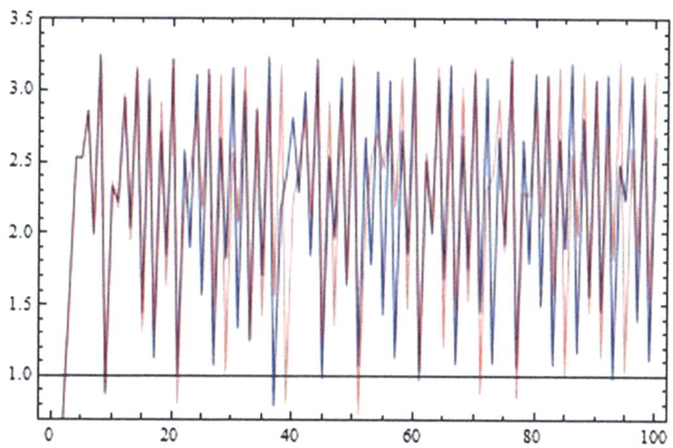

Fig. 10 Two previous graphs of Fig. 9 in one

4 Conclusions

In this paper, through a discrete dynamical system the dynamics of a nonlinear discrete-time duopoly game with a convex and log-linear demand and quadratic cost functions are analyzed. We suppose that the first player is a public company and cares about the maximization of social welfare (W) and the rival player is a private company that cares about the maximization of his profit function only. The stability of equilibria, bifurcation and chaotic behavior are investigated. We showed that the speed of adjustment (k) and the parameter of product differentiation (d) may change the stability of equilibrium and cause a structure to behave chaotically. The main result is that a lower degree of product differentiation and a higher value of the speed of adjustment may destabilize the Cournot–Nash equilibrium. Outside of the stability space for each one of these parameters more complex attractors are appeared making the system's behavior chaotic and unpredictable. Finally, we showed that for very small values of the speed of adjustment the equilibrium is stable for each value of the differentiation parameter.

References

1. G. Sarafopoulos, K. Papadopoulos, On a Cournot duopoly game with differentiated goods, heterogeneous expectations and a cost function including emission costs. Scientif. Bullet. Economic Sci. **1**(1), 11–22 (2017)
2. G. Sarafopoulos, K. Papadopoulos, Chaos in oligopoly models. Int. J. Product. Manage. Assessment Technologies (IJPMAT) **7**(1), Article 4 (2018)
3. G. Sarafopoulos, K. Papadopoulos, On the stability of a Cournot dynamic game under the influence of information, Knowledge E Publishing, ISSN 2518-668X, 131–144 (2018)
4. G. Sarafopoulos, On a duopoly game with homogeneous players and a quadratic demand function. East-West J. Econom. Business **XX**(2), 39–48 (2017)
5. G. Sarafopoulos, K. Papadopoulos, Complexity in a Bertrand Duopoly Game with heterogeneous players and differentiated goods. Springer Proc. Business Econ. 15–26 (2019)
6. H.N. Agiza, On the analysis of stability, bifurcation, chaos and chaos control of Kopel map. Chaos, Solitons Fractals **10**, 1909–1916 (1999)
7. H.N. Agiza, On the stability, bifurcations, chaos and chaos control of Kopel map. Chaos Solitons Fract. **11**, 1909–1916 (2004)
8. H.N. Agiza, A.S. Hegazi, A.A. Elsadany, Complex dynamics and synchronization of duopoly game with bounded rationality. Math. Comput. Simulat. **58**, 133–146 (2002)
9. A. Agliari, L. Gardini, T. Puu, Some global bifurcations related to the appearance of closed invariant curves. Math Comput Simulation **68**, 201–219 (2005)
10. A. Agliari, L. Gardini, T. Puu, Global bifurcations in duopoly when the Cournot point is destabilized via a subcritical Neimark bifurcation. International Game Theory Review **8**, 1–20 (2006)
11. G.I. Bischi, M. Kopel, Equilibrium selection in a nonlinear duopoly game with adaptive expectations. J. Econ. Behav. Organ. **46**, 73–100 (2001)
12. M. Kopel, Simple and complex adjustment dynamics in Cournot duopoly models. Chaos Solitons Fract. **12**, 2031–2048 (1996)
13. T. Puu, The chaotic duopolists revisited. J Econ. Behav. Org. **37**, 385–394 (1998)
14. G. Sarafopoulos, Complexity in a duopoly game with homogeneous players, convex, log linear demand and quadratic cost functions. Proc. Econ. Finance **33**, 358–366 (2015)
15. H.N. Agiza, A.A. Elsadany, Chaotic dynamics in nonlinear duopoly game with heterogeneous players. Appl. Math. Comput. **149**, 843–860 (2004)
16. H.N. Agiza, A.A. Elsadany, Nonlinear dynamics in the Cournot duopoly game with heterogeneous players. Phys. A **320**, 512–524 (2003)
17. W.J. Den Haan, The importance of the number of different agents in a heterogeneous asset—pricing model. J. Econ. Dynam. Control **25**, 721–746 (2001)
18. L. Fanti, L. Gori, The dynamics of a differentiated duopoly with quantity competition. Econ. Model. **29**, 421–427 (2012)
19. F. Tramontana, Heterogeneous duopoly with isoelastic demand function. Econ. Model. **27**, 350–357 (2010)
20. J. Zhang, Q. Da, Y. Wang, The dynamics of Bertrand model with bounded rationality. Chaos, Solitons Fractals **39**, 2048–2055 (2009)
21. A.K. Naimzada, G. Ricchiuti, Complex dynamics in a monopoly with a rule of thumb. Appl. Math. Comput. **203**, 921–925 (2008)
22. S.S. Askar, Complex dynamic properties of Cournot duopoly games with convex and log-concave demand function. Operations Research Letters **42**, 85–90 (2014)

23. A. Naimzada, L. Sbragia, Oligopoly games with nonlinear demand and cost functions: two boundedly rational adjustment processes. Chaos, Solitons Fractals **29**, 707–722 (2006)
24. Y. Tanaka, Irrelevance of the choice of strategic variables in duopoly under relative profit maximization. Econ. Business Lett. **2**(2), 75–83 (2013)
25. A.A. Elsadany, Dynamics of a Cournot duopoly game with bounded rationality based on relative profit maximization. Appl. Math. Comput. **294**, 253–263 (2017)

Influence of the Heart Rate on Dynamics of Cardiorespiratory System

Evgeniy D. Pechuk, Tatyana P. Konovalyuk, Tatyana V. Sobol and Tatyana S. Krasnopolskaya

Abstract Laws of dynamics of the DeBoer's model of the cardiovascular and respiratory systems are studied taking into account dependence of parameters on a variation of the heartbeat. The linear approximation of unknown functional dependences of feedback influence of a cardiosystem on respiratory are investigated. By methods of the modern theory of dynamical systems the steady-state regimes of the modified models are studied. It has been shown that a reverse influence of parameters of a cardiosystem frequency on frequency of respiratory oscillations results in irregularity of the dynamic mode of the combined system, i.e. a cause of chaos appearances in the cordially-respiratory system is the internal interaction of subsystems: cardiovascular and respiratory. A decrease in heart rate causes a decrease in pressure.

Keywords Heart rate · Cardiovascular system · Respiratory system · Feedback · Chaos

1 Introduction

In the paper of DeBoer and coauthors [1] the model of interaction of the cardiovascular and respiratory systems was developed, which well agrees with experimental information. The DeBoer's model was sub-stantially developed in future. In the work of Seidel and Herzel in 1998 [2] the so-called SH-model, different from the DeBoer's

E. D. Pechuk · T. P. Konovalyuk · T. V. Sobol · T. S. Krasnopolskaya (✉)
Institute of Hydromechanics NASU, Kiev, Ukraine
e-mail: t.krasnopolskaya@tue.nl

E. D. Pechuk
e-mail: evgdmp@gmail.com

T. P. Konovalyuk
e-mail: tpk_8_4@ukr.net

T. V. Sobol
e-mail: sobol2017liza@gmail.com

C. H. Skiadas and Y. Dimotikalis (eds.), *12th Chaotic Modeling and Simulation International Conference*, Springer Proceedings in Complexity,
https://doi.org/10.1007/978-3-030-39515-5_17

model the account of some elements of activity of sine knot, is probed. Thus found out existence in this model of chaotic dynamics. A SH-model got further development [3], where the reverse affecting of heart system is taken into account respiratory activity. In the Grinchenko-Rudnitsky model [4], within the framework of DeBoer's model and in accordance with principles of optimum management, adjusting and interaction of blood pressure and amplitudes of breathing oscillations is investigated. The model built at such approach allowed, in particular, to explain appearance of peak on frequency of Meyer in the spectrums of respiratory oscillations and synchronization of cardiac and respirator rhythms [5]. In the present research we construct a linear approximation of unknown functional dependences of feedback influence of a cardiosystem on respiratory and investigate a variation of the heartbeat.

2 The Model

A DeBoer's model includes the followings main characteristics of the heartbeat: systolic pressure S, diastolic pressure D, R-R interval I and arterial time constant T (in a state of rest for a healthy man S = 120 mmHg, D = 80 mmHg, I = 800 ms, T = 1500 ms). A mathematical model is system of five discrete nonlinear mappings. This model contains direct mechanical influence of the respirator system only:

$$
\begin{aligned}
D_i' &= S_{i-1}' \exp\left(-\frac{2}{3}\frac{I_{i-1}'}{T_{i-1}'}\right),\\
S_i' &= D_i' + \gamma\frac{T_0}{S_0}I_{i-1}' + \frac{A}{S_0}\sin(2\pi f T_0 t_i) + \frac{c_2}{S_0},\\
I_i' &= G_v\frac{S_0}{T_0}\widehat{S}_{i-\tau_v}' + G_\beta\frac{S_0}{T_0}F(\widehat{S}', \tau_\beta) + \frac{c_3}{T_0},\\
T_i' &= 1 + G_\alpha\frac{S_0}{T_0} - G_\alpha\frac{S_0}{T_0}F(\widehat{S}', \tau_\alpha)\\
\widehat{S}_i' &= 1 + \frac{18}{S_0}\arctan\frac{S_0(S_i'-1)}{18},
\end{aligned}
\tag{1}
$$

where $i \geq 1, D' = D/S_0, S' = S/S_0, \widehat{S}' = \widehat{S}/S_0, I' = I/T_0, T' = T/T_0$, $F(\widehat{S}, \tau) = 1/9(\widehat{S}_{i-\tau-2} + 2\widehat{S}_{i-\tau-1} + 3\widehat{S}_{i-\tau} + 2\widehat{S}_{i-\tau+1} + \widehat{S}_{i-\tau+2})$, $t_i = \sum_{k=0}^{i-1} I_k'$—the moment of time of heartbeat, A = 3 mmHg—breathing amplitude, f = 0.25 Hz—breathing frequency, $c_2 = S_0 - D_0 - \gamma I_0$, $c_3 = I_0 - S_0(G_v + G_\beta)$, $\gamma = 0.016$ mmHg, $G_\alpha = 18$ ms/mmHg, $G_\beta = 9$ ms/mmHg, $G_v = 9$ ms/mmHg, $\tau_\alpha = \tau_\beta = 4$, $\tau_v = 0$, if frequency of heart rhythmic less then 75 beat/min, and $\tau_v = 1$, if frequency >75 beat/min.

This system of equations after the deletion of dependent variables can be reduced to a system of three discrete mappings with six delays in the variable S', of the following form:

$$\begin{cases} S'_i = \varphi_1(I'_{i-1}, S'_{i-1}, S'_{i-3}, S'_{i-4}, S'_{i-5}, S'_{i-6}, S'_{i-7}, t_{i-1}) \\ I'_i = \varphi_2(I'_{i-1}, S'_{i-1}, S'_{i-2}, S'_{i-3}, S'_{i-4}, S'_{i-5}, S'_{i-6}) \\ t_i = t_{i-1} + I'_{i-1} \end{cases}, \qquad (2)$$

where φ_1 and φ_2 are nonlinear functions. The resulting system (2) could be written by replacing the variables as a nonlinear system of nine discrete maps without delay. Thus, the presence of such an inoperable property of functioning of the cardiovascular system of man as a delay, leads to a significant increase in the order of the resulting system. Influence of the respiratory system on the cardiovascular system, within the framework of this model is taken into account by parameters A and f. Therefore, the reverse effect of the cardiosystem can be realized by influencing the activity of the heart on these two characteristics. In this paper, this feedback is modeled by the simplest linear approximations. Thus, the amplitude of the change in systolic pressure due to movement of the chest during breathing is assumed to be dependent on the value of systole pressure on the previous cardiointerval according to the law:

$$A^i = A_0 - A_1(S_{i-1} - S_0),$$

where $A_0 = A$, $A_1 \geq 0$, which simulates the decrease in the amplitude of respiration with increasing systolic pressure. The frequency of breathing is assumed to depend on the frequency of cardiac contractions according to the dependence:

$$f^i = f_0 + f_1(1/I_{i-1} - 1/I_0),$$

where $f_0 = f$, $f_1 \geq 0$.

Thus, we will study the dynamics of the modified model of cardiorespiratory system, which consists of DeBoer's model, with direct respirator influence

$$(A + r_i) \sin \varphi_i,$$

and with reverse influence modeled by the linear functions.

3 Simulations

For a simulation in accordance with physiology of healthy man, the followings values of variables and constants are used: $I'[0] = 0.53$, $S'[-j] = 1.08$, $j = 0, \ldots, 6$, $r'[0] = 0$, $\varphi'[0] = 0$, $\kappa = 0.001$ 1/ms, $\nu = 0.001$ 1/ms mmHg.

Now we start to study influence of heartbeat on dynamics of cardiorespiratory system. First, we study the system solutions for the high heart rate, namely, when the pulse is 75 beats in minute and:

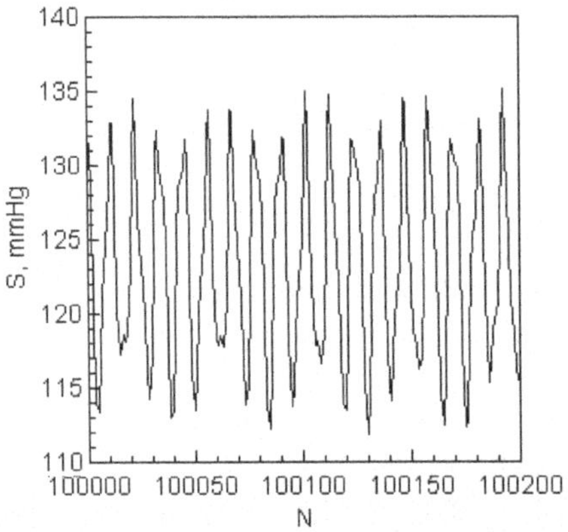

(a) $I_0 = 800$ ms

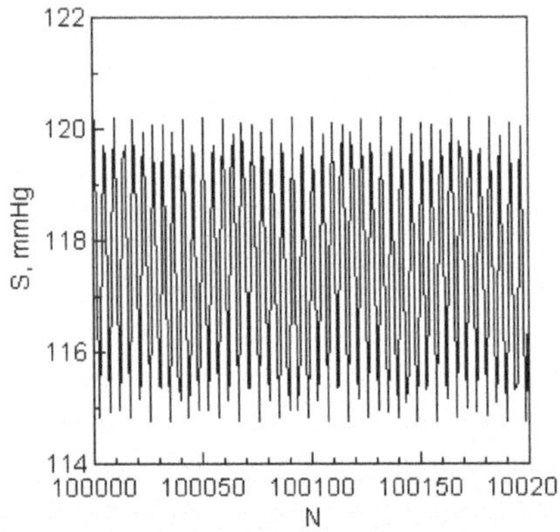

(b) $I_0 = 1000$ ms

Fig. 1 Systolic pressure at $I_0 = 800$ ms (**a**) and at $I_0 = 1000$ ms (**b**)

(a) $I_0 = 800\,\text{ms}, A_0 = 3\,\text{mmHg}, A_1 = 0, f_0 = 0.25\,\text{Hz}, f_1 = 0.22.$

Then we compare the solutions with the results of simulation of the system when the heart rate is lower, the pulse is 60 beats in minute, and:

(b) $I_0 = 1000\,\text{ms}, A_0 = 3\,\text{mmHg}, A_1 = 0, f_0 = 0.22\,\text{Hz}, f_1 = 0.21.$

The graphs of simulated systolic pressures are shown in Fig. 1a at $I_0 = 800$ ms and in Fig. 1b at $I_0 = 1000$ ms. As could be seen, a decrease in heart rate causes a decrease in systolic pressure: the mean value of systolic pressure is around 125 mmHg at the heart rate 75 beats in minute and is 117.5 mmHg at the heart rate 60 beats in minute. Also the amplitudes of systolic pressure oscillations are decreased, but the frequencies are increased with decreasing of the heart rate. The dependences of values of R-R interval on the heart rate are shown in Fig. 2. Obviously, for bigger cardio interval at $I_0 = 1000$ ms larger is R-R interval. And finally, the graphs of the respiratory frequency are shown in Fig. 3a at $I_0 = 800$ ms and in Fig. 3b at $I_0 = 1000$ ms, from which could be seen, that a decrease in heart rate causes a decrease in the respiratory frequency: the mean value of a respiratory frequency is around 0.235 at the heart rate 75 beats in minute and is 0.23 at the heart rate 60 beats in minute.

4 Conclusions

On the basis of the DeBoer's model of interaction of the cardio and respirator systems dependences of solutions of systems on variation of the heartbeat are studied. The

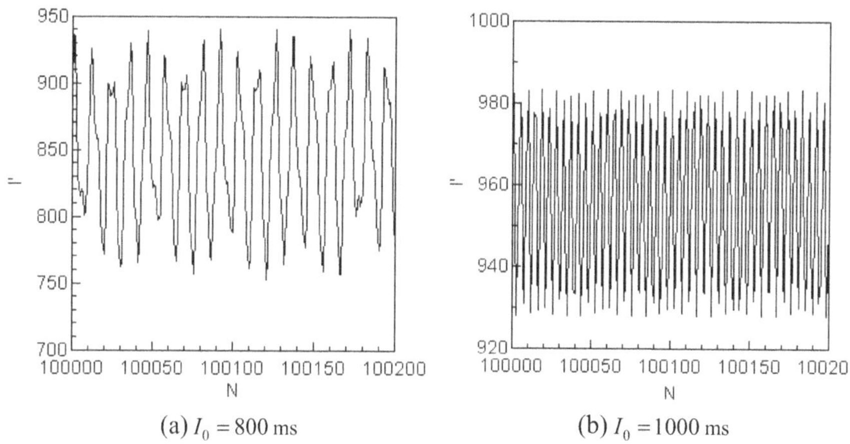

Fig. 2 Simulated R-R interval data at $I_0 = 800$ ms (**a**) and at $I_0 = 1000$ ms (**b**)

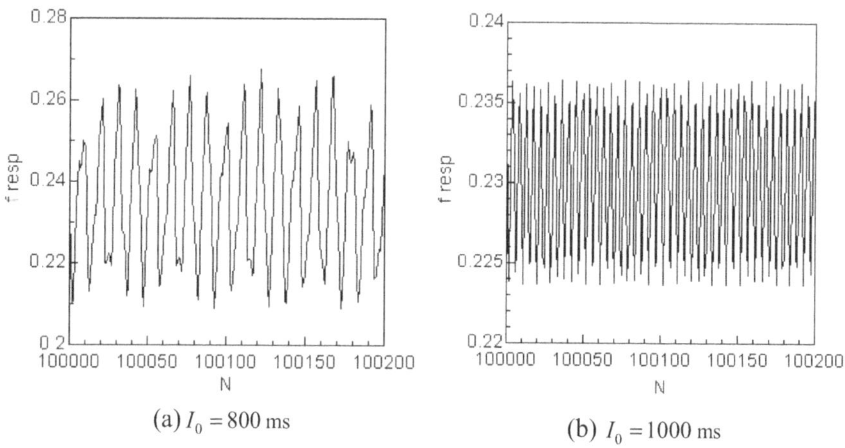

Fig. 3 Respiratory frequency at $I_0 = 800$ ms (**a**) and at $I_0 = 1000$ ms (**b**)

new model takes into account both direct and reverse influence of subsystems—cardiovascular and respiratory. The linear approximation of unknown functional dependences of feedback influence are investigated. As was found a decrease in human pulse leads to a decrease in systolic pressure and respiratory frequency.

Acknowledgements We gratefully acknowledge funding from the National Academy of Sciences of Ukraine under the grant 0118U005412.

References

1. R.W. DeBoer, J.M. Karemaker, J. Strakee, Hemodynamic fluctuations and baroreflex sensitivity in humans: A beat-to-beat model. Amer. J. Physiol. **253**, H680–H689 (1987)
2. H. Seidel, H. Herzel, Bifurcations in a nonlinear model of the baroreceptor-cardiac reflex. Acoust. J. **115D**, 145–160 (1998)
3. K. Kotani et al., Model for cardiorespiratory synchronization in humans. Phys. Rev. E. **65**, 051923–051932 (2002)
4. V.T. Grinchenko, A.G. Rudnitskiy, A model of intersection of the cardiovascular and respiratory systems. Acoust. J. **9**(3), 16–26 (2006)
5. T.S. Krasnopolskaya, E.D. Pechuk, Chaos In a modified cardiorespiratory model. CMSIM **4**, 563–570 (2013)

Threshold Method for Control of Chaotic Oscillations

Volodymyr Rusyn and Christos H. Skiadas

Abstract The classical Chua's circuit that realizes chaotic behavior is presented. This circuit having a simple nonlinear element designed to be accurately piecewise-linear modelled. The circuit was modelled by using MultiSim software environment. The system's behavior is investigated through numerical simulations, by using well-known tools of nonlinear theory, such as chaotic attractor and time distributions of the chaotic coordinates. Using threshold method was practical realization of the control of chaotic attractor. This classical Chua's circuit that generates a chaotic and controlled attractor with a fixed period can be used in modern systems transmitting and receiving information. Number of periodic (controlled) attractor can be used as a key for masking of information carrier.

Keywords Chaos · Control · Threshold · MultiSim

1 Introduction

Chaos is the most interdisciplinary thematic areas; it includes very interesting, complex, nonlinear phenomena that have been intensively studied and regard many different areas ranging from sciences, mathematics and engineering to social systems [1–4].

In the area of engineering, chaos has been found to be very useful and has great potential in many different technological disciplines, such as in information and computer sciences, power systems protection, liquid mixing, economics, magnetism, biomedical systems analysis etc. [5–9].

V. Rusyn (✉)
Department of Radio Engineering and Information Security,
Yuriy Fedkovych Chernivtsi National University, Chernivtsi, Ukraine
e-mail: rusyn_v@ukr.net

C. H. Skiadas
ManLab, Technical University of Crete, Chania, Greece

C. H. Skiadas and Y. Dimotikalis (eds.), *12th Chaotic Modeling and Simulation International Conference*, Springer Proceedings in Complexity,
https://doi.org/10.1007/978-3-030-39515-5_18

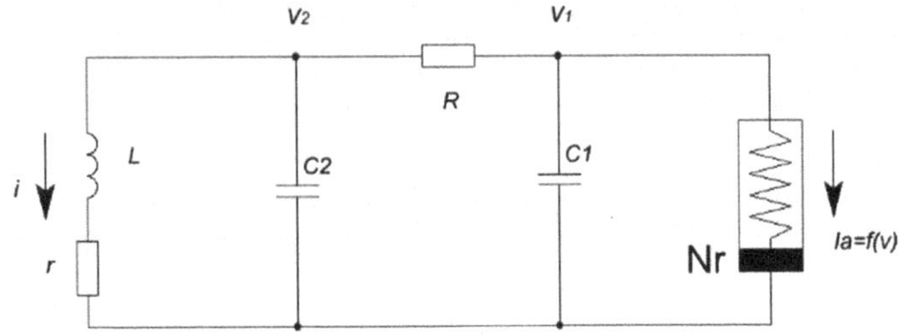

Fig. 1 Chua circuit

Chaotic signals can be generated by electronic circuits [10–15]; memristors [16–20]; simple [21, 22] and more complex [23]; analog, digital or mixed signal [24]. These signals depend on the system's initial conditions and this dependence is very sensitive; thus, they demonstrate the feature of being unpredictable. At the same time, chaotic signals are wide-band signals. Although they seem to be random, they are fully deterministic, highly sensitive to the system parameters, as well.

For chaos control have been proposed many different approaches or techniques, such as linear feedback control, OGY, inverse optimal control, etc. [25–29]. The theoretical basis of most known methods for control chaos is stabilizing the unstable periodic orbits via parameter perturbation.

The Chua circuit is the simplest electronic circuit exhibiting chaos that was invented in 1983 by Leon Chua [30, 31]. In Fig. 1 is shown one of the classical Chua's circuit.

The circuit consists of one nonlinear element with characteristics $f(v)$, called Chua diode, one inductor L, two capacitors (C_1 and C_2) and passive resistors (R and r). It is described by the following equations

$$\begin{cases} \dfrac{dv_1}{dt} = \dfrac{1}{C_1}\left(\dfrac{1}{R}(v_2 - v_1) - f(v)\right), \\ \dfrac{dv_2}{dt} = \dfrac{1}{C_2}\left(\dfrac{1}{R}(v_1 - v_2) + i_L\right), \\ \dfrac{di_L}{dt} = -\dfrac{1}{L}(v_2 + ri_L), \end{cases} \tag{1}$$

$$f(v) = \begin{cases} G_b v_1 + (G_b - G_a)E_1, & if\ v_1 \le -E_1, \\ G_b v_1, & if\ |v_1| < E_1, \\ G_b v_1 + (G_a - G_b)E_1, & if\ v_1 \ge E_1. \end{cases}$$

Here function $f(v)$ is volt-ampere characteristics of Chua's diode, i_L is a current through inductor L, v_1 and v_2 are voltages across capacitors C_1 and C_2, respectively.

2 Modelling and Analysis of Non-linear Element

Nonlinear elements—these are elements in which the relation between voltage and current is a nonlinear function. For example, it is a diode, in which the current is an exponential function of the voltage. Circuits that consist of nonlinear elements are harder to analyze and design.

The circuit realization for modelling and analysis of the non-linear element is displayed in Fig. 2, with component: one operational amplifier TL082; resistors $R_1 = R_2 = 290\ \Omega$, $R_3 = 1.2\ k\Omega$, $R_4 = 47\ k\Omega$, $R_5 = R_6 = 3.3\ k\Omega$, $R_7 = 47\ k\Omega$; two diodes 1N4148; voltage—±9 V.

The nonlinear characteristic was modelled by the following parameters: = 6.5 V, = 2 kHz, = 3 kΩ. Figure 3 shows result of modelling of nonlinear elements using MultiSim. The simulation parameters: $_1$ = 2 V/div, $_2$ = 2 V/div.

Figure 4 shows V/I characteristic of nonlinear element that was realized practically.

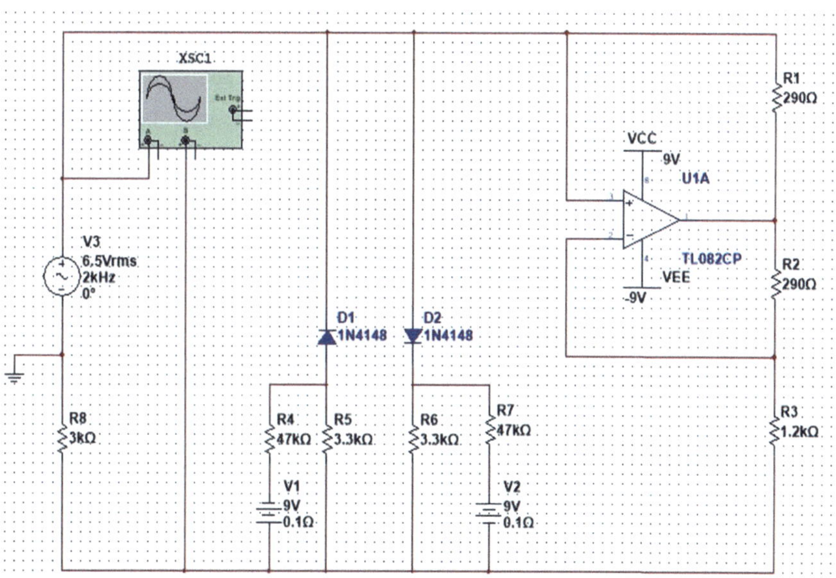

Fig. 2 Circuit realization for modelling and analysis of nonlinear characteristic

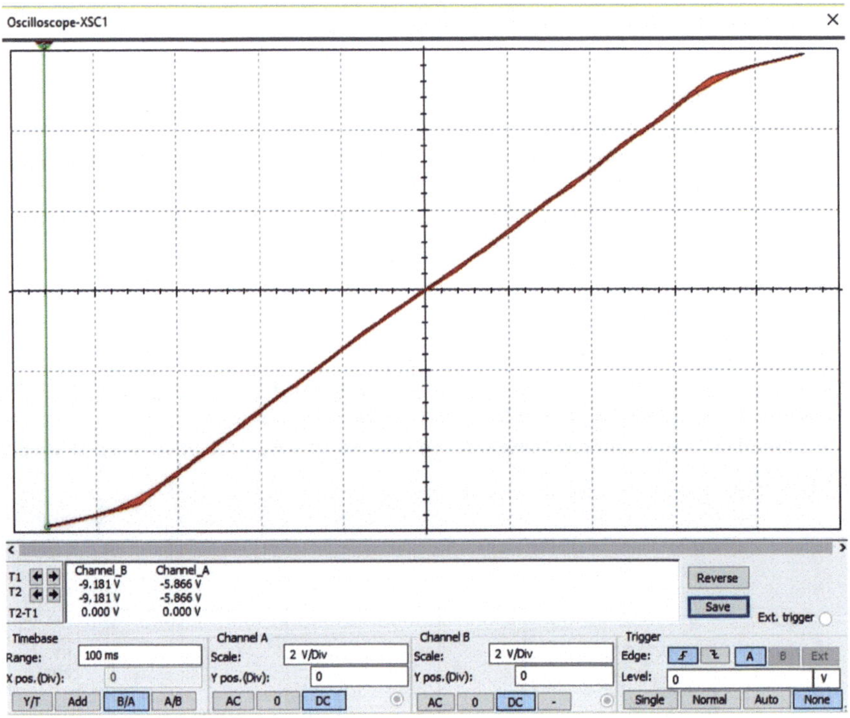

Fig. 3 V/I characteristic of a nonlinear element

3 Modelling and Analysis of the Classical Chaotic Chua's Generator

Figure 5 shows simulated scheme of the classical chaotic Chua's generator by using MultiSim.

Circuit was realized on the one operational amplifier TL082, powered by a 9 V, two diodes 1N4148, resistors $R_1 = R_2 = 290\ \Omega$, $R_3 = 1.2\ k\Omega$, $R_4 = 47\ k\Omega$, $R_5 = R_6 = 3.3\ k\Omega$, $R_7 = 2$ k (1.7 k), $R_8 = 47\ k\Omega$, two capacitors $C_1 = 10$ nF, $C_2 = 100$ nF, inductor $L_1 = 18$ mH.

Figure 6 shows the result of computer circuit simulation. On the virtual oscilloscope is presented generated chaotic signal in the plane XY. Coordinate X in the circuit correspond voltage V_{C2}, coordinate Y—voltage V_{C1}. The simulation parameters: $U_1 = 500$ mV/div, $U_2 = 1$ V/div.

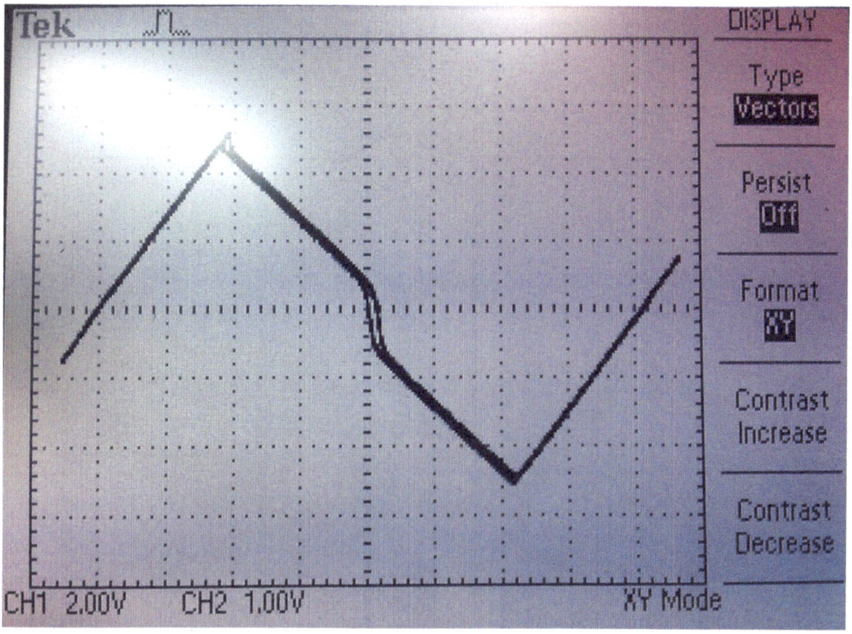

Fig. 4 V/I characteristic of nonlinear element realized practically

In Fig. 7 the time-series of both x- and y-signals appear. Their non-periodic nature is evident. Shows time dependences of the coordinates X (top) and Y (bottom) respectively (the channels' settings were for channel A, 1 V/div and for channel B, 5 V/div.

Figures 8 and 9 shows chaotic attractor and time-series that was realized practically.

4 Practical Realization of the Threshold Method for Control of Chaotic Oscillations

Threshold formalism for multidimensional systems was described in [32]. Now to experimentally demonstrate the range and efficacy of the method, we implement it on the classical chaotic Chua's circuit. We consider a realization of the double scroll chaotic Chua's attractor given by the following set of (rescaled) three coupled ODEs:

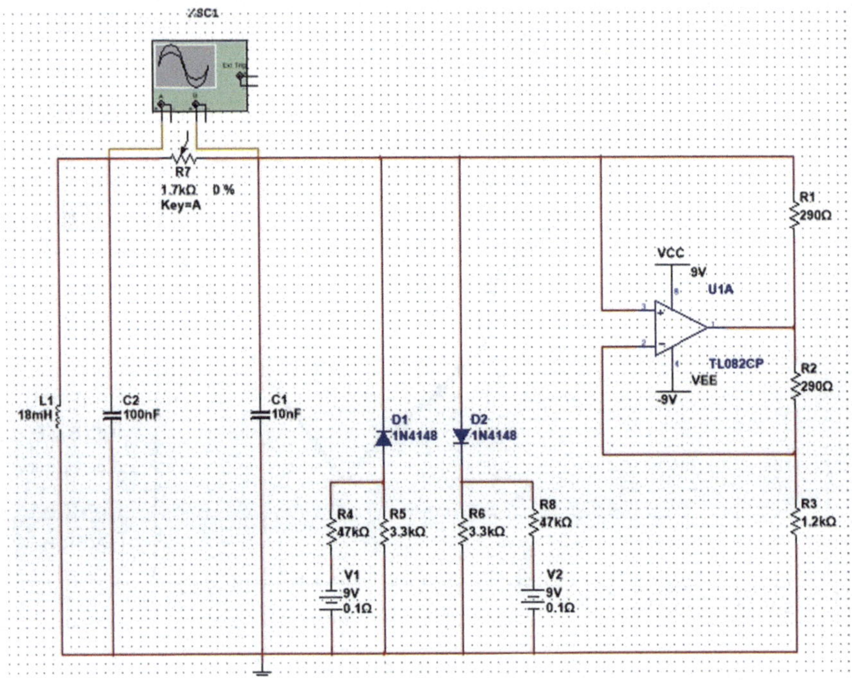

Fig. 5 The simulated circuit of the classical chaotic Chua's generator

$$\begin{cases} \dfrac{dx}{dt} = \alpha(y - x - g(x)), \\ \dfrac{dy}{dt} = x - y + z, \\ \dfrac{di_L}{dt} = -\beta y, \end{cases} \tag{2}$$

where $\alpha = 10$, $\beta = 14.87$, $g(x)$—nonlinear function.

The circuit realization of the above is displayed in Fig. 10 with component values: capacitors $C_1 = 100$ nF, $C_2 = 10$ nF, DA_1–DA_4—operational amplifier TL082, powered by a 9 V, GB_1—threshold reference voltage, inductor $L_1 = 18$ mH, resistors $R_1 = R_2 = 1.71$ k, $R_3 = 47$ kΩ, $R_4 = R_5 = 3.3$ kΩ, $R_6 = 47$ kΩ, $R_7 = R_8 = 290$ Ω, $R_9 = 1.2$ kΩ, $R_{10} = 1$ kΩ, potentiometer $R_{11} = 100$ kΩ, diodes VD_1–VD_3—1N4148.

In the circuit, the voltage V_T corresponds to x^*. Figures 11, 12 and 13 shows experimental results of the control of chaotic oscillations.

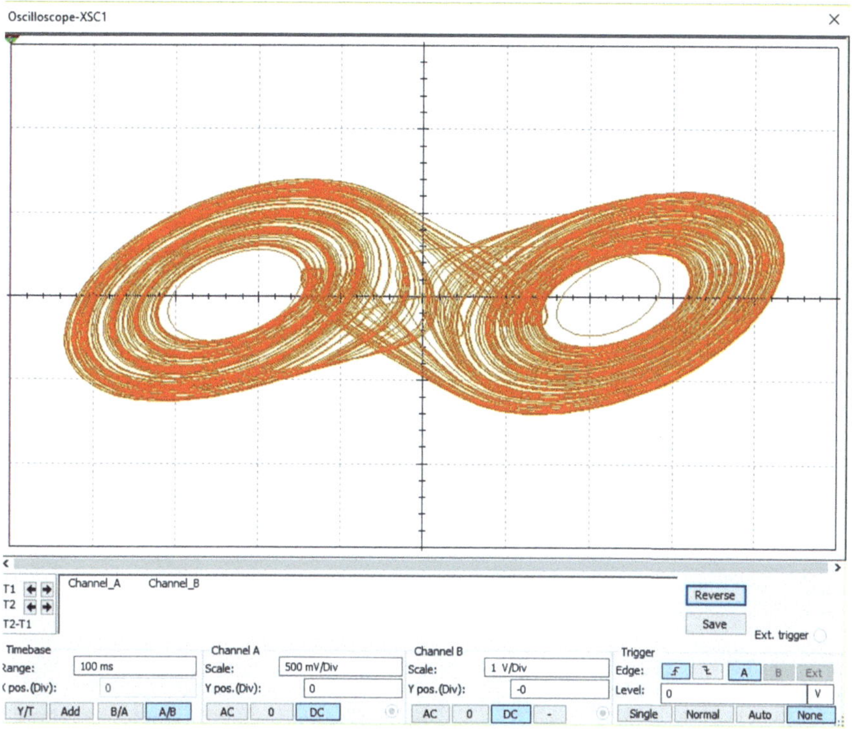

Fig. 6 Chaotic attractor

We implement more minimal thresholding. Instead of demanding that the x variable be reset to x^* if it exceeds x^*, we only demand this in (2). This has easy implementation, as it avoids modifying the value of x in the nonlinear element (x). So, then all we do is to implement $dy/dt = x^* - y + z$ instead of (2), when $x > x^*$, and there is no controlling action if $x < x^*$.

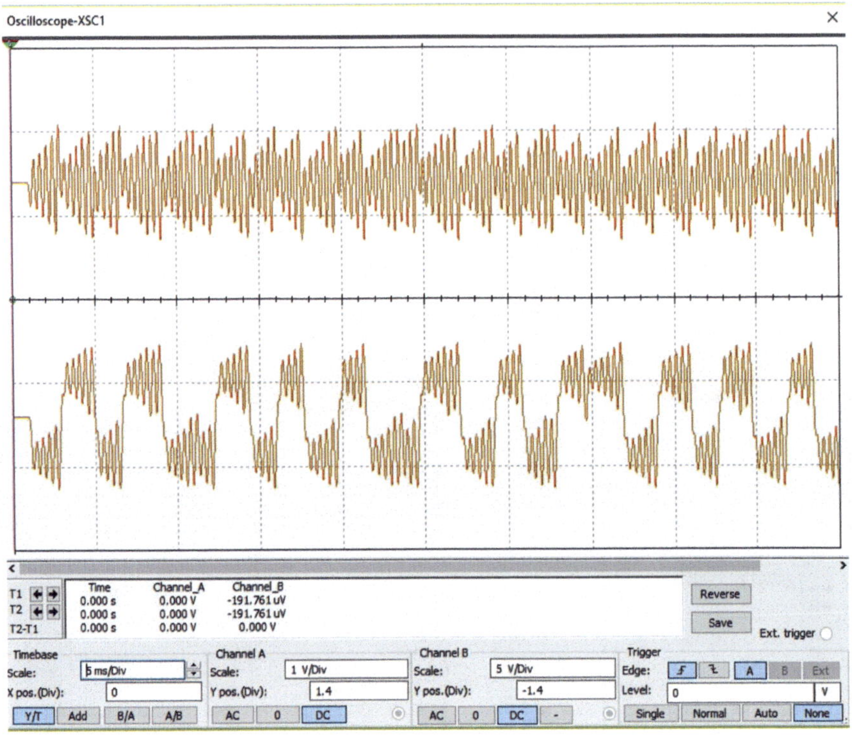

Fig. 7 The x-signal (upper) and the y-signal (lower) timeseries. Their non-periodic nature is evident

5 Conclusions

For the first time was used threshold method for control of chaotic oscillations for classical Chua's chaotic generator. This classical Chua's circuit that generate a chaotic and controlled attractor with a fixed period can be used in modern systems transmitting and receiving information. Number of periodic (controlled) attractor can be used as a keys for masking of information carrier.

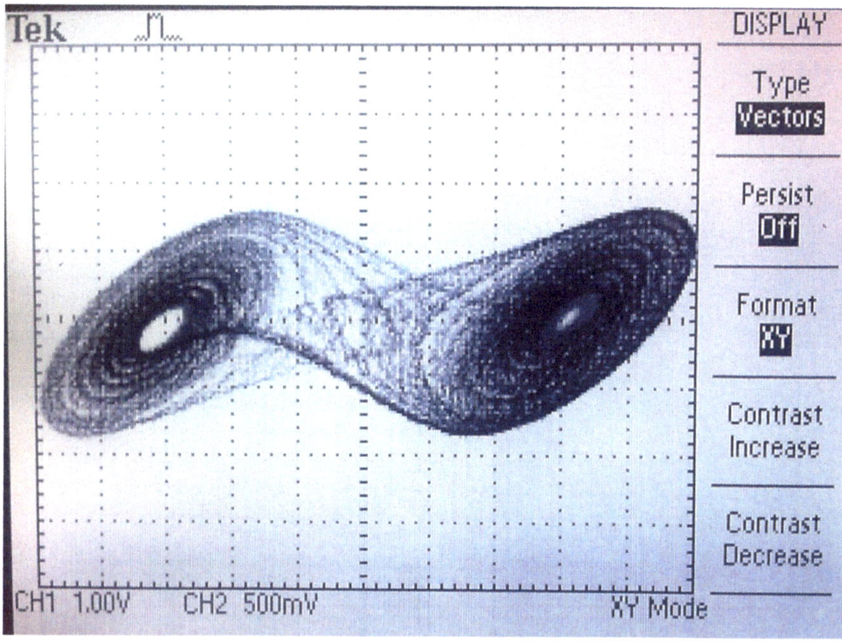

Fig. 8 Chaotic attractor realized practically

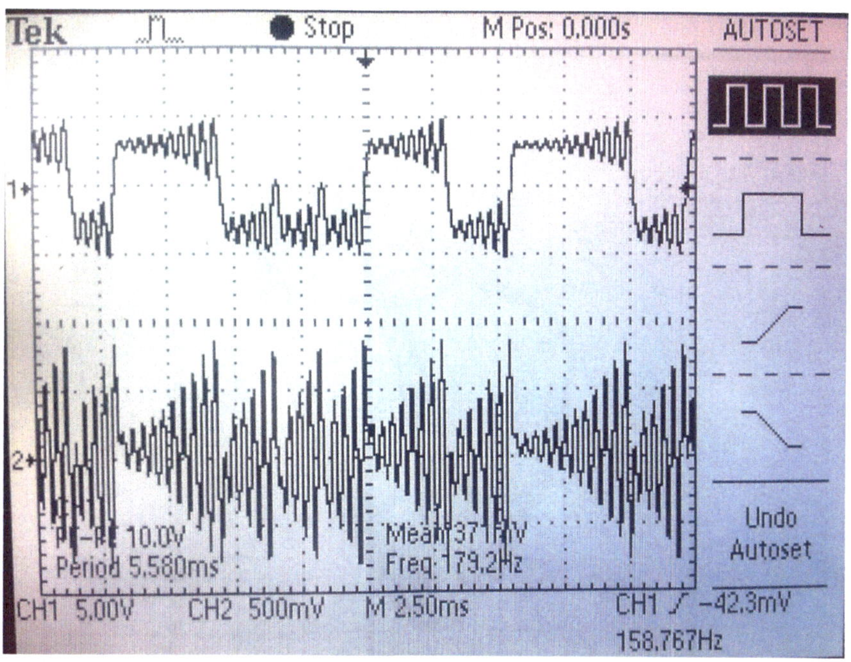

Fig. 9 The x-signal (upper) and the y-signal (lower) timeseries realized practically. Their non-periodic nature is evident

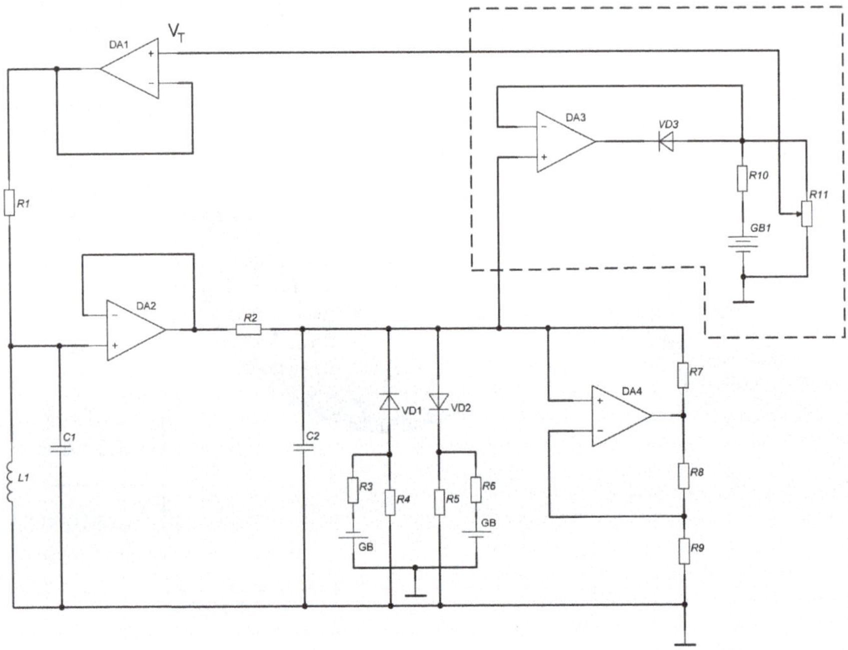

Fig. 10 The x-signal (upper) and the y-signal (lower) time-series realized practically. Their non-periodic nature is evident

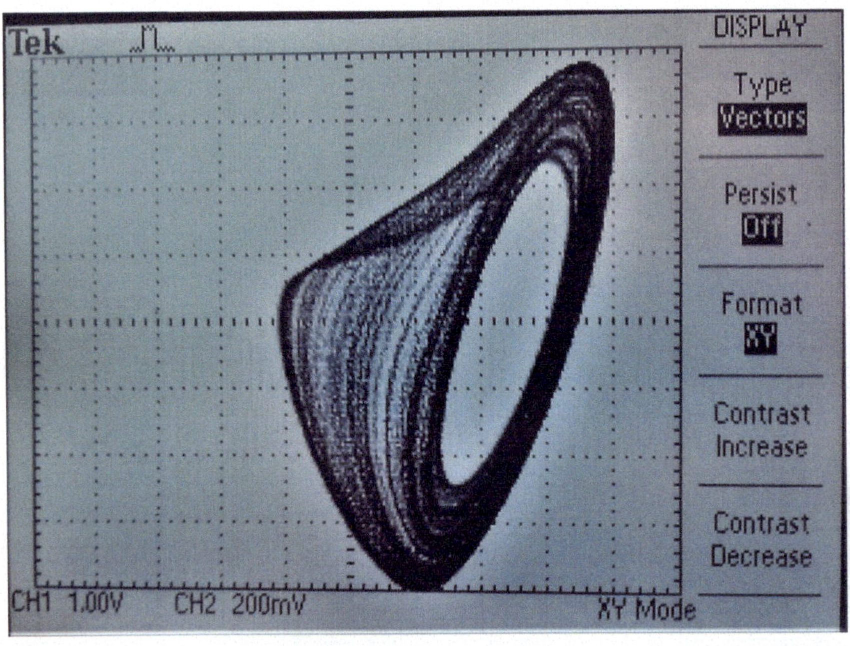

Fig. 11 Uncontrolled chaotic attractor in the $V_1 - V_2$ plane

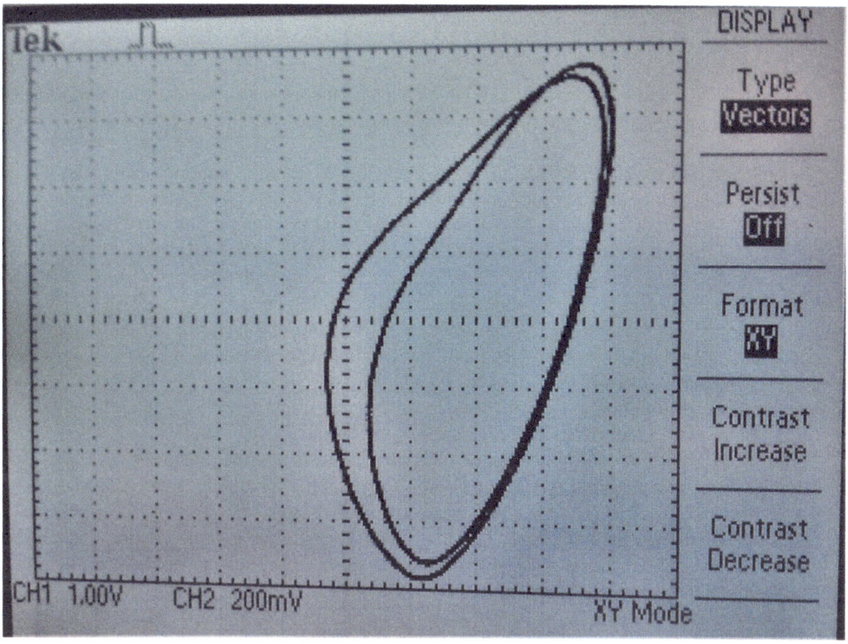

Fig. 12 2-period controlled attractor obtained when $x^* = 2.7$ V in the $V_1 - V_2$ plane

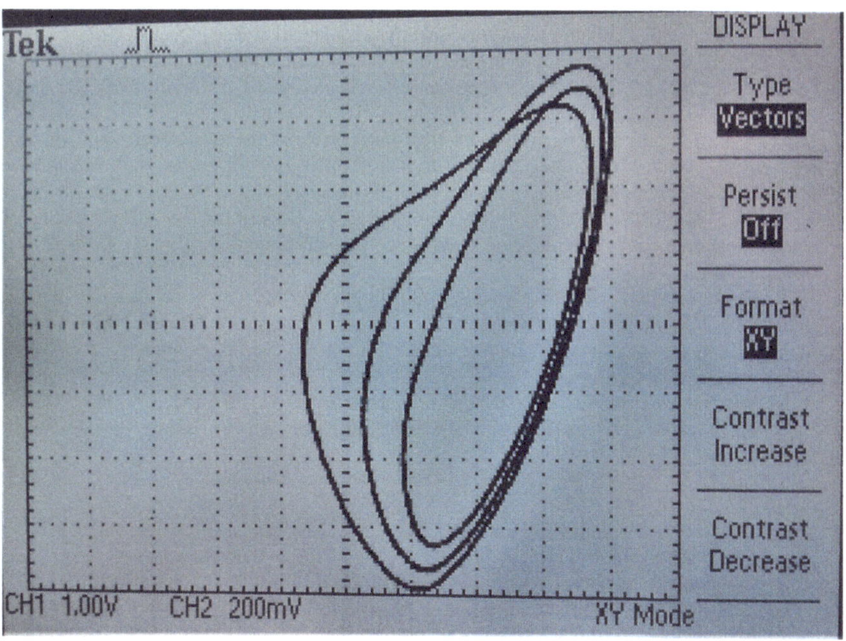

Fig. 13 3-period controlled attractor obtained when $x^* = 2.71$ V in the $V_1 - V_2$ plane

References

1. N. Lingala, N.S. Namachchivaya, I. Pavlyukevich, Random perturbations of a periodically driven nonlinear oscillator: escape from a resonance zone. Nonlinearity **30**(4), 1376–1404 (2017)
2. N. Lingala, N.S. Namachchivaya, I. Pavlyukevich, W. Wedig, Random perturbations of periodically driven nonlinear oscillators. Procedia IUTAM **19**, 91–100 (2016)
3. C.H. Skiadas, Exact solutions of stochastic differential equations: Gompertz, generalized logistic and revised exponential. Methodol. Comput. Appl. Probab. **12**(2), 261–270 (2010)
4. C.H. Skiadas, C. Skiadas, *Chaotic Modelling and Simulation: Analysis of Chaotic Models, Attractors and Forms* (Taylor & Francis Group, LLC, 2008), pp. 1–345
5. P.P. Horley, M.Ya. Kushnir, M. Morales-Meza, A. Sukhov, V. Rusyn, Period-doubling bifurcation cascade observed in a ferromagnetic nanoparticle under the action of a spin-polarized current. Physica B **486**, 60–63 (2016)
6. L. Pribylova, Bifurcation routes to chaos in an extended Van der Pol's equation applied to economic models. Electron. J. Differ. Equ. **53**, 1–21 (2009)
7. V. Rusyn, O. Savko, Modeling of chaotic behavior in the economic model, in *CHAOS 2015—8th Chaotic Modeling and Simulation International Conference, Proceedings 2015* (2015), pp. 705–712
8. S. Vaidyanathan, A. Sambas, S. Kacar, U. Cavusoglu, A new finance chaotic system, its electronic circuit realization, passivity based synchronization and an application to voice encryption. Nonlinear Eng. **8**(1), 193–205 (2019)
9. V. Hajnova, L. Pribylova, Two-parameter bifurcations in LPA model. J. Math. Biol. **75**(5), 1235–1251 (2017)
10. V. Rusyn, Modeling and research information properties of Rucklidge chaotic system using LabView, in *CHAOS 2017—Proceedings: 10th Chaotic Modeling and Simulation International Conference* (2017), pp. 739–744
11. C.-H. Lien, S. Vaidyanathan, A. Sambas, Sukono, M. Mamat, W.S.M. Sanjaya, Subiyanto, A new two-scroll chaotic attractor with three quadratic nonlinearities, its adaptive control and circuit design, in *IOP Conference Series: Materials Science and Engineering* (Vol. 332, Issue 1, Article no. 012010, 2018)
12. V.B. Rusyn, Modeling and research of Chaotic Rossler system with LabView and Multisim software environments, in *Bulletin of National Technical University of Ukraine «Kyiv Polytechnic Institute»*, Series Radiotechnique Radioapparatus Building (Eddition 59, 2014), pp. 21–28
13. A. Sambas, W.S.M. Sanjaya, M. Mamat, P.R. Putra, A.T. Azar, Mathematical modelling of chaotic jerk circuit and its application in secure communication system. Stud. Fuzziness Soft Comput. **337**, 133–153 (2016)
14. V. Rusyn, A. Stancu, L. Stoleriu, Modeling and control of chaotic multi-scroll Jerk system in LabView, in *Bulletin of National Technical University of Ukraine «Kyiv Polytechnic Institute»*, Series Radiotechnique Radioapparatus Building (Edition 62, 2015), pp. 94–99
15. A. Sambas, S. Vaidyanathan, M. Mamat, W.S. Mada Sanjaya, A six-term novel chaotic system with hidden attractor and its circuit design. Stud. Syst. Decis. Control **133**, 365–373 (2018)
16. L. Chua, Memristor—the missing circuit element. IEEE Trans. Circuit Theory **18**(5), 507–519 (1971)
17. Z.F. Wang, L. Shi, H. Wu, N. Helian, O.L. Chua, Fractional memristor. Appl. Phys. Lett. **111**, 243502 (2017)
18. M. Itoh, L. Chua, Dynamics of hamiltonian systems and memristor circuits. Int. J. Bifurc. Chaos **27**(2), 1730005 (2017)
19. S.G. Stavrinides, How to teach memristor in school. Phys. Educ. **52**(3), 033008 (2017)
20. V. Rusyn, S. Khrapko, Memristor: modeling and research of information properties, in *Springer Proceedings in Complexity* (2019), pp. 229–238
21. J. Kalomiros, S.G. Stavrinides, A.N. Miliou, I.P. Antoniades, L. Magafas, A.N. Anagnostopoulos, The nonlinear current behaviour of a driven R-L-Varactor in the low frequency range. Nonlinear Anal. Real World Appl. **10**(2), 691–701 (2009)

22. V.B. Rusyn, I. Pavlyukevich, L. Pribylova, H.C. Skiadas, Design, modeling and research of the new non-autonomous chaotic generator. Visn. NTUU KPI, Ser. Radioteh. Radioaparatobuduv. **77**, 13–16 (2019)
23. A.N. Miliou, S.G. Stavrinides, A.P. Valaristos, A.N. Anagnostopoulos, Nonlinear electronic circuit—Part II: Synchronization in a chaotic MODEM scheme (Review Paper). Nonlinear Anal. Theory, Methods Appl. **71**, e21–e31 (2009)
24. S.G. Stavrinides, N.F. Karagiorgos, K. Papathanasiou, S. Nikolaidis, A.N. Anagnostopoulos, A digital non-autonomous chaotic oscillator suitable for information transmission. IEEE TCAS-II: Express Briefs **60**(12), 887–891 (2013)
25. E. Ott, C. Grebogi, J.A. Yorke, Controlling chaos. Phys. Rev. Lett. **64**, 1196–1199 (1990)
26. V. Rusyn, M. Kushnir, O. Galameiko, Hyperchaotic control by thresholding method, in *Modern Problems of Radio Engineering, Telecommunications and Computer Science—Proceedings of the 11th International Conference, TCSET'2012*, art. No. 6192785 (2012), p. 67
27. V.B. Rusyn, A. Stancu, L. Stoleriu, Modeling and control of chaotic multi-scroll Jerk system in LabView. Visn. NTUU KPI, Ser. Radioteh. Radioaparatobuduv. **63**, 94–99 (2015)
28. V.B. Rusyn, L. Pribylova, D.-G. Dimitriu, Control of the modified chaotic Chua's circuit using threshold method. Visn. NTUU KPI, Ser. Radioteh. Radioaparatobuduv. **75**, 61–65 (2018)
29. V. Rusyn, Modeling, analysis and control of chaotic Rucklidge system. J. Telecommun. Electron. Comput. Eng. (JTEC), Tom **11**(1), 43–47 (2019)
30. T. Matsumoto, A chaotic attractor from Chua's circuit. IEEE Trans. Circuits Syst. **31**, 1055–1058 (1990)
31. L. Chua, A zoo of strange attractors from the canonical Chua's circuits, in *Proceedings of the IEEE 35th Midwest Symposium on Circuits and Systems* (Cat. No. 92CH3099-9, Vol. 2, 1992), pp. 916–926
32. K. Murali, S. Sinha, Experimental realization of chaos control by thresholding. Phys. Rev. E **68**(1), 016210 (2003)

Some Implications of Invariant Model of Boltzmann Statistical Mechanics to the Gap Between Physics and Mathematics

Siavash H. Sohrab

Abstract Some implications of a scale-invariant model of Boltzmann statistical mechanics to physical foundation of the gap between physics and mathematics, Riemann hypothesis, analytic number theory, Cantor uncountability theorem, continuum hypothesis, Goldbach conjecture, and Russell paradox are studied. Quantum nature of space and time is described by introduction of dependent *internal measures* of space and time called spacetime and independent *external measures* of space and time. Because of its *hyperbolic geometry*, its discrete fabric, and its stochastic atomic motions, physical space is called *Lobachevsky-Poincaré-Dirac-Space*.

Keywords Riemann hypothesis · Continuum hypothesis · Analytical number theory · Infinitesimals · Spacetime · Goldbach conjecture · Russell paradox

1 Introduction

Boltzmann statistical mechanics is known to be universally valid for all systems composed of very large numbers of weakly interacting particles considered as ideal gas. The similarities between stochastic quantum fields [1–17] and classical hydrodynamic fields [18–30] resulted in recent introduction of a scale-invariant model of statistical mechanics and its application to thermodynamics [31], fluid mechanics [32, 33], and quantum mechanics [34, 35].

The present study focuses on further implications of the model to the physical foundation of the gap between physics and mathematics. In particular, the impact of the model on Riemann hypothesis, analytical number theory, Cantor uncountability theorem, continuum hypothesis, infinitesimals, Goldbach conjecture, and Russell's paradox are examined. Thermodynamic temperature is shown to provide *internal*

12th CHAOS Conference Proceedings, 18 - 22 June 2019, Chania, Crete, Greece.

S. H. Sohrab (✉)
McCormick School of Engineering and Applied Science, Department of Mechanical Engineering, Northwestern University, 2145 Sheridan Road, Evanston, IL 60208-3111, USA
e-mail: s-sohrab@northwestern.edu

C. H. Skiadas and Y. Dimotikalis (eds.), *12th Chaotic Modeling and Simulation International Conference*, Springer Proceedings in Complexity,
https://doi.org/10.1007/978-3-030-39515-5_19

measures of space and time called spacetime of a manifold called *Lobachevsky-Poincaré-Dirac-Space*. Cantor continuum hypothesis is disproved by showing that cardinality of real numbers is the same as that of natural numbers namely $\aleph_0$.

2 A Scale Invariant Model of Statistical Mechanics

The scale-invariant model of statistical mechanics for equilibrium galactic-, planetary-, hydro-system-, fluid-element-, eddy-, cluster-, molecular-, atomic-, subatomic-, kromo-, and tachyon-dynamics at the scale β = g, p, h, f, e, c, m, a, s, k, and t is schematically shown on the left hand side and the corresponding laminar flows on the right hand side of Fig. 1.

For each statistical field, one defines particles that form the background fluid and are viewed as point-mass or "*atom*" of the field. Next, the *elements* of the field are defined as finite-sized composite entities composed of an ensemble of "*atoms*" as shown in Fig. 1. Finally, the ensemble of a large number of "*elements*" is defined as the statistical "*system*" at that particular scale.

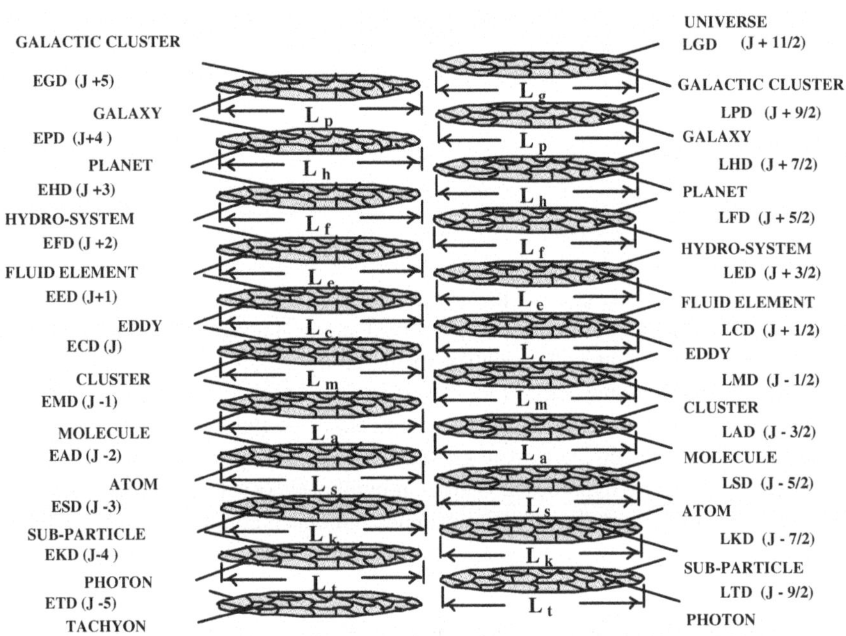

Fig. 1 Scale invariant model of statistical mechanics. Equilibrium-β-Dynamics on the left-hand-side and non-equilibrium Laminar-β-Dynamics on the right-hand-side for scales β = g, p, h, f, e, c, m, a, s, k, and t as defined in Sect. 2. Characteristic lengths of (system, element, "atom") are $(L_\beta, \lambda_\beta, \ell_\beta)$ and λ_β is the mean-free-path [34]

Following the classical methods [19, 36–38] the invariant definitions of density ρ_β, and velocity of *atom* $\mathbf{u}_\beta$, *element* $\mathbf{v}_\beta$, and *system* $\mathbf{w}_\beta$ at the scale β are given as [33, 34]

$$\rho_\beta = n_\beta m_\beta = m_\beta \int f_\beta du_\beta,\ \mathbf{u}_\beta = \mathbf{v}_{w\beta-1} \tag{1}$$

$$\mathbf{v}_\beta = \rho_\beta^{-1} m_\beta \int \mathbf{u}_\beta f_\beta d\mathbf{u}_\beta,\ \mathbf{w}_\beta = \mathbf{v}_{w\beta+1} \tag{2}$$

The most probable element of the lower scale β is identified as the "atom" of the next higher scale β + 1 such that $\mathbf{v}_{mp\beta} = \mathbf{u}_{\beta+1}$, resulting in the hierarchy of embedded statistical fields. At thermodynamic equilibrium velocity, energy, and speed of "particles" or Heisenberg-Kramers [39] *virtual oscillators* are governed by Gaussian (Maxwell), Planck [40], and Maxwell-Boltzmann distribution functions [34].

3 Invariant Definitions of External Space and Time Versus Internal Spacetime

According to the model in Fig. 1, physical space, Aristotle fifth element, de Broglie hidden thermostat [3], Casimir vacuum [41], or Dirac stochastic ether [42] is identified as a tachyon fluid [34, 43]. It is important to emphasize that *space is the tachyonic fluid itself and not the container that is occupied by this fluid*, as in the classical theories of ether [44], in harmony with perceptions of Euler [45] on reality of space and time.

To describe the nature of physical space and time, we consider the large scale of cosmology β = g with (atom, element, system) velocities $(\mathbf{u}_g, \mathbf{v}_g, \mathbf{w}_g)$ and the adjacent lower scale of astrophysics β = s with the velocities $(\mathbf{u}_s, \mathbf{v}_s, \mathbf{w}_s)$. According to Eq. (1) the "atom" of cosmic field is defined as the most probable element of astrophysical field $\mathbf{u}_g \equiv \mathbf{v}_{ws}$. Also, by Eq. (2) the most-probable element of cosmic field is identified as "system" of astrophysical field $\mathbf{v}_{wg} \equiv \mathbf{w}_s$. Therefore, every "point" constituting the "space" of cosmic field will be composed of either galaxy (g) or Casmir vacuum (∅) [41, 43] as shown in Fig. 2.

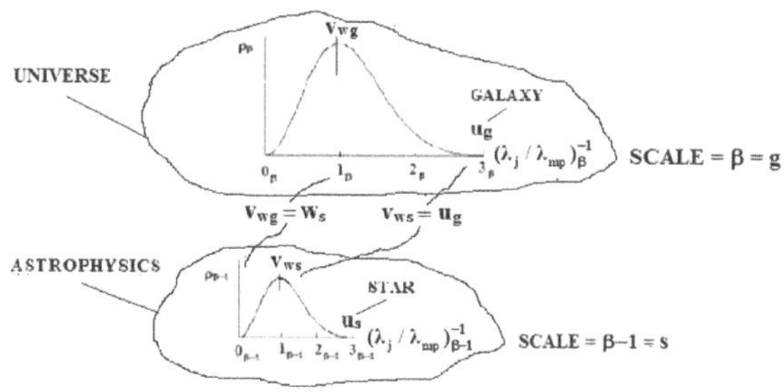

Fig. 2 Normalized Maxwell-Boltzmann speed distribution at cosmic and astrophysical scales [43]

Because of the definition, $\mathbf{u}_g \equiv \mathbf{v}_{ws}$ the atomic energy at cosmic scale is identical to the most-probable energy of astrophysical field, under local thermodynamic equilibrium and cosmic field will have a homogenous constant temperature [31]

$$m_g u_g^2 = m_{ws} v_{ws}^2 = kT_{\beta=s} = k\lambda_{ws} \quad (3)$$

Since $v_{ws} = \lambda_{ws}\nu_{ws} = \lambda_{ws}/\tau_{ws}$ by Eq. (3), one associates constant internal measures of (extension λ_{ws}, duration τ_{ws}) with every "point" of space in a universe at temperature $T_g = T_s$. Thus, temperature of physical space or Casimir vacuum [41] fixes local measures of *spacetime* $(\lambda_{ws}, \tau_{ws})$ that are not *independent* since $v_{ws} = \lambda_{ws}/\tau_{ws}$ must satisfy the vacuum temperature $T_g = T_s$ in Eq. (3) [31]. Since Wien displacement law $\lambda_w T = 0.29$ cm.K = 0.0029[m^2] suggests the change of units [m/cm] = 100, the classical temperature conversion formula is expressed as T [m] = °C [m] + 2.731 with 2.731 close to Penzias-Wilson [46] cosmic microwave background temperature $T_{CMB} \simeq 2.73$ [m].

When expressed as $(\lambda_{wxs}, \lambda_{wys}, \lambda_{wzs}, \tau_{ws})$ the internal coordinates are in harmony with the four-dimensional spacetime of Poincaré [47] and Minkowski [48]. Hence, Poincaré [49] anticipated the importance of hyperbolic geometry of physical space that parallel to his anticipation of an external stress (Poincaré stress) [34], is another testimony to the true genius of this great mathematician, physicist, and philosopher. Because of its *hyperbolic geometry*, its discrete or quantum fabric, and its stochastic atomic motions, physical space is called *Lobachevsky-Poincaré-Dirac-Space*.

At cosmic scale $\beta = g$ one employs internal (ruler, clock) of the lower scale of astrophysics $\beta = s$ to define *external* space and time coordinates (x_β, t_β) as $(x_\beta = N_x\lambda_{w\beta-1}, t_\beta = N_t\tau_{w\beta-1})$. Because of *independence* of numbers N_x and N_t, coordinates (x_β, t_β) could be employed to study the dynamics of universe.

4 Connections Between Riemann Hypothesis, Quantum Mechanics, and Kinetic Theory of Ideal Gas

At the most fundamental level, the science of mathematics is concerned with the relationships between *numbers* according to certain assigned axiomatic rules. The numbers could be natural $\mathbb{N}$, integer $\mathbb{Z}$, rational $\mathbb{Q}$, irrational $\mathbb{I}$, transcendental $\mathbb{T}$, real $\mathbb{R}$, complex $\mathbb{C}$, p-adic $\mathbb{Q}_p$, … numbers. What is of central significance is that no *physical dimension* need be attributed to such numbers in the ideal Platonic world of mathematics. In theoretical study of physical sciences on the other hand, one aims at describing physical phenomena through mathematical models involving numbers with *physical dimensions* such as length, time and mass (meter, second, kg). It is only when physical variables are made *dimensionless* that mathematical physics assumes the same degree of universality as mathematics.

The scale invariant model of statistical mechanics (Fig. 1) has an impact on Hilbert's number eight problem namely Riemann hypothesis. A recent study [34]

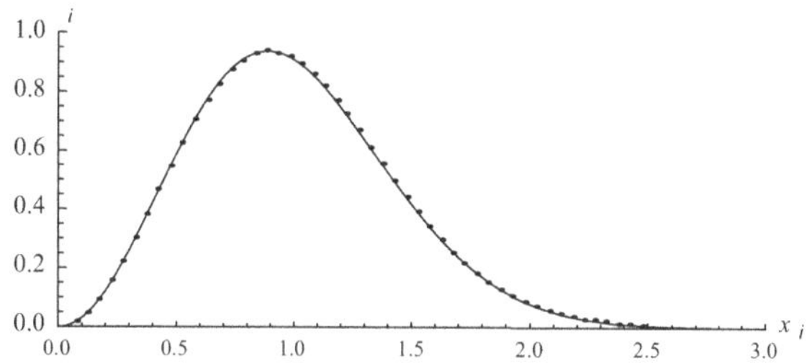

Fig. 3 Probability density of normalized spacing between zeros of *Riemann* zeta function [51] γ_n, $10^{12} \leq n \leq 10^{12}+10^5$, normalized spacing between eigenvalues of GUE [51], and the NMB distribution function in Eq. (4) [34]

was focused on possible connections between kinetic theory of gas and the theoretical findings of Montgomery [50] and Odlyzko [51] on analytical number theory and what is known as Montgomery-Odlyzko law [52, 53]. The pair correlation of Montgomery [50] was subsequently recognized by Dyson to correspond to that between the energy levels of heavy elements [52, 53] and thus to the pair correlations between eigenvalues of *Hermitian* matrices [54]. Hence, a connection was established between quantum mechanics on the one hand and *quantum chaos* [55] on the other hand.

When Maxwell-Boltzmann distribution is renormalized with respect to the most-probable speed it leads to *Normalized Maxwell-Boltzmann* (NMB) distribution function [34]

$$\rho_j = (8/\pi_\beta)[(2/\sqrt{\pi_\beta})x_{j\beta}]^2 e^{-[(2/\sqrt{\pi_\beta})x_{j\beta}]^2} \tag{4}$$

that can be viewed as distribution of sizes of "*clusters of numbers*" or Hilbert "*condensations*" [52]. Normalized spacing between non-trivial zeros of *Riemann* zeta function and the eigenvalues of GUE calculated by Odlyzko [51], are compared with calculation from Eq. (4) in Fig. 3.

As discussed in [34], Maxwell-Boltzmann distribution function in Fig. 3 could be viewed as spectrum of sizes of probabilities of "number clusters" that are identified as "*energy levels*" corresponding to *stationary-states* of Bohr [39] in quantum mechanics. The distribution of cluster sizes shown in Fig. 3 is called *unitary distribution* $(\text{NMB})_u$ because all clusters are composed of identical units as "atoms".

One next considers an infinite number of $(\text{NMB})_p$ distribution functions constructed from different prime numbers $p = 2, 3, 5, 7, 11, \ldots.$ as "atomic specie" [34]. Hence, the elements of these statistical fields are Gauss's *clock-calculators* or Hensel's *p-adic numbers* $\mathbb{Q}_p = p^N$ [52]. For example, the prime "specie" p distribution contains spectrum of clusters.

$$p, p^2, p^3, p^4, \ldots.p^N \ldots = \mathbb{Q}_p, p = 2, 3, 5, 7, 11, \ldots. \tag{5}$$

Finally, following Connes [56, 57], one constructs an Adele or *mixture* distribution function $(\text{NMB})_A$ combining unitary $(\text{NMB})_u$ and all suitably normalized prime "*species*" distribution functions

$$\mathbb{A} = (\mathrm{NMB})_{\mathrm{A}} = (\mathrm{NMB})_{\mathrm{u}} + \sum_{\mathrm{p}} (\mathrm{NMB})_{\mathrm{p}} = \mathbb{R} + \sum_{\mathrm{p}} \mathbb{Q}_{\mathrm{p}} \tag{6}$$

Because $(\mathrm{NMB})_{\mathrm{u}}$ and $(\mathrm{NMB})_{\mathrm{p}}$ are statistically independent and follow identical distribution functions, by Gauss central limit theorem their ensemble namely Adele distribution $(\mathrm{NMB})_{\mathrm{A}}$ in Eq. (6) will also follow a Normalized Maxwell-Boltzmann distribution function shown in Fig. 3. Because of renormalization, the infinite hierarchy of distribution functions in Eq. (6) will all collapse on a single universal curve (Fig. 3) covering the interval $(0 \to 4/\sqrt{\pi} = 2.567)$.

With homogeneous $(p_1 + p_1)$ and heterogeneous $(p_2 - p_1)$ even numbers, and under constant most-probable "measure" $\mathrm{T} = \lambda_{\mathrm{w}}$ [34], an even number of Adele field in Eq. (6) could be expressed as $\mathbb{A}_{\mathrm{E}} = 2\mathbb{R}_1 + 2p_1 = 2\mathbb{R}_2 - (p_2 - p_1)$ leading to $2(\mathbb{R}_2 - \mathbb{R}_1) = 2\mathbb{R} = p_2 + p_1$ that is in accordance with Goldbach conjecture [58].

Since by Eq. (6) all $(\mathrm{NMB})_{\mathrm{p}}$ are embedded in $(\mathrm{NMB})_{\mathrm{A}}$, the distribution of primes will be random in accordance with the perceptions of Cramér [59]. Notwithstanding the statement by Euler [60].

> Mathematicians have tried in vain to discover some order in the sequence of prime numbers but we have every reason to believe that there are some mysteries which the human mind will never penetrate.

The model described above may provide a physical foundation for the random nature of distribution of prime numbers. Therefore, Boltzmann kinetic theory of gas provides the *physical dynamic system* needed to bridge the gap between *Adele space* of noncommutative geometry of Connes [56] and *physical space* as described earlier [34, 43]. In other words, when *physical space* or Casimir vacuum [41] is identified as a *fluid* governed by a statistical field, it will have an energy spectrum given by Schrödinger equation of quantum mechanics [34] that in view of Heisenberg [61] *matrix mechanics* will be described by noncommutative spectral geometry [56, 57]. Although the exact connection between noncommutative geometry and Riemann hypothesis is yet to be understood according to Connes [56]

> The process of verification can be very painful: one's terribly afraid of being wrong…it involves the most anxiety, for one never knows if one's intuition is right- a bit as in dreams, where intuition very often proves mistaken.

the ideas suggested above and further described in [34] may help in the construction of the physical foundation of such a mathematical theory.

By Euler golden key [52] the zeros of zeta function $\prod_{p} p^{s} = 0$ and Planck [40] formula for energy quantum $\varepsilon_{\mathrm{j}} = \mathrm{h}\nu_{\mathrm{j}} = \mathrm{h}p^{s} = mu_j^2$ lead to zeros of particle velocity at *stationary states* [34].

$$\mathrm{u}_{\mathrm{j}} = \sqrt{\mathrm{h}/m}\, \nu_{\mathrm{j}}^{1/2} = \sqrt{\mathrm{h}/m}\, p^{s/2} = \sqrt{\mathrm{h}/m}\, p_j^{a+\mathrm{i}b} = 0 \tag{7}$$

By symmetry, *stationary states* corresponding to zeros of particle translation, rotational, and vibration velocity with the normalized value between (0, 1) will occur only at the mean value hence on the critical line a = 1/2 in accordance with Riemann hypothesis [34].

5 Implications to Cantor Uncountability Theorem, Continuum Hypothesis, Infinitesimals, and Russell Paradox

There are no physical or mathematical reasons for the hierarchy of statistical fields in Fig. 1 not to continue to larger than cosmic $\beta > g$ or smaller than tachyon $\beta < t$ scales ad infinitum in harmony with axiom of infinity. This is because infinite divisibility of matter as eloquently stated by Anaxagoras [62]

> Neither is there a smallest part of what is small, but there is always a smaller, for it is impossible that what is should ever cease to be.

is a logical requirement imposed by criteria of reversibility of mathematical operations. This is because if infinite divisibility of a body leads to nothing i.e. absolute zero, then such a process could not be reversed in order to return to the original body. Therefore, both absolute zero and absolute infinite are singularities as ideal Aristotle potential limits never actualized. Hence, according to the theory of numbers described herein Leibnitz house of *Monad* will always have a window.

When normalized invariant Maxwell-Boltzmann distribution is *re-normalized* with respect to the *most-probable state,* it leads to transformation ($x_{mp} \to 1$, $x_{\infty} \to 4/\sqrt{\pi} = 2.567$) as shown in Fig. 3. An example with ($0_\beta = 10^{-120}, 1_\beta = (4/\sqrt{\pi})10^{-100}, \infty_\beta = (4/\sqrt{\pi})^2 10^{-80}$) and the size of zero and infinity relative to unity taken as (10^{-20}, 1, 10^{20}) for three consecutive scales is shown in Fig. 4.

In Fig. 4 instead of the unit $1_{\beta-1} = 10^{-120}$ [m], any arbitrary value such as $1_\beta = 10^{-120^{10}}$[m] could be chosen thus requiring many more statistical fields before arriving at the scales shown in Fig. 4. Therefore, as suspected by ancient mathematicians and philosophers, the number *unity* plays a unique and fundamental role in mathematics just like the number π. This is because it is the dimension of unity (…,

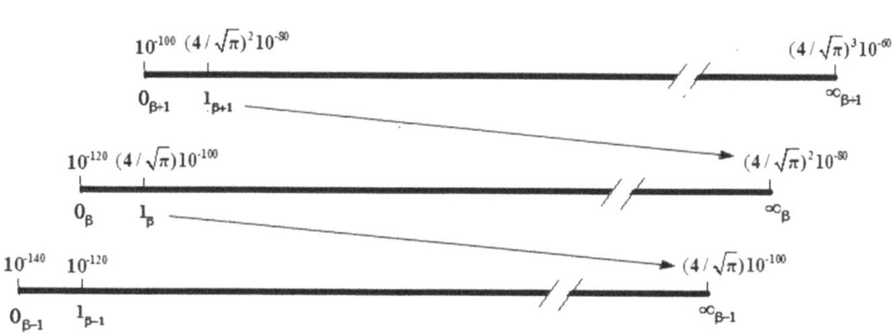

Fig. 4 Hierarchy of coordinates for cascades of embedded statistical fields

mm, m, km, ...) that defines the limits in Fig. 4 involving $4/\sqrt{\pi}$ as actual physical extensions (mm, m, km). Any finite value of unity naturally leads to a paradox since *infinite number of any finite quantity leads to infinite* a fact that may be referred to as "*ultraviolet catastrophe of mathematics*". However, in the present model this paradox is not encountered because $(\mathrm{NMB})_{\mathrm{A}}$ given in Eq. (6) is based on "quantum" units analogous to the theory of equilibrium radiation of photon gas by Planck [34, 40]. Avoidance of "*ultraviolet catastrophe*" is of central importance to the theory of numbers. This is because without *quantum foundation of theory of numbers* discussed in previous Section, one cannot confine the *infinite* to the realm of the *finite* in order to construct a real line.

Some implications of scale-invariant model of Boltzmann statistical mechanics to nonstandard analysis of Robinson [63] and Nelson [64] discussed earlier [65] are now further examined. First, it is reasonable to expect that all numbers $\mathbb{N}$, $\mathbb{Q}$, $\mathbb{I}$, $\mathbb{T}$, $\mathbb{R}$, and $\mathbb{Q}_{\mathrm{p}}$ be expressible in unitary field $(\mathrm{NMB})_{\mathrm{u}}$. This is because unitary field $(\mathrm{NMB})_{\mathrm{u}}$ has *maximum resolution* since it is constructed from the most elementary "atom" constituting all number fields of larger scales. Next, all numbers in $0_{\beta-1} < \mathbb{N} < 1_{\beta-1}$ are defined as *nonstandard numbers* at scale β since they are $\mathbb{N} < 0_{\beta}$ and hence *unobservable* at this scale [65]. Finally, all numbers $\mathbb{N} < 1_{\beta-2}$ are identified as *infinitesimals* for the scale β and above because even an infinite number of them will be $\infty_{\beta-2} = 1_{\beta-1}$ that is $1_{\beta-1} \le 0_{\beta}$.

The invariant model of statistical mechanic shown in Fig. 1 involving hierarchies of (atom, element, system) is most harmonious with Cantor theory of sets [66]. Clearly, leaving Everett's many universe aside, our universe at scale $\beta = g$ could be considered as the set of all sets (SOAS). According to Figs. 1 and 2 the universe U is defined as the largest supercluster SC (cluster of cluster of galaxies)

$$U = SC = \sum_{j} (CG_j) = \sum_{j} \overline{N}_{GC} = N_{CU}\overline{N}_{GC} = N_{GU} \tag{8}$$

that is the "limiting" SC subject to chaotic boundary condition due to random motion of galaxies. Hence, negative response to the question "is SOAS a member of itself" is also not contradictory since SOAS as the limiting set and a *quantum mechanical system* is subject to random and chaotic boundary condition.

A related problem known as Russell's paradox is about the barber who shaves all men who do not shave themselves. The paradox arises by noting that if the barber shaves himself he should not, and if he does not shave himself he should thus leading to contradiction either way. The set of all men $\{\mathbf{M}\}$ either can shave themselves ST $\{\overline{\mathbf{M}}^*, \overline{\mathbf{B}}^*\}$ or be shaved by someone else SE $\{\widetilde{\mathbf{M}}^*, \widetilde{\mathbf{B}}^*\}$. The shaving process is viewed as chemical reactions through steps (1)–(4) in Eq. (9). The case $\widetilde{\mathbf{B}}^*$ leads to $\{\mathbf{M}\} = \{\widetilde{\mathbf{M}}^*, \widetilde{\mathbf{B}}^*\} = \{\widetilde{\mathbf{M}}^*\}$ and the barber cannot self-shave and could follow reactions (1)-(3) without contradiction.

$$\begin{array}{ll} \widetilde{\mathbf{B}}^* + \widetilde{\mathbf{M}}^* \to \widetilde{\mathbf{B}}^* + \widetilde{\mathbf{M}} + * & (1) \\ \widetilde{\mathbf{B}}^* + \widetilde{\mathbf{M}} \to \widetilde{\mathbf{B}} + \widetilde{\mathbf{M}} + * & (2) \\ \widetilde{\mathbf{B}}^* + \widetilde{\mathbf{M}}^* \to \widetilde{\mathbf{B}} + \widetilde{\mathbf{M}} + ** & (3) \\ \overline{\mathbf{B}}^* \to \overline{\mathbf{B}} + * & (4) \end{array} \tag{9}$$

The case $\overline{\mathbf{B}}^*$ leads to $\{\mathbf{M}\} = \{\overline{\mathbf{M}}^*, \overline{\mathbf{B}}^*\} = \{\widetilde{\mathbf{M}}^*\}$ and the barber can self-shave by reaction (4) above. However, like SOAS problem, the paradox is only resolved if one assumes that the barber is on the border and oscillates between two complimentary sets. Hence, in harmony with Heisenberg uncertainty principle of quantum mechanics, Russell's barber must locate on the border and oscillate $\{\widetilde{\mathbf{B}}^* \leftrightarrow \overline{\mathbf{B}}^*\}$ between the complimentary sets $\{\widetilde{\mathbf{M}}^*\}$ and $\{\overline{\mathbf{M}}^*\}$ to avoid the paradox.

Cantor's uncountability theorem [58, 66] proving that the interval of real numbers (0, 1) is not countable is discussed next. This theory plays a central role in Cantor theory of transfinite numbers and Continuum Hypothesis CH [66] and is eloquently described by Casti [58]

> Since each real number as an infinite sequence of the ten natural numbers 0, 1, 2, ..., we can express a real number between 0 and 1as $r = .z_{r_1} z_{r_2} z_{r_3} z_{r_4} \ldots$, where each z_{r_i} is one of the ten integers between 0 and 9. By assumption, we can put any such number into one-to-one correspondence with a natural number. Suppose the following table is such a correspondence
>
> $$\begin{aligned} r_1 &= .\underline{z_{11}} z_{12} z_{13} z_{14} \ldots \\ r_2 &= .z_{21} \underline{z_{22}} z_{23} z_{24} \ldots \\ r_3 &= .z_{31} z_{32} \underline{z_{33}} z_{34} \ldots \\ &\vdots \end{aligned} \tag{10}$$
>
> Now consider the number represented by taking the diagonal elements in this array, which are underlined, changing each diagonal element to be a digit different from the digit z_{ii} that is there initially (with proviso that the new digit is not either 0 or 9).

It is argued that since the new number differs from all old ones in Eq. (10) due to new diagonal elements, the cardinality of reals must be greater than natural numbers. However, Cantor's uncountability theorem is not consistent with the model of real line schematically shown in Fig. 4. In fact, it can be shown that real numbers on the interval (0, 1) can be counted by natural numbers and hence have the same cardinality $\aleph_0$. To show this, assuming any arbitrary small unit $\varepsilon = 10^{-100}$ to define zero measure of scale β as $0_\beta = 10^{-100} = 1_{\beta-1} = \infty_{\beta-2}$, one begins counting $\infty_\beta = 1/0_\beta = 10^{100}$ numbers each time adding $1_{\beta-1} = 10^{-100}$ to the previous number constructing the list in Eq. (11).

$$\begin{array}{lr} 0.0000 & 00000001 \\ 0.0000 & 00000002 \\ 0.2862239 & 2865598 \\ 0.70782246021 & 667310 \\ 0.999999999999 & 999999999 \end{array} \tag{11}$$

The list of numbers given in Eq. (11) covers *all possible numbers* within the interval (0, 1) given the zero measure $0_\beta = 10^{-100} = 1_{\beta-1} = \infty_{\beta-2}$ defining the unity of the lower scale $\beta-1$. Since all possible real numbers on (0, 1) in Eq. (10) must be in the list of Eq. (11), the *diagonal method* of Cantor cannot produce any new numbers by changing any digits of the list in Eq. (11) since all possible numbers are presented by *construction*. Therefore, one must conclude that the cardinality of $\mathbb{R}$ is the same as that of $\mathbb{N}$ namely $\aleph_0$ and Cantor *general continuum hypothesis* GCH is disproved as suspected by both Gödel and Poincaré [62]. In this regards we quote Gödel [58].

> I believe that adding up all that has been said one has good reason to suspect that the role of the continuum problem in set theory will be to lead to the discovery of new axioms which will make it possible to disprove Cantor's conjecture.

Even if higher accuracy of $\mathbb{R}$ numbers were demanded, the list in Eq. (11) could be enlarged without limit by starting from a much smaller unit for zero at the lower scale of $\beta-1$ for example by choosing a new unit as $0_{\beta-1} = 10^{-100^{100}} = 1_{\beta-2} = \infty_{\beta-3}$.

6 Concluding Remarks

Some implications of a scale-invariant model of Boltzmann statistical mechanics to physical foundation of mathematics namely Riemann hypothesis, Cantor uncountability theorem, continuum hypothesis, Goldbach conjecture, and Russel paradox were examined. Cantor continuum hypothesis was disproved by showing that cardinality of real numbers is the same as that of natural numbers namely $\aleph_0$.

In addition, quantum nature of space and time as well as *internal* spacetime versus *external* space and time were described. Because of its *hyperbolic geometry*, its discrete fabric, and its stochastic atomic motions, physical space is called *Lobachevsky-Poincar -Dirac-Space*. The results may help in addressing the warning signs emphasized by Nelson [67].

Acknowledgements This research was in part supported by NASA grant No. NAG3-1863.

References

1. L. de Broglie, Interference and corpuscular light. Nature **118**, 441–442 (1926); Sur la Possibilité de Relier les Phénomènes d'Interférence et de Diffraction à la Théorie des Quanta de Lumière. Acad. Sci. Paris **183**, 447–448 (1927); La Structure Atomique de la Matière et du Rayonnement et la Mécanique Ondulatoire. **184**, 273–274 (1927); Sur le Rôle des Ondes Continues en Mécanique Ondulatoire. **185**, 380–382 (1927)
2. L. de Broglie, *Non-linear wave mechanics: a causal interpretation* (Elsevier, New York, 1960)
3. L. de Broglie, The reinterpretation of wave mechanics. Found. Phys. **1**(5), 5–15 (1970)
4. E. Madelung, Quantentheorie in hydrodynamischer form. Z. Physik. **40**, 332–326 (1926)

5. E. Schrödinger, Über die Umkehrung der Naturgesetze. Sitzber Preuss Akad Wiss Phys-Math Kl **193**, 144–153 (1931)
6. R. Fürth, Über Einige Beziehungen zwischen klassischer Staristik und Quantenmechanik. Z. Phys. **81**, 143–162 (1933)
7. D. Bohm, A suggested interpretation of the quantum theory in terms of "Hidden Variables" I. Phys. Rev. **85**(2), 166–179 (1952)
8. T. Takabayasi, On the foundation of quantum mechanics associated with classical pictures. Prog. Theor. Phys. **8**(2), 143–182 (1952)
9. D. Bohm, J.P. Vigier, Model of the causal interpretation of quantum theory in terms of a fluid with irregular fluctuations. Phys. Rev. **96**(1), 208–217 (1954)
10. E. Nelson, Derivation of the schrödinger equation from newtonian mechanics. Phys. Rev. **150**(4), 1079–1085 (1966)
11. E. Nelson, *Quantum Fluctuations* (Princeton University Press, Princeton, New Jersey, 1985)
12. L. de la Peña, New foundation of stochastic theory of quantum mechanics. J. Math. Phys. **10**(9), 1620–1630 (1969)
13. L. de la Peña, A.M. Cetto, Does quantum mechanics accept a stochastic support? Found. Phys. **12**(10), 1017–1037 (1982)
14. A.O. Barut, Schrödinger's interpretation of ψ as a continuous charge distribution. Ann. Phys. **7**(4–5), 31–36 (1988)
15. A.O. Barut, A.J. Bracken, Zitterbewegung and the internal geometry of the electron. Phys. Rev. D **23**(10), 2454–2463 (1981)
16. J.P. Vigier, De Broglie waves on dirac Aether: a testable experimental assumption. Lett. Nuvo Cim. **29**(14), 467–475 (1980); C. Cufaro Petroni, J.P. Vigier, Dirac's Aether in relativistic quantum mechanics, found. Physics **13**(2), 253–286 (1983); J.P. Vigier, Derivation of inertia forces from the einstein-de broglie-bohm (E.d.B.B.) causal stochastic interpretation of quantum mechanics, Found. Phys. **25**(10), 1461–1494 (1995)
17. F.T. Arecchi, R.G. Harrison, *Instabilities and Chaos in Quantum Optics* (Springer, Berlin, 1987)
18. O. Reynolds, On the dynamical theory of incompressible viscous fluid and the determination of the criterion. Phil. Trans. R. Soc. A **186**(1), 123–164 (1895)
19. D. Enskog, *Kinetische Theorie der Vorgange in Massig Verdunnten Gasen* (Almqvist and Wiksells Boktryckeri-A.B., Uppsala, 1917). English translation: G. S. Brush, *Kinetic Theory* (Pergamon Press, New York, 1965), pp. 125–225
20. G.I. Taylor, Statistical theory of turbulence-parts I-IV. Proc. R. Soc. A **151**(873), 421–478 (1935)
21. T. Kármán, L. Howarth, On the statistical theory of isotropic turbulence. Proc. R. Soc. A **164**(917), 192–215 (1938)
22. H.P. Robertson, The invariant theory of isotropic turbulence. Proc. Camb. Phil. Soc. **36**, 209–223 (1940)
23. A N. Kolmogoroff, Local structure on turbulence in incompressible fluid. Acad. Sci. URSS **30**, 301–305 (1941); A refinement of previous hypothesis concerning the local structure of turbulence in a viscous incompressible fluid at high reynolds number. J. Fluid Mech. **13**, 82–85 (1962)
24. A.M. Obukhov, On the distribution of energy in the spectrum of turbulent flow. Bull. Acad. Sci. USSR **32**, 19–22 (1941); Some specific features of atmospheric turbulence. J. Fluid Mech. **13**, 77–81 (1962)
25. S. Chandrasekhar, Stochastic problems in physics and astronomy. Rev. Mod. Phys. **15**(1), 1–89 (1943)
26. S. Chandrasekhar, *Stochastic, Statistical, and Hydrodynamic Problems in Physics and Astronomy, Selected Papers*, vol. 3 (University of Chicago Press, Chicago, 1989), 199–206
27. W. Heisenberg, On the theory of statistical and isotropic turbulence, Proc. R. Soc. A **195**, 402–406 (1948); Zur Statistischen Theorie der Turbulenz. Z. Phys. **124**, 7–12, 628–657 (1948)
28. G.K. Batchelor, *The Theory of Homogeneous Turbulence* (Cambridge University Press, Cambridge, 1953)

29. L.D. Landau, E.M. Lifshitz, *Fluid Dynamics* (Pergamon Press, New York, 1959)
30. H. Tennekes, J.L. Lumley, *A First Course in Turbulence* (MIT Press, 1972)
31. S.H. Sohrab, On a scale-invariant model of statistical mechanics and the laws of thermodynamics. J. Energy Resour. Technol. **138**(3), 1–12 (2016)
32. S.H. Sohrab, Invariant forms of conservation equations and some examples of their exact solutions. J. Energy Resour. Technol. **136**, 1–9 (2014)
33. S.H. Sohrab, Solutions of modified equation of motion for laminar flow across (Within) Rigid (Liquid) sphere and cylinder and resolution of stokes paradox. AIP Conf. Proc. **1896**, 130004 (2017)
34. S.H. Sohrab, On a scale invariant model of statistical mechanics, kinetic theory of ideal gas, and riemann hypothesis. Int. J. Mod. Commun. Tech. Res. **3**(6), 7–37 (2015)
35. S.H. Sohrab, Scale Invariant Model of Statistical Mechanics and Quantum Nature of Space, Time, and dimension. Chaotic Model. Simul (CMSIM) **3**, 231–245 (2016)
36. R.S. de Groot, P. Mazur, *Nonequilibrium Thermodynamics* (North-Holland, 1962)
37. H. Schlichting, *Boundary-Layer Theory* (McGraw Hill, New York, 1968)
38. F.A. Williams, *Combustion Theory*, 2nd edn. (Addison Wesley, New York, 1985)
39. B.L. van der Waerden (ed.), Towards quantum mechanics. In: *Sources of Quantum Mechanics* (Dover, New York, 1967), pp. 1–59
40. M. Planck, On the law of the energy distribution in the normal spectrum. Ann. Phys. **4**, 553–558 (1901)
41. H.B.G. Casimir, On the attraction between two perfectly conducting plates. Proc. K. Ned. Akad. Wet. **51**, 793–795 (1948)
42. P.A.M. Dirac, Is there an Aether? Nature **168**, 906 (1951)
43. S.H. Sohrab, Invariant Model of Statistical Mechanics, Quantum Mechanics, and Physical Nature of Space and Time. In: *Proceedings of the International Conference on 8th CHAOS* (Henri Poincaré Institute, Paris, France, 2015), pp. 769–801, 26–29
44. E.T. Whittaker, *A history of the theories of aether and electricity*, vol. 2 (Tomash Publishers, New York, 1954)
45. L. Euler, Réflexions sur l' Espace et le Temps, Histoire de l'Academie Royale des Sciences et Belles Letters. **4**, 324–33 (1748)
46. A.A. Penzias, R.W. Wilson, A measurement of excess antenna temperature at 4080 Mc/s. Astrophys. J. **142**, 419–421 (1965)
47. H. Poincaré, *Sur la Dynamique de l'Electron*, vol. 21 (Rendiconti del Circolo matematico di Palermo, 1906), pp. 9–175
48. H. Minkowski, Space and time. In: *Theory of Relativity* (Dover, New York, 1952), p. 75
49. H. Poincaré, *Science and Hypothesis* (Dover, New York, 1952), p. 65
50. H.L. Montgomery, The pair correlation of zeroes of the zeta function, analytic number theory. In: *Proceedings of the Symposium on Pure Mathematics*, vol. 24 (American Mathematical Society, Providence, RI, 1973), pp. 181–193
51. A.M. Odlyzko, On the distribution of spacings between zeroes of the zeta function. Math. Comp. **48**, 273–308 (1987)
52. J. Derbyshire, *Prime Obsession* (Joseph Henry Press, Washington, D.C, 2003)
53. S. du Marcus, *The Music of Primes* (Harper Collins, New York, 2003)
54. M.L. Mehra, *Random Matrices*, 3rd edn. (Elsevier, Amsterdam, 2004)
55. M.V. Berry, Quantum chaology. Proc. R. Soc. London A **413**, 183–198 (1990)
56. A. Connes, Geometry from the spectral point of view. Lett. Math. Phys. **34**(3), 203–238 (1995); *Trace Formula in Noncommutative Geometry and Zeroes of the Riemann Zeta Function*, Preprint ESI 620 (Vienna, 1998), pp. 1–88
57. A. Connes, *On the Fine Structure of Spacetime, on Space and Time*, ed. S. Majid (Cambridge University Press, 2008), pp. 196–237
58. J.L. Casti, *Mathematical Mountaintops* (Oxford University Press, 2001)
59. H. Cramér, Prime numbers and probability. Skand. Mat. Kongr. **8**, 107–115 (1936)
60. A. Granville, Harald Cramér and the distribution of prime numbers. Scand. Actuarial J. **1**, 1995 (1995)

61. W. Heisenberg, *The Physical Principles of Quantum Theory* (Dover, New York, 1949)
62. J.S. Bell, *The Continuous and the Infinitesimal in Mathematics and Philosophy* (Polimetrica, Milan, Italy, 2006)
63. A. Robinson, *Non-Standard Analysis* (North-Holland Publishing Company, Amsterdam, 1966)
64. E. Nelson, Internal set theory: a new approach to nonstandard analysis. Bull. Am. Math. Soc. **83**(6), 1165–1198 (1977)
65. S.H. Sohrab, Implications of a scale invariant model of statistical mechanics to nonstandard analysis and the wave equation. WSEAS Trans. Math. **7**(3), 95–103 (2008)
66. G. Cantor, *Contributions to the Founding of the Theory of Transfinite Numbers* (Dover, New York, 1955)
67. E. Nelson, Warning signs of a possible collapse of contemporary mathematics. In Infinity, New Research Frontiers, eds. M. Heller, W. Hugh Woodin (Cambridge University Press, 2011), pp. 76–85

Deterministic Irreversibility Mechanism and Basic Element of Matter

V. M. Somsikov

Abstract The role of the deterministic mechanism of irreversibility (DMI) in the organisation of matter is studied. The nature of DMI is explained. The connection of the DMI with the nature of entropy, chaos and the breaking of the space-time symmetry is discussed. How the evolution of the system can be described in the dual phase space is shown. The fact that the open nonequilibrium dynamic system is a basic element of matter is shown. The nature of supersymmetry and the principle of uncertainty, which can be entered into classical mechanics, were analysed. How expansion of classical mechanics, which follows from the accounting of structure of a body in its dynamics, leads to solutions of the existing problems of physics and to creation of the uniform theory is discussed.

Keywords Classical mechanics · Irreversibility · Entropy · Chaos · Phase space · Evolution

1 Introduction

There are important problems on the way to creation the physical picture of the world: how mass arise; what are elementary particles of the matter; how complex systems from simple elements arise; whether there are such a formalisms that allows building a theory of everything based on dynamics laws of its elements. One of the great obstacles on the way to solution of these problems was a fact that fundamental physics can describe the existing world, but do not describe the processes of its evolution, that is, the emergence, development and transformation of the nature systems [1–3]. This is explained by that that evolution in nature is irreversible but the fundamental physics laws are reversible [4–6].

In the process of solution of the irreversibility problem, at first its probabilistic mechanism was found [5, 6]. This mechanism of irreversibility is based on the

V. M. Somsikov (✉)
Institute of Ionosphere, Almaty 050020, Kazakhstan
e-mail: vmsoms@rambler.ru

Al-Farabi, Kazakh National University, Almaty 050040, Kazakhstan

C. H. Skiadas and Y. Dimotikalis (eds.), *12th Chaotic Modeling and Simulation International Conference*, Springer Proceedings in Complexity,
https://doi.org/10.1007/978-3-030-39515-5_20

hypothesis of the existence of a random external influence on the Hamiltonian system. According to the Poincaré theorem, Hamiltonian systems are reversible. However, they are exponentially unstable and for an arbitrarily small random disturbance, they can deviates from the deterministic trajectory on any predetermined distance within the available phase space. It means irreversibility. But since this irreversibility mechanism is probabilistic, it is impossible to answer on the next questions: how in the nature the "order" can arise from "chaos"; how to build of the world's picture, if the theory of evolution is based on the probabilistic mechanism of irreversibility etc.

The existence of a probabilistic mechanism of irreversibility does not exclude existence of the ***deterministic mechanism of irreversibility*** (DMI). Indeed, DMI was relatively recently found [7, 8]. DMI not only eliminated the problems associated with a probabilistic explanation of irreversibility, but also opened up opportunities to build a deterministic evolutionary picture of the world within the framework of the fundamental laws of physics [9, 10].

Here, we will explain the nature of DMI. We will show that in according with the DMI nature, the basic element of matter should be an ***open nonequilibrium dynamic system*** (ONDS). We will show also how ONDS can be described in the dual orthogonal phase space with taking into account of the DMI. We will discuss the nature of supersymmetry and the principle of uncertainty, which can be entered into classical mechanics.

2 The Nature of DMI

The main obstacles on the way to the DMI explanation were as follows. As is known, the Newton's motion equation for a material point (MP) is reversible. The motion equations for a system from potentially interacting MPs, which are derived from the Lagrange equation or by summation of the motion equations for MPs, are also reversible [11, 12], although in nature we observe the irreversibility of systems. The reason for the discrepancy due to the fact that the Lagrange equations are derived under the condition of the holonomic restrictions [6]. This condition is inapplicable for systems with non-holonomic restriction when the work of external forces not only go on the body's motion, but also on increase its internal energy, for example, due to the work of friction forces.

It has been shown that for taking into account the change of the internal energy it was necessary to present energy of a body in according with the principle of duality of symmetry (PDS) in the form of the sum of the motion energy and internal energy. For this purpose it is necessary to take body model in the form of the equilibrium structured particle (SP), which is a system consisting from sufficiently large number of potentially interacting MPs. It was found, that the motion equation of SP is asymmetric with respect to time and is applicable for systems with non-holonomic restrictions [7].

Let us explain shortly the key ideas and concepts that were used in the substantiation of the DMI on the example a body sliding along inclined rough surface under the action of gravity. At the initial moment, its energy is equal to potential energy. During motion the part of this energy goes on increase in kinetic energy of the body. This energy determined by the velocity of the body's ***centre of mass*** (CM). Another part of the energy goes on the heat of the body as a result of the conversion of the energy of motion into internal energy due to the work of the friction force. Therefore, the motion energy at each subsequent point of the position of the body is less than the consumed potential energy on the value of change of the internal energy (the absorption of the energy of body motion by the external environment is not taken into account, since this does not fundamentally change anything in our discussion).

The body motion in a classical mechanics is described by the empirical equation, in which the force of friction is proportional to the velocity [11, 12]:

$$M\dot{V}_0 = -F_0 - \mu V, \tag{1}$$

Here M-is a mass of a body, V_0-is a velocity of the CM of a body, F_0-is a force acted on CM (here and after the forces are expressed through the gradient of the potential energy), μ-is empirical friction coefficient.

The Eq. (1) empirically takes into account the transformation of the motion energy into internal energy, but does not describe the process of such a transformation itself. However, the motion equation for SP makes such a description possible [6, 7]. Let us show why it so.

For describing the body's motion with taking into account of the friction, we will use the molecular-kinetic theory. In this case, the body can be represented as SP. Without loss of generality, we will assume that the masses of all MPs are equal to one. Let us use the fact that any scalar sum of square functions of a vector can be presented so [6]:

$$\sum_{i=1}^{N} r_i^2 = NR_0^2 + \sum_{\substack{i=1 \\ i \neq j}}^{N} (r_{ij})^2/N \tag{2}$$

where the vector r_i is a position of MP in the laboratory system of coordinate O; $r_{ij} = r_i - r_j; i, j = 1, 2, 3 \ldots N$, i,j-is a number of MP, $i \neq j$, $v_i - v_j = v_{ij} = \dot{r}_{ij}$-is a relative velocity of two MP; $R_0 = \left(\sum_{i=1}^{N} r_i\right)/N$-is velocity of the CM of SP.

Let us: $r_i = R_0 + \tilde{r}_i$, where $\tilde{r}_i$-is a coordinate of MP in relative of the CM. Therefore, we have: $v_i = V_0 + \tilde{v}_i$ and the next equality has a place [6]:

$$\sum_{i=1}^{N} r_i^2 = NR_0^2 + \sum_{\substack{i=1 \\ i \neq j}}^{N} (\tilde{r}_i)^2 \tag{3}$$

The motion of everyone MP is a mix of two motions: the motion MP with the CM and motion MP in relative to the CM. The coordinate system in which the coordinate and velocity vectors are represented as the sum of the corresponding motion vectors of the CM of the system and motion MP in relative to the CM is called as the dual coordinate of the system [6]. Only in dual coordinate of the system, the work of external forces can be divided on two works: the work on the change of the CM velocity and the work, on change of MPs velocities in relative to the CM. The variables that determine the change of MPs velocities in relative to the CM are micro-variables. The variables that determine the work on change of the CM velocity are macro-variables. The micro—and macro-variables are independent and belong to different symmetry groups [7]. Thus, we have two invariants corresponding to the two symmetry groups that determine the motion of bodies. The body's motion energy is connected with the space symmetry group. The body's internal energy is connected with the body's symmetry groups. The statement that the motion of the system determined by the two type of symmetry: the internal symmetry of the system and the space symmetry called as the principle of duality of the symmetry (PDS) [10].

The energy of the system in micro-and macro-variables has the form [7–10]:

$$E_N = E_N^{tr} + E_N^{ins} = const \tag{4}$$

where $E_N^{ins} = T_N^{ins} + U_N^{ins}$ is SP internal energy; $T_N^{ins} = \sum_{i=1}^N m\tilde{v}_i^2/2$ is a kinetics part of SP internal energy; $U_N^{ins}(r_{ij}) = \sum_{i=1}^{N-1}\sum_{j=i+1}^N U_{ij}(r_{ij})$ is a potential part of the internal energy due to MP interaction; $E_N^{tr} = T_N^{tr} + U_N^{tr}$ is the SP motion energy; T_N^{tr} is a kinetic part of the motion energy, U_N^{tr}-is a potential part of the motion energy in the external forces.

The energy's conservation law for SP is: the sum of the motion energy and the internal energy of the system is conserved along its trajectory.

The SP's motion equation was obtained from the energy's expression (4) [13]:

$$M_N \dot{V}_N = -F_N^0 - \mu V_N, \tag{5}$$

where $F_N^0 = \sum_{i=1}^N F_i^0$; F_i^0-is external force acted on i-th MP; F_{ij} is interacted force i and j MP; $F_{ij}^0 = F_i^0 - F_j^0$; $\dot{E}_N^{\text{int}} = \sum_{i=1}^{N-1}\sum_{j=i+1}^N v_{ij}(m\dot{v}_{ij} + F_{ij}^0 + NF_{ij})$; $\mu = \dot{E}_N^{\text{int}}/(V_N^{max})^2$.

Equation (5) was obtained in dual system of coordinate by differentiation on time of the energy under the condition of constancy of it. But the classical motion equation for the system of MP was derived by summation of the motion equations for each MP, therefore the change of internal energy was not taken into account here, since the sum of the changes in the MP momentum in relative to the CM is zero [11, 12].

The Eq. (5) is strongly different from empirical Eq. (1) due to the friction coefficient, which has acquired an analytical form in Eq. (5). The first term in the right hand side defines the external force acted on the CM. The second term is nonlinear bisymmetrical term. This term was called evolutionary nonlinearity because it is responsible for irreversibility and origin of new structures of a substance [19]. It depends on both micro—and macro-variables that belong to different groups of symmetry. Thus, the Eq. (5) takes into account that the work of external forces divided into the work of moving the system, and the work changing its internal energy.

As it following from the Eq. (5), the acceleration of the system is zero when the system's velocity is equal to V_N^{max}. In this case all work of external forces go into internal energy and we have Aristotle equation: $V_N^{max} F_N^0 = -\dot{E}_N^{\text{int}}$. So, knowing the external force acting on the body and its steady speed, we determine the energy absorbed by the body.

In according with Aristotle equation, the system has a constant speed if the friction force is equal to the external force. But in according with the principle of Galileo, the body maintains a state of rest or moves with a constant speed, if no force acts on it. It would seem that there is a contradiction. However, it is not so. The Galilean principle is valid for Hamiltonian systems when friction is equal to zero and reversibility has a place. However, in the general case of non-holonomic systems, we have: body maintains a state of rest or rectilinear motion with the constant velocity if the external force is less or equal to friction force respectively. If $\mu = 0$, then we have Newton's equation: $M_N \dot{V}_N = -F_N^0$.

The occurrence of attractor is possible only for dissipative systems [14]. However, according to Eq. (5) the dissipation is possible only for systems. **Therefore, the matter is diviseble infinitely. That is, all natural objects are systems** [13].

Let us to summarize these results. MPs do not have structure. Therefore, its total energy is the motion energy consisting from of the potential and kinetic components. These types of energy can be transformed into each other only and nothing else. Hence, the MP's motion energy is constant and the time is reversible. The canonical motion equations for a system of MP are following from the Lagrange or Hamilton equations, applicable only for systems with holonomic constraints. These constraints are equivalent to the fact that external forces are potential and the system's motion energy is a constant. Therefore, these motion equations are reversible.

The expansion of the classical mechanics to dissipative mechanics arises by replacement of the body's model from MP on the structural body's model in the form of a SP and because of use of PDS.

According to the Eq. (5), trajectory of the body depends on change of internal energy. It means that the trajectory, defined on the basis of canonical formalism of classical mechanics, has uncertainty. It can be written as: $\Delta E_N^{tr} \Delta t > 0$. Because any matter has a structure, this uncertainty has to take place always. It is similar to Heisenberg's uncertainty principle in quantum mechanics although have another explanation.

3 Evolutionary Non-linearity and Symmetry Breaking

In 1937, Landau proposed explanation for spontaneous symmetry breaking during phase transitions [15, 16]. He introduced the so-called scalar order parameter for a quantitative description of spontaneous symmetry breaking and empirically determined the type of free energy decomposition in the phase transition region. By using order parameter, he writes of the thermodynamic potential, $F(\varphi)$ so [15, 16]:

$$F(\varphi) = F_0 + V(a_2\varphi^2/2 + a_4\varphi^4/4 - h\varphi) \tag{6}$$

where φ-is a scalar order parameter, a_2, a_4, h—are empirical coefficients.

The Eq. (6), according to its graphic form, was called later as the "Mexican hat." It turned out later, the similar potential describes not only the symmetry breaking during phase transitions of the second kind. A similar type of decomposition of thermodynamic potentials or scalar functions that determine the violation of space-time symmetry exists when describing the violation of supersymmetry [17, 18].

The invariance of the MP motion energy corresponds to the symmetry of time. For the MP of the motion equation the motion energy is conserved. However, according to Eq. (5), the SP motion energy in a non-homogeneous field of external forces is not saved. This means a time symmetry breaking, which has a place due to transformation of the body's motion energy into the internal energy. Nonlinearity, responsible for the breaking of space-time symmetry, was called evolutionary nonlinearity [9, 19]. These nonlinearities are absent in the framework of the canonical formalism of the classical mechanics. The physical essence of evolutionary nonlinearity in that, that the direct flow of the motion energy into the internal energy is described by term of the second order of smallness but the reverse flows of the motion energy is described by a fourth-order term of smallness [19]. So, for evolutionary nonlinearity we can write:

$$\Phi(\chi) \approx H_0 + H(b_2\chi^2 - b_4\chi^4) \tag{7}$$

where χ—is a smallness parameter, which characterized effectiveness exchange between motion energy and internal energy; b_2, b_4-are coefficients.

We can assume, basing on the Noether's theorem, that the similarity of Eqs. (6) and (7) indicates that the violation of any type of symmetry is connected with the evolutionary nonlinearity, which describe transformations of one type of energy to another. These two types of energy, corresponding for two types of the symmetry groups. Thus we can see that violations of symmetry is nonlinear transformation one type of energy into another. Universality of function of breaking of symmetry in various physical processes allows to assume that the nature of breaking of symmetry in all cases is defined by evolutionary nonlinearity [19].

4 On the Nature of Chaos and Order

As it known, in nature there are two types of chaos. The first type is chaos, which has a purely probabilistic nature. The state of a system with such chaos is unpredictable and cannot be defined by deterministic equations. Its typical example is white noise

[14]. Probability laws describe this chaos. The dimension of such chaos is infinite. We will call it stochastic chaos, random or probabilistic chaos.

The second type is deterministic chaos, since it characterizes the motion of deterministic systems, defined by strict equations of dynamics. It was discovered much later than stochastic chaos [9, 20]. As it turned out, even the oscillator dynamics at certain points in the phase space has a solution almost indistinguishable from stochastic chaos. But unlike stochastic chaos, deterministic chaos has a finite dimension [14].

According to the Poincaré theorem, the dynamical chaos is reversible. However, in nature any chaos is irreversible. This discrepancy for a long time was explained by the probabilistic fluctuations of the boundary conditions. That is, the explanation of irreversibility was based on its probabilistic mechanism. However, the existence of DMI radically changes this explanation of irreversible chaos. Indeed, DMI is deterministic and does not require the presence of random external influences on the system. This means that the motion of the system to equilibrium is described by deterministic equations. Leaving aside the extensive literature on the discovery of deterministic chaos, problems in the development of the mathematical apparatus for its description [14], we briefly explain the nature of deterministic chaos, which follows from the DMI.

The body's dynamics is determined by its symmetry and the symmetry of space. It follows that in order to determine the dynamics of a body, its energy must be represented as the sum of the motion energy and internal energy. The key property of the internal energy is the equality to zero of the sum of the momentums of its elements. The implementation of this condition occurs through of the chaotic motion of the system's elements. DMI is associated with an increase of the chaotic motion energy of the system's elements due to its motion energy and impossibility to returning of the internal energy into motion energy.

In according with the condition of the infinite divisibility of matter and the mechanics of SP in wide limits of generality all bodies are ONDS [13]. Let us to consider a case when the external influences on the ONDS are absent. In this case, ONDS is possible to consider as the closed nonequilibrium system, which consists from a set of SP, which in relative motion to each other. The momentum vectors for all SP are isotropic and therefore the sum of the momentum of SP is equal to zero. The equilibration of NS is associated with the aspiration of the relative speeds of SP to zero. Hence, the concepts of absolute chaos or equilibrium correspond to the equality to zero of the energies of the relative motions of such subsystems. That is, the ideal equilibrium or absolute chaos corresponds to the case when the motion energy for any sufficiently large subsystem, which is chosen from system in arbitrary way, is equal to zero. The process of establishing equilibrium is described by the Eq. (5).

Let us now consider the case when external influence is exists. Only in this case, the ONDS steady state as a set of attractors is possible. Such a set of attractors can be considered as a hierarchical ladder of a set ONDS, when simpler ONDSs, are the constituent elements of the upper step of the hierarchical ladder ONDS. The steady state of ONDS is ensured by the equality of incoming and leaving flows of energy, matter, and entropy for each step of the hierarchical ladder [13]. Thus, in nature, nonequilibrium, or order is correspond to ONDS hierarchical ladder. New attractors occur through chaos with the destruction of the old attractors as a result of changes in external conditions or the evolution of the system. Such a process is observed, for example, during phase transitions, when the state of a ONDS with one type of symmetry changes to its state with other types of symmetry.

It should be noted that absolute or perfect chaos or perfect order does not occur in nature, since all natural systems are ONDS. Absolute chaos is an approximation when dissipative processes can be neglected.

Thus, the key conclusion that follows from the existence of DMI is that that the matter is infinitely divisible [10]. From this fact and from the fact of the existence of DMI is follows that all natural systems should be hierarchy ONDS [14].

The unity of the laws of evolution and the forms of matter implies the necessity of the interrelation of the field and corpuscular representation of matter. These two representations must satisfy the requirement that matter is a hierarchical system of ONDS. Then we should take an oscillator as a base ONDS. From the condition of infinite divisibility of matter, it follows that it must have a field shape, for example, be a pulsating vortex of the electromagnetic field, which has a characteristic scale, frequency and momentum. All these properties are satisfied by a photon [2, 23].

5 Dual Phase Space

A characteristic feature of the hierarchical ladder of matter is that the preceding step of the ONDS serves as a structural element of the subsequent step. The existence of a connection between the hierarchical steps of an ONDS follows from the deterministic relationship between the laws of dynamics of MP and SP. Therefore, to study the processes of establishing order, all systems must be considered as ONDS, since all evolutionary processes in systems are associated with changes in external constraints. For example, the phase transitions can occur when the ambient temperature changes. In accordance with it the next question arises: what order of the matter may arise under certain changes of the external restrictions. Obviously, for the answer to this question, we need to know which stable states exist in the accessible parameter space of external constraints for the systems. Moreover, it is necessary to know the possible paths for the system from one to another attractor, that is, what is the scenario of the space-time changes of external constraints, so that the system arrives at a given attractor. The study of these questions can be executed based on methods of phase portraits. For systems of MP, it is convenient to take the dual phase space [9, 21]. In contrast to the usual phase space, this space allows us to take into account the PDS. Let us explain the essence of the dual phase space.

In the simplest cases, but taking into account all the main evolutionary laws, ONDS can be presented as the nonequilibrium system consisting from a number of SP. The motion energy of everyone MP is a mix of two energies: the motion energy of MP with velocity of the CM and motion energy of MP in relative to the CM. Consequently, the energy of the ONDS in the general case should be represented as follows [9]: $E^{ONDS} = \sum_{i=1}^{R} (E_i^{tr} + E_i^{ins})$. The stationary state of ONDS can have a place when the total energy of the system is conserved: $E^{ONDSs} = const$. Thus the phase space, which allows analyzing the processes of the evolution of the ONDS, when ONDS is a non-equilibrium system from a set of SP, must be represented in the form of two subspaces corresponding to the SPs internal energy and the motion energy. In the case when the external influence on ONDS can be neglected, the following condition has a place:

$$\left(\sum_{i=1}^{R} E_i^{tr}/E_i^{ins}\right)_{t\to\infty} \to 0, \qquad (8)$$

where $i = 1, 2, 3 \ldots R$, R-is a number of SP into ONDS, E_i^{tr}, E_i^{ins}—are energy of the relative motion and the internal energy for i-th SP. The value $\left(\sum_{i=1}^{R} E_i^{tr}/E_i^{ins}\right) = \Theta$ is determined the degree of non-equilibrium of ONDS.

Since ONDS is a set of SPs moving relative to each other, then the state of ONDS can be associated with a point in the phase space $(6R - 1)$ of dimensions, where R is the number of SP included in the ONDS. That is, three coordinates and three components of momentum define the position of each SP. This space is determined in macro-variables, which determine the coordinates and momentum of the SP. It was called the *S—space* [1, 11]. In this space, a one-to-one correspondence between the state and the point of the phase space is exists, when dissipation can be neglected.

In the presence of dissipation, the internal energy of the ONDS will increase due to the energy of relative motion of the SP. That is, each point of *S*—space will already correspond to a state with different values of internal energy. This ambiguity can be eliminated if the *S*-space is supplemented by the *D*-space of micro-variables, which determine the internal energy of the SP. Such dual space is called as *SD*—space. The main difference between SD-space and ordinary phase space is that that the SD-space was built by dividing the energy of each MP into two parts. One part corresponds to the contribution of MP's motion energy to the energy of relative motion of the SP, and the other part corresponds to the contribution of MP's motion energy to the internal energy of the SP.

The ONDS state is determined by two points in the SD-space. One point is belonging to *S*, and another belonging to *D* subspaces. These two points are defined by two $\boldsymbol{S}$ and $\boldsymbol{D}$ vectors. The modulus of the $\boldsymbol{S}$—vector is proportional to the sum of the energies of the relative motion of the SP. The modulus of the $\boldsymbol{D}$—vector is proportional to the sum of the internal energies of the SPs.

When ONDS equilibrates, the **S**—vector module tends go to zero, because the SPs energies of the relative motion tends go to zero, and the **D**—vector module increases.

If ONDS is conservative, then the condition $\boldsymbol{S}^2 + \boldsymbol{D}^2 = const$ has a place. This condition is following from the law of conservation of energy and can be written as: $ZZ^* = const$, where $Z = S + iD$, i-is imaginary unit. Since, for any transformations of **S** and **D** vectors, the sum of the SP motion energy and their internal energies is preserved, the state of the ONDS corresponds to the point of the complex plane, defined by: $Z = S + iD$. We can written for vectors **S**, **D**: $\varphi = arctg(D/S)$. In equilibrium we have: $\mathbf{S} = 0$, and $\varphi = \pi/2$.

The energy of SP is represented by the sum of the motion energy and the internal energy. Then for this phase space can be written:

$$\int_{\Delta_0^S} dq_0^S dp_0^S + \int_{\Delta_0^D} dq_0^D dp_0^D = \int_{\Delta_t^S} dq_t^S dp_t^S + \int_{\Delta_t^D} dq_t^D dp_t^D \tag{9}$$

Here q_0^S, p_0^S—are coordinates and momentums of the SP forming the phase volume element Δ_0^S at the initial moment of time; q_0^D, p_0^D—are coordinates and momentums of MP relative to the CM of the SP, forming the phase volume element Δ_0^D at the initial moment of time; q_t^S, p_t^S—are coordinates and momentums of the SP forming the phase volume element Δ_t^S at the moment of time t; q_t^D, p_t^D-are coordinates and momentums of MP relative to the CM of the SP, forming the phase volume element Δ_t^D at the moment of $\boldsymbol{t}$.

The evolution of the phase volume can be written so: $\Delta_0^S + \Delta_0^D = \Delta_t^S + \Delta_t^D$. It is equivalent to the law of energy conservation. Due to the transformation of the energy of the relative motions of the SP into their internal energy, we will have:

$$(\Delta_t^S/\Delta_t^D)_{t\to\infty} \to 0 \tag{10}$$

Rate of change Δ_t^S is determined by D-entropy—S^d [9, 10, 22]. D-entropy is the ratio of the change of the internal energy of the system to the full internal energy. In this case we can write: $\frac{d}{dt}\sum_{i=1}^{R} E_i^{tr} = T\frac{d}{dt}S^d$, where T-is temperature of SP, which determined by the component of the internal average kinetic energy. If we have non-dissipative system, then: $dS^d/dt = 0$, and S—space coincides with the usual phase space.

In general case, the evolution of ONDS determined by external constraints. In this case, also, its evolution can be determined in SD—space. But in general the phase trajectory will be determined not only by dissipative processes, as in the case of a closed ONDS, but also by the nature of external constraints, which should be set by the functional: $\Phi(\alpha, \beta, \ldots \zeta, t)$. In general case the $\Phi(\alpha, \beta, \ldots \zeta, t)$ will depend both from the system's internal properties and from the characteristics of environment.

The main advantage of the SD-phase space is that it allows tracing the nature of the change in the internal energy, which determines the system's motion to the attractor in connection with the change of the external constraints.

6 Discussion

The natural bodies have a structure. Therefore, their evolution is determined by the PDS. In according with PDS, the body's energy is a sum of the motion energy and the internal energy. In the non-homogeneous field of external forces, the motion energy can transfer into the internal energy. This transformation is determined by the evolutionary nonlinearity. That is, taking into account the body's structure and the non-holonomic restrictions in their dynamics leads to taking into account the violation of time symmetry. DMI is connected with the transition of the motion energy into the chaotic motions of the elements.

The key conclusion that follows from the existence of DMI is that that the matter is infinitely divisible [10]. From this fact and from the fact of the existence of DMI is follows, that the basic element of matter should be an open non-equilibrium dynamic system (ONDS). Therefore, all natural systems should be a hierarchy of the ONDS [13].

For graphic display the body's motion in accordance with the PDS, the phase space is divided into two subspaces: the S-space of macro-variables, and the D-space of micro-variables. S—space reflects the body's motion, and D—space reflects the motion of body's elements relative to the CM.

As a result of the expansion of mechanics arising because of taking into account of the body's structure, new approaches to the solution of many problems of physics appear. For example, generalisation of potential of classical mechanics which defines symmetry breaking of time because of evolutionary nonlinearity is possible. The Noether's theorem about existence of invariants for dynamic groups of symmetry confirms possibility of such generalisation. It is possible to assume that violation of any symmetry is connected with the evolutionary nonlinearity, meshing of variable from various groups of symmetry. It speaks about a possibility of the analytical description of spontaneous violation of symmetry and about necessity of creation of a formalism for the description of various types of bifurcation.

Other task is connected with emergence of the principle of uncertainty in mechanics. This principle is connected with uncertainty of a trajectory of a body because of neglect a role of absorption of the motion energy. The quantum mechanics was created basing on the canonic formalism of classical mechanics. Therefore, it is quite natural that Heisenberg's uncertainty can be related to the uncertainty due to the structures of the quantum particles.

And of course, the important are the questions: to what physics conclusions will lead the accounting of infinite divisibility of matter, and also the existence of the principles of connection between its hierarchical levels, which allowing to build the laws of the upper hierarchical levels of matter, if to know laws of the lower hierarchical levels.

Whether tells us the infinite divisibility of matter about the field nature of matter whether it will return ideas of ether in any new form, it is not known yet.

The accounting of the body's structure in the form of set of MP introduces evolution into physics and establishes of unity of physics within its fundamental laws. Is this accounting of infinite divisibility of matter lead to the new stage of physics development? Anyway, it is obvious that the solution of these and other tasks will exert deep impact on development of physics of evolution.

References

1. P.W. Anderson, More is different. Science, New Series **177**(4047) (1972)
2. G.W. Hooft, Light is heavy. arXiv:1508.06478v1 [physics.hist-ph] 26 Aug 2015
3. G.W't. Hooft, Free will in the theory of everything. arXiv:1709.02874v1[quant-ph]8
4. H.G. Callaway, Fundamental physics, partial models and time's arrow (2016). https://www.researchgate.net/publication/296327588
5. I. Prigogine, *From Being to Becoming.* (Nauka, Moscow, 1980)
6. G.M. Zaslavsky, *Stochasticity of Dynamical Systems.* (Nauka, Moscow, 1984)
7. V.M. Somsikov, Irreversibility and physics of evolution, in *Chaotic Modeling and Simulation (CMSIM)*, vol. 1 (2018)
8. V.M. Somsikov, Deterministic mechanism of irreversibility. JAP **14**(3) (2018)
9. V.M. Somsikov, *To the basics of the physics of evolution.* (Science, Almaty, 2016)
10. V.M. Somsikov, Problems of evolution of open systems. PEOS **9**(2) (2007)
11. C. Lanczos, *The Variational Principles of Mechanics.* (Peace, Moscow, 1962)
12. H. Goldstein, *Classical mechanics.* (Science, Moscow, 1975)
13. V. Somsikov, Deterministic irreversibility and the matter structure. JAP, **16**(1) (2019)
14. A.Y. Loskutov, A.S. Mikhailov, *Introduction to synergetics.* (Science, Moscow, 1990)
15. L.D. Landau, E.M. Lifshitz, *Statistical Physics.* (Nauka, Moscow, 1976)
16. L.D. Landau, To the theory of phase transitions. I. ЖЭТФ **7**, 19 (1937)
17. J. Bernstein, A question of mass. AJP **79**(1) (2011)
18. F. Wilczek, The origin of mass. Modern Phys. Let. A **21**(9) (2006)
19. V.M. Somsikov, Non-linearity of dynamics of the non-equilibrium systems. World J. Mech. **2**(7) (2017)
20. B.V. Chirikov, Resonance processes in magnetic traps. Atom. Energy **6**(6) (1959)
21. V.M. Somsikov, A.B. Andreev, A.I. Mokhnatkin, V.I. Kapytin. Dual phase space of a nonequilibrium system. PEOS **20**(1) (2018)
22. V.M. Somsikov, The dynamical entropy. Int. J. Sci. **4**(36) (2015)
23. P.S. Baranov, Photon structure (2006). x4u.lebedev.ru/personalia/baranov/Baranov_Structura_fotona.html

Global Indeterminacy and Invariant Manifolds Near Homoclinic Orbit to a Real Saddle in a Resource Optimal System

Beatrice Venturi

Abstract The paper investigates the dynamic properties of a resource optimal system with externality derived by [1]. To this end, we determine the whole set of conditions which lead to global indeterminacy and the existence of a homoclinic orbit that converges in both forward and backward time to a real saddle equilibrium point. The dynamics near this homoclinic orbit have been investigated. If the exponent of the externality is varied the homoclinic is broken and it bifurcates in a stable periodic orbit. The economic implications are discussed in the conclusions.

Keywords Externality · Homoclinic orbits to real saddle · Global indeterminacy

1 Introduction

In this paper, we find and analyze the homoclinic bifurcation (or connecting) to a saddle equilibrium point with purely real eigenvalues in a resource optimal system. The model concerns an endogenous growth two-sector model with externalities (see [1–3]). Following standard change variables, related with the growth rate of the fundamental variables, we may put the system in a reduced form: a first-order nonlinear three-dimensional differential system, (see [2–4]). We use steady-state analysis to obtain information related to the existence of hyperbolic equilibrium point. We show that the model posses multiple equilibria.

By using the method of undetermined coefficients (see [5]), we prove the existence of a homoclinic orbit that converges both forward and backward to the equilibrium point whose linearisation matrix admits two negative and one positive real eigenvalues The stable and unstable manifolds are locally governed by real eigenvalues. As well known, the mathematical concept of homoclinic bifurcations are very important from a dynamical point of view. Infact, such phenomena cause global re-arrangements in phase space, including changes to basins of attractions and

B. Venturi (✉)
Department of Economics and Business, University of Cagliari, Cagliari, Italy
e-mail: venturi@unica.it

C. H. Skiadas and Y. Dimotikalis (eds.), *12th Chaotic Modeling and Simulation International Conference*, Springer Proceedings in Complexity,
https://doi.org/10.1007/978-3-030-39515-5_21

generation of chaotic dynamics (see [6]). We introduce the mechanism that gives rise to global invariant manifolds near homoclinic orbit of a real saddle equilibrium point (with purely real eigenvalues) (see Aguera). The dynamics near these homoclinic orbits have been discovered and investigated. In our formulations, we consider a homoclinic bifurcation of dimension one, that bifurcates in a cycle. It means that only a parameter is allowed to varied. In our memory, in the economic contest this phenomenon has been studied only by [7] for the Lucas model.

Our goal is to show that such a cycle gives rise to global indeterminacy in a parameter set commonly investigated by means of the instruments of the local analysis (see [8]).

The paper develops as follows. In the second Section, we analyse the optimal control model and we introduce the equivalent three-dimensional continuous time abstract stationary system with cubic nonlinearity. The third Section is devoted to the steady-state analysis of the model in reduced form. In section four we apply the procedure developed by [5], and we show that a homoclinic loop emerges as a solution trajectory. In the last Section, we consider a homoclinic bifurcation of dimension one, that bifurcates in a stable periodic orbit. The economic implications are discussed in the conclusions.

2 The Model

We consider a natural resource optimal control model. It deals with the maximisation of a standard utility function with constraints

$$\underset{c(t),\, u(t)}{Max} \int_0^\infty \frac{c^{1-\sigma}-1}{1-\sigma} e^{-\rho t} dt \qquad (\mathcal{P})$$

$$\begin{aligned} \textit{subject to}\ :& \\ \dot{k} &= Ak^{\beta}\,(nr)^{1-\beta}\, r_a^{\gamma} - c \\ \dot{r} &= \delta r\left[1-(nr)^2\right] \\ k(0) &= k_0 \\ r(0) &= r_0 \end{aligned}$$

where: c is per-capita consumption; $\rho \in \mathbb{R}_{++}$ is a positive discount factor; σ is the inverse of the intertemporal elasticity of substitution; k is the physical capital; $A \in \mathbb{R}_{++}$ is a measure of the stock of existing technology (without loss of generality, we put $A = 1$); r is a disposal natural resource; n is a fraction of r, $n \in [0, 1]$; $\delta \in \mathbb{R}_{++}$ is the growth rate of the variable r; r_a represents an external effect due to the presence of a common pool natural resource; $\gamma \in \mathbb{R}$ is a parameter, the exponent of r_a. The set of parameters $\omega \equiv (\beta, \delta, \gamma, \rho, \sigma)$ lives inside an significant economic parameter set $\Omega \equiv \{(0, 1) \times \mathbb{R}_{++} \times \mathbb{R} \times \mathbb{R}_{++} \times \mathbb{R}_{++} \setminus \{1\}\}$.

The state variables of optimal control model are k and r. The control variables are c and n.

The solution candidates for the problem $\mathcal{P}$ can be obtained by means of the Pontryagin Maximum Principle with the usual transversality condition:

$$\lim_{t\to\infty}\left[e^{-\rho t}\left(\lambda_1 k+\lambda_2 r\right)\right]=0. \tag{1}$$

We consider only the competitive equilibrium solution (As well known, it follows from the presence of the externality that the competitive solution differs from the planner's solution[1]) in equilibrium $r = r_a$ for all t.

In standard way, we introduce the discounted Hamiltonian:

$$\mathcal{H}=\frac{c^{1-\sigma}-1}{1-\sigma}+\lambda_1\left[k^{\beta}\left(nr\right)^{1-\beta}r_a^{\gamma}-c\right]+\lambda_2\delta r(1-(nr)^2) \tag{2}$$

where λ_1 and λ_2 are the costate variables. The necessary first order conditions are:

$$\frac{\partial\mathcal{H}}{\partial c}=0\ c^{-\sigma}=\lambda_1 \tag{2a}$$

$$\frac{\partial\mathcal{H}}{\partial n}=0\Longrightarrow 2\delta n^2r^2\lambda_2=\lambda_1(1-\beta)k^{\beta}n^{1-\beta}r^{1-\beta+\gamma-1} \tag{2b}$$

$$-\frac{\partial\mathcal{H}}{\partial k}=\dot{\lambda}_1-\rho\lambda_1\Longrightarrow\dot{\lambda}_1=\rho\lambda_1-\lambda_1\beta k^{\beta-1}n^{1-\beta}r^{1-\beta+\gamma} \tag{2c}$$

$$-\frac{\partial\mathcal{H}}{\partial r}=\dot{\lambda}_2-\rho\lambda_2\Longrightarrow\dot{\lambda}_2=\rho\lambda_2-\lambda_1(1-\beta+\gamma)k^{\beta}n^{1-\beta}r^{-\beta+\gamma}-\lambda_2\delta\left[1-3\left(rn\right)^2\right] \tag{2d}$$

Since the (not maximized) Hamiltonian is jointly concave in both its state and control variables, by Mangasarian's condition, the first order conditions are also sufficient for the existence of an interior solution of the problem $\mathcal{P}$.

By eliminating the costate variables λ_1 and λ_2, we can re-write the equations of the optimal control problem in terms of a system of four non-linear differential equations in the state (k and r) and control variables (c and n). We consider only the competitive equilibrium solution (As well known, it follows from the presence of the externality that the competitive solution differs from the planner's solution) in equilibrium $r = r_a$ for all t.

---

[1]The planner's solution involves a choice of k, r, c, n, and r_a which maximizes the control optimal model ($\mathcal{P}$) and to $r = r_a$ for all t.
In the other hand, the path for r coincides with the given path r_a in the competive solution then the system is in equilibrium.
The equilibrium solution takes r_a as exogenously determined.

$$
\begin{aligned}
\dot{k} &= k\left(k^{\beta-1}(nr)^{1-\beta}r^{\gamma} - \frac{c}{k}\right)\\
\dot{r} &= r\delta\left(1-(nr)\right)^2\\
\dot{c} &= c\left(-\frac{\rho}{\sigma} + \frac{\beta}{\sigma}k^{\beta-1}n^{1-\beta}r^{1-\beta+\gamma}\right)\\
\dot{n} &= n\left[-\frac{\beta}{(\beta+1)}\frac{c}{k} + \frac{\delta(-\beta+\gamma-1)}{(\beta+1)} + \delta(nr)^2\left(\frac{(1-\beta+\gamma)}{(1-\beta)}\right)\right]
\end{aligned}
\tag{4}
$$

Remark We call a Balance Growth Path (BGP) a solution of (4) in which the growth rate of k, r, c and n are constant.

By using the following substitution:

$$
x_1 = \frac{c}{k};\; x_2 = nr;\; x_3 = \frac{y}{k}
$$

where $y = k^{\beta}(nr)^{1-\beta}r_a^{\gamma}$ is the production function, we can rewrite the above conditions in term of variables' growth rates (see [2, 3]).

We get a deterministic system of three differential equations with two non-predetermined variables and one predetermined variable. At the end with some algebra, we find the simpler form:

$$
\begin{aligned}
\dot{x}_1 &= x_1\left[x_1 + \left(\frac{1}{\sigma} - 1\right)x_3 - \frac{\rho}{\sigma}\right]\\
\dot{x}_2 &= x_2\left[-\frac{\beta}{(\beta+1)}x_1 + \frac{\delta\gamma}{1-\beta}x_2^2 + \frac{\delta\gamma}{\beta+1}\right]\\
\dot{x}_3 &= x_3\left[\frac{1-\beta}{\beta+1}x_1 - (1-\beta)x_3 + \frac{2\delta\gamma}{\beta+1}\right]
\end{aligned}
\tag{5}
$$

2.1 Steady States Analysis

A stationary (equilibrium) point $P^* = P^*(x_1^*, x_2^*, x_3^*)$ of the system (5) is any solution of the following system:

$$
\begin{aligned}
&x_1\left[x_1 + \left(\frac{1}{\sigma} - 1\right)x_3 - \frac{\rho}{\sigma}\right] = 0\\
&x_2\left[-\frac{\beta}{(\beta+1)}x_1 + \frac{\delta\gamma}{1-\beta}x_2^2 + \frac{\delta\gamma}{\beta+1}\right] = 0\\
&x_3\left[\frac{1-\beta}{\beta+1}x_1 - (1-\beta)x_3 + \frac{2\delta\gamma}{\beta+1}\right] = 0
\end{aligned}
\tag{6}
$$

We get four admissible steady states:

$$
\begin{aligned}
&i. \quad P^*(0,0,0) = P_0^*\\
&ii. \quad P^*\left(0,0,\frac{2\delta\gamma}{1-\beta^2}\right) = P_1^*\\
&iii. \quad P^*\left(\frac{\rho}{\sigma},0,0\right) = P_2^*\\
&iv. \quad P^*\left[\frac{1}{\sigma}\left(\rho - \frac{(1-\sigma)(1-\beta+2\sigma\delta\gamma)}{(1-\beta)\left(1-\sigma\beta^2\right)}\right), 0, \frac{1-\beta+2\sigma\delta\gamma}{(1-\beta)\left(1-\sigma\beta^2\right)}\right] = P_3^*
\end{aligned}
\tag{7}
$$

The Jacobian matrix J of the system (5) is:

$$
J(P) = \begin{bmatrix}
2x_1 + \frac{1-\sigma}{\sigma}x_3 - \frac{\rho}{\sigma} & 0 & \frac{1-\sigma}{\sigma}x_1\\
-\frac{\beta}{\beta+1}x_2 & -\frac{\beta}{\beta+1}x_1 + \frac{3\gamma\delta}{1-\beta}x_2^2 + \frac{\delta\gamma}{\beta+1} & 0\\
\frac{1-\beta}{\beta+1}x_3 & 0 & \frac{1-\beta}{\beta+1}x_1 - 2(1-\beta)x_3 + \frac{2\delta\gamma}{\beta+1}
\end{bmatrix}
\tag{8}
$$

where $P = P(x_1, x_2, x_3)$.

We denote $J = J(P_i^*)$ with $i = 0, 1, 2, 3$ the Jacobian matrix J evaluated at one of the above four equilibrium point.

3 The Existence of a Saddle with Three Purely Real Eigenvalues

We are interested in the special case which $J = J(P^*)$ has three real eigenvalues, one positive and two negative.

To this end, we analyze the dynamics of the model around the equilibrium point: $P_1^*\left(0,0,\frac{2\delta\gamma}{1-\beta^2}\right)$.

Lemma 1 *We consider the following subsets of the parameters space Ω:*

$$
\Omega_1 = \{\omega \in \Omega : \rho \in R_{++};\ \ 0 < \sigma < 1; \delta \in\ R_{++}; \gamma < 0\}
$$

Let $\omega \in \Omega_1$ be. Then $J(P_1^*)$ has one positive and two negative purely real eigenvalues.

Proof Let $J(P_1^*)$ be the Jacobian matrix evalueted at the equilibrium point P_1^* $\left(0,0,\frac{2\delta\gamma}{1-\beta^2}\right)$, we get :

$$J(P_1^*) = \begin{bmatrix} +\frac{1-\sigma}{\sigma}\left(\frac{2\delta\gamma}{1-\beta^2}\right) - \frac{\rho}{\sigma} & 0 & 0 \\ 0 & +\frac{\delta\gamma}{\beta+1} & 0 \\ \left(\frac{2\delta\gamma}{(\beta+1)^2}\right) & 0 & -\frac{2\delta\gamma}{\beta+1} \end{bmatrix} \tag{9}$$

The eigenvalues of $J(P_1^*)$ are the solutions of its characteristic equation:

$$\det(\lambda \mathbf{I} - J) = \lambda^3 - Tr(J)\lambda^2 + B(J)\lambda - Det(J) \tag{10}$$

where $\mathbf{I}$ is the identity matrix. $Tr(J)$ and $Det(J)$ are Trace and Determinant of $J(P_1)$, respectively. $B(J)$ is the sum of principal minors of order 2.

We apply the Cardano's formula to (10) and we get the result.

Formally, the eigenvalues are: purely real eigenvalues:

$$\lambda_u = -\frac{2\delta\gamma}{\beta+1} > 0 > \lambda_s = \frac{\delta\gamma}{\beta+1} \geq \lambda_{ss} = +\frac{1-\sigma}{\sigma}\left(\frac{2\delta\gamma}{1-\beta^2}\right) - \frac{\rho}{\sigma}$$

■

Example 1 Set $(\beta, \gamma, \delta, \rho, \sigma) = (0.33, -0.4, 0.04, 0.01, 0.5)$. This economy has $P^* \equiv (x_1^*, x_2^*, x_3^*) \simeq (0, 0, 0.03591067220)$. The computation of the eigenvalues of $\mathbf{J}$ leads to $\lambda_u \simeq 0.02406015038$, $\lambda_s \simeq -0.01203007519$, $\lambda_{ss} \simeq -0.02897766805$.

Example 2 Set $(\beta, \gamma, \delta, \rho, \sigma) = (0.33, -0.1, 0.04, 0.01, 0.188)$. This economy has $P^* \equiv (x_1^*, x_2^*, x_3^*) \simeq (0, 0, 0.008977668050)$. The computation of the eigenvalues of $\mathbf{J}$ leads to $\lambda_u \simeq 0.006015037592$, $\lambda_s \simeq -0.03007518797$, $\lambda_{ss} \simeq -0.05456198426$.

4 The Existence of a Homoclinic Orbit

The second step of our calculations is the explicit calculus of the homoclinic orbit in $J(P_1^*)$.

Lemma 2 (*Existence of homoclinic orbits to the real saddle* P_1^*) *Let* $\omega \in \Omega_1$. *Then*

$$\Omega_1{}^H = \left\{\omega \in \Omega_1 \ : \ (2.5) \ \textit{possesses a homoclinic orbit} \ \Gamma_{P_1^*}\right\} \neq \varnothing$$

In order to construct the homoclinic orbit analytically, we apply the procedure developed by [5]. We compute the stable and unstable manifolds, of the saddle equilibrium point $J(P_1^*)$, respectivaly: $W^s$and W^u with the undetermineted coefficients method. We show that a homoclinic loop emerges as a solution trajectory of system (5) for parameter values belonging to the set $\Omega_1^H \subset \Omega_1$ The application of the method leads to the following relationship

$$\varphi \equiv \xi - \frac{+F_{1e}\psi^2}{(2\lambda_3 + \lambda_1)} - \frac{F_{1d}}{-\lambda_1}(\xi^2 \frac{F_{3d}}{(\lambda_3 + 2\lambda_1)} + \frac{1}{\lambda_3}\psi^2 F_{3f}) \frac{(4\lambda_1^2 + 4\lambda_1\lambda_3)}{(+F_{3d})(+\lambda_3 - 2\lambda_1)} = 0 \tag{11}$$

In (11), ξ and $\psi$are arbitrary constants with $(\xi, \psi) \in (0, 1)^2$, the $F_{i,j}$ coefficients, $i = 1, 2, 3$ and $j = e, d, f$, F_{ij} are intricate combinations of the original parameters of the model and of three scaling factors $(\zeta_1, \zeta_2, \zeta_3)$ associated with the choice of the eigenvectors.

We now introduce a normal topological form for homoclinic bifurcation (see [9]).

We consider a two-dimensional cross-section Σ with coordinates (ξ, ψ) of W^u, the unidimensional unstable manifold. Suppose that $\xi = 0$ corresponds to the intersection of Σ with the stable manifold W^s of P_1^*. Let conversely the point with coordinates (ξ^u, ψ^u) correspond to the intersection of W^u with Σ. Then, the following occurs.

Definition 1 A split function can be defined as $\varphi = \xi^u$. Its zero $\varphi = 0$ gives a condition for the homoclinic bifurcation in $\mathbb{R}^3$.

It might be impossible to characterise the system for a full set of parameter spaces, and the boundary of the homoclinic orbit region. Using γ as a bifurcation parameter, we observe that parameters remain inside Ω_1^H for $-0.9 < \gamma < -0.001$ and $0 < \sigma < 1$.

Example 3 Therefore, If we set $\gamma \simeq -0.2435913142$ and $(\beta, \delta, \rho, \sigma) = (0.33, 0.04, 0.01, 0.5)$ a homoclinic orbit can emerge as solution trajectories of system (5).

We found this result after many maple simulations.

5 Existence of a Unique Limit Cycle In the Neighborhood of a Homoclinic Orbit

In order to introduce the main result of the paper we recall some basic definitions (see [9]).

Definition 2 Real negative eigenvalues that are closest to the imaginary axis are called leading (or principal) eigenvalues, while the corresponding eigenspace is called a leading (or principal) eigenspace.

Definition 3 A saddle quantity S_{x_0} of a saddle equilibrium point x_0 with real eigenvalues, is the sum of the positive eigenvalue and one of the leading eigenvalue.

Lemma 3 *Let* $S_{P_1^*} = \lambda_u(P_1^*) + \lambda_{ss}(P_1^*)$ *be in* $\omega \in \Omega_1^H$ *then* $S_{P_1^*} > 0$.

Proof It follows from of direct calculus.

Theorem 1 *Consider a three-dimensional system*

$$\dot{x} = f(x, \alpha), \qquad x \in \mathbb{R}^3, \quad \alpha \in \mathbb{R}^1$$

with smooth f, having at $\alpha = 0$ a saddle equilibrium point $x_0 = 0$ with real eigenvalues $\lambda_u(0) > 0 > \lambda_s(0) \geq \lambda_{ss}(0)$ and a homoclinic orbit Γ_0. Assume the following genericity conditions also hold:
$(H.1)$ $S_0 = \lambda_u(0) + \lambda_s(0) > 0$;
$(H.2)$ $\lambda_s(0) \neq \lambda_{ss}(0)$;
$(H.3)$ Γ_0 *returns to* $x_0 = 0$ *along the leading eigenspace*;
$(H.3)$ Γ_0 *is simple or twisted;*
$(H.5)$ $\varphi'(0) \neq 0$, *where* $\varphi(\alpha)$ *is a split function.*
Then, for all sufficiently small $|\alpha|$ there exists a neighborhood U_0 of $\Gamma_0 \cap x_0$ in which a unique saddle limit cycle L_ϕ bifurcates from Γ_0. The cycle exists for $\varphi < 0$ if Γ_0 is nontwisted, and for $\varphi > 0$ if Γ_0 is twisted. They are locally topological equivalents in a neighborhood U_0 of $\Gamma_0 \cap x_0$ for sufficiently small $|\alpha|$.

Proposition 1 *(Existence of a limit cycle in the neighborhood of the homoclinic orbit). Let $\omega \in \Omega_1^H$. Then γ is the bifurcation parameter such that a homoclinic orbit emerges as a solution trajectory of system (5). Since there exist values of γ sufficiently close to $\hat{\gamma}$ such that $\varphi(\gamma) > 0$ and $\varphi(\hat{\gamma}) = 0$, a unique stable limit cycle around the homoclinic orbit emerges.*

Proof To prove the statement, we need to show that conditions in Theorem 1 are satisfied. By Lemma 1 and Lemma 2, we already know that since H1 is satisfied, there exist regions in the parameters space such that P_1^* is a real saddle with $S_0 > 0$. Furthermore, there is a critical value of the parameter $\gamma = \hat{\gamma}$ at which an orbit connecting P_1^* to itself in backward and forward time emerges as a solution trajectory, then H3 is satisfied. Our examples also show that, generically, $\lambda_s(P_1^*) \neq \lambda_{ss}(P_1^*)$ so that (H.2) is also satisfied. By construction, the homoclinic orbit returns to P_1^* along the leading eigenspace. Finally equation (11) represents the split function of the homoclinic orbit, as stated and H5 is satisfied. ■

6 Conclusions

In this paper we have focused on the parameter regions around a saddle equilibrium point with purely real eigenvalues.

From an economic point of view we show that the small negativity of the exponent of the externality due to over-exploitment of natural resources combined with a low inverse intertemporal elasticity of substitution plays a crucial role in determining global indeterminacy We have applied the procedure developed by [5] and we have shown that a homoclinic loop emerges as a solution trajectory of the reduced system for a economic set of parameter values.

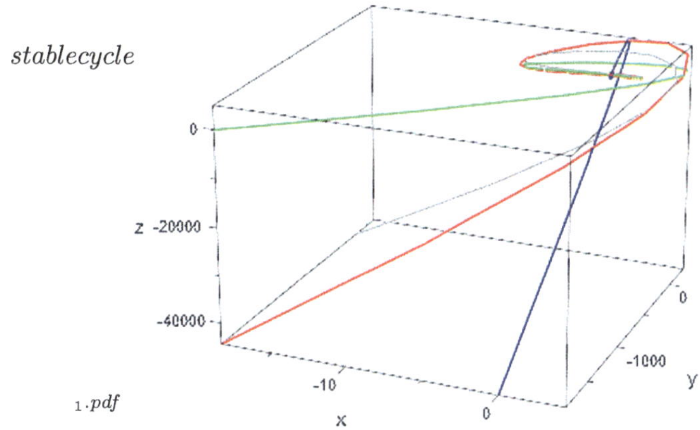

Fig. 1 In Fig. 1. A unique stable limit cycle in the neighborhood of the homoclinic orbit

In order to get a homoclinic bifurcation we have introduced a normal topological form. By varing the exponent of the externality we consider a homoclinic bifurcation of dimension one, that bifurcates a unique stable cycle.

As clearly pointed out in the literature [10], the rupture of a homoclinic orbit connecting the unique steady state to itself implies the existence of a tubular neighborhood of the original homoclinic orbit, such that any initial condition starting inside this tubular neighborhood gives rise to a perfect-foresight equilibrium which is allowed to converge to the limit cycle. Finally, with similar arguments introduceted in [6–8, 11], we are able to show global indeterminacy of the equilibrium for the model, since the result is valid beyond the small neighborhood relevant for the local analysis.

References

1. F. Wirl, Sustainable growth, renewable resources and pollution: thresholds and cycles. J. Econ. Dyn. Control **28**, 1149 –1157 (2004)
2. A. Pirisinu, *Stability and Sunspots in Endogenous Growth Models* (Department of Economic and Business, University of Cagliari, Italy, 2018). PhD Thesis
3. B. Venturi, A. Pirisinu, Bifurcations and sunspots in continuous time optimal models with externalities. In *Classification, (Big) Data Analysis and Statistical Learning*. vol. 53 (Springer International Publishing, Berlin, 2018), pp. 235–242
4. C.B. Mulligan, I. Sala, X. Martin, Transitional dynamics in two-sector models of endogenous growth. Q. J. Econ. **108**(3), 739–773 (1993)
5. D. Shang, M. Han, The existence of homoclinic orbits to saddle-focus. Appl. Math. Comput. **163**, 621–631 (2005)
6. G. Bella, P. Mattana, B. Venturi, Shilnikov chaos in the Lucas model of endogenous growth. J. Econ. Theory (2017). https://doi.org/10.1016/j.jet.2017.09.010. ISSN: 0022-0531
7. G. Bella, P. Mattana, B. Venturi, Globally indeterminate growth paths in the lucas model of endogenous growth. Macroecon. Dyn. 1–12 (2019)

8. A. Antoci, M. Galeotti, P. Russu, Global analysis and indeterminacy in a two-sector growth model with human capital. Int. J. Econ. Theory **10**, 313–338 (2014)
9. Y.A. Kuznetsov, *Elements of Applied Bifurcation Theory*, 3rd edn. (Springer, New York, 2004)
10. P. Aguirre, B. Krauskopf, H.M. Osinga, Global invariant manifolds near homoclinic orbits to a real saddle: (Non)orientability and flip bifurcation. SIAM J. Appl. Dyn. Syst. **12**, 1803–1846 (2013)
11. P. Mattana, B. Venturi, Existence and stability of periodic solutions in the dynamics of endogenous growth. Int. Rev. Econ. Bus. **46**, 259–284 (1999)

The Method of Singular Integral Equations in the Theory of Microstrip Antennas Based on Chiral Metamaterials

Dmitriy S. Klyuev, Anatoly M. Neshcheret, Oleg V. Osipov, Alexander A. Potapov and Julia V. Sokolova

Abstract The present chapter is devoted to the study of microstrip antennas (MSA) based on chiral metamaterials. The method of electrodynamic analysis of similar antennas with chiral substrates is considered, ensuring high accuracy of calculations and at the same time being less exacting to computational resources in comparison with known complexes of electrodynamic modeling. A mathematical model of MSA with a biisotropic chiral substrate is shown on the basis of the mathematical apparatus of singular integral equations (SIE), whose numerical solution belongs to the class of well-posed mathematical problems. The results of calculation of the impedance, directional and polarization characteristics of MSA with chiral substrates are presented. The problem of diffraction of a plane electromagnetic wave of linear polarization on a re-radiating structure based on a chiral metamaterial is considered.

Abbreviations

EMF	Electromagnetic field
EMW	Electromagnetic wave
DD	Directionality diagrams

D. S. Klyuev · A. M. Neshcheret · O. V. Osipov · J. V. Sokolova
L. Tolstoy str., 23, Samara 443010, Russia
e-mail: klyuevd@yandex.ru

A. M. Neshcheret
e-mail: neshchereta@gmail.com

O. V. Osipov
e-mail: nanometa@yandex.ru

J. V. Sokolova
e-mail: sokolova-yu-v@yandex.ru

A. A. Potapov (✉)
V. A. Kotelnikov Institute of Radio Engineering and Electronics,
Russian Academy of Sciences, Moscow, Russia
e-mail: potapov@cplire.ru

C. H. Skiadas and Y. Dimotikalis (eds.), *12th Chaotic Modeling and Simulation International Conference*, Springer Proceedings in Complexity,
https://doi.org/10.1007/978-3-030-39515-5_22

MSA Microstrip antenna
SIE Singular integral equation
SIR Singular integral representation
UHF Ultra-high frequency

The expansion of the range of tasks solved by modern radioelectronics, as well as their complication, has stimulated intensive development of antenna theory and technology in recent decades. The main areas of radioelectronics using are communication, television, radiolocation, radio control, radio astronomy and others are impossible without the use of different antenna systems. Therefore, the problem of improving their electrical characteristics and miniaturization is extremely important.

Currently, the most common type of antennas used in portable electronic devices are antennas in microstrip performance [1, 2]. A lot of scientific articles, monographs have been published about these antennas, hundreds of constructive and functional varieties have been described and patented. Such close interest is primarily due to the advantages of this type of antennas: improved mass-size characteristics, the possibility of using modern technologies for serial production of both emitters and excitation devices, matching and controlling the radiation characteristics of such antennas. However, the MSA technology, which was so promising about two decades ago, has now reached its limits with respect to reducing the dimensions of microwave devices. Therefore, the search for new approaches to the development of microwave technology has recently become much more active. One of the promising directions in creating antennas of the new generation is associated with the use in their design of artificial composite metamaterials [3], which have proved at the moment, their effectiveness.

Among the metamaterials, a special place is occupied by chiral media, which are a dielectric container in which the conductive inclusions of the mirror-asymmetric shape are evenly distributed [4, 5]. These media have a number of unique properties, the main ones being the cross-polarization of the incident field and the circular dichroism, thanks to which the bifurcation of normal waves to waves with left- and right-circular polarizations occurs in the medium. The two main types of chiral media are biisotropic (isotropic chiral) materials, in which the mirror-asymmetric elements are oriented arbitrarily, thereby providing an isotropy of the medium for waves with right- and left-handed polarizations and bi-anisotropic (anisotropic chiral) media in which the elements are oriented identically, creating an anisotropy axis.

Theoretical and experimental studies show that the use of metamaterials makes it possible to significantly improve the electrical and mass-dimensional characteristics of MSA [5–9], so the development of MSA technique is directly related to their use. In particular, their use in antenna technology can significantly reduce the dimensions, compensate for the reactivity of electrically small antennas, significantly improve the directivity, improve coordination, reduce the mutual influence between radiators in antenna arrays, etc.

On the other hand, at the moment there is no uniform strict theory of MSA on their basis, there are no strict self-consistent physical and mathematical models of such antennas, as well as methods for their analysis and synthesis. In addition, the processes, phenomena and regularities of the radiation of electromagnetic waves by such antennas have not been studied sufficiently, and their characteristics have not been sufficiently studied. In this regard, there is an urgent scientific problem of the development of the theory of MSA with substrates from metamaterial, as well as the development of adequate methods for their electrodynamic analysis that ensure the correctness, accuracy and stability of solutions.

At present, two main approaches to the research of MSA on the basis of chiral metamaterials can be distinguished. The first approach, which is most popular, is the use of software complexes of electrodynamic modeling, such as CST Microwave Studio, Feko, HFSS, etc. As you know, in the basis of these packages, numerical algorithms are laid, presupposing the partitioning (discretization) of the entire space or its part into unit cells. Here it should be noted that discretization violates the original structure of the object in such a way that its properties can be distorted. Undoubtedly, as the size of the cells decreases and their number increases, the field calculated with the help of these packages tends to the true field in the structure, that is, there is the convergence of the iterative process, but the solution of the problem with a large number of cells requires the use of high-speed processors and large amounts of RAM. Therefore, it is necessary to seek a compromise between the accuracy of the solution, the time and the computational resources necessary for its implementation. In addition, very often in the solution of electrodynamic problems even small distortions in the structure can cause significant deviations in the calculated field. This happens, for example, in resonant structures, to which the chiral media also belong. Therefore, such programs often produce results that simply do not correspond to physical meaning. And the question of "trust" to the results is always open.

Another approach to the analysis of such microstrip structures is associated with the development of electrodynamic models of MSA with substrates from metamaterial based on Fredholm integral equations of the first kind when representing surface currents by an equivalent distribution [10, 11]. An essential shortcoming of these models is the difficulty of obtaining a stable solution, due to the incorrectness of the mathematical problem of Hadamard [12].

To eliminate these drawbacks, it is proposed to use an approach based on chiral metamaterials for the correct electrodynamic analysis of MSA based on chiral metamaterials, the essence of which is to obtain a singular integral representation of the electromagnetic field that, when the appropriate boundary conditions is substituted, is transformed into a singular integral equation (SIE) whose solution is a correct mathematical task.

In connection with the foregoing, in this chapter the method of electrodynamic analysis of MSA with chiral substrates will be considered in detail, which ensures high accuracy of calculations and is thus less demanding for computational resources in comparison with known complexes of electrodynamic modeling.

1 Statement of the Problem. Surface Impedance Method

Let us consider a microstrip antenna in which the substrate is made of chiral metamaterial, of thickness d, and metallized from the lower side (Fig. 1). The macroscopic parameters of the chiral substrate are equal to ε_1, μ_1 and χ, where ε_1, μ_1—relative permittivity and magnetic permeability, and χ—the chirality parameter, the physical meaning of which, is to determine the degree of interrelation between the processes of magnetic and electrical polarizations in the medium. On this substrate there is an infinitely thin and ideally conducting rectangular symmetrical radiator of length $2l$ and width $2a$. Above the radiator is a dielectric half-space, in which the relative dielectric and magnetic permeability's are equal to ε_2, μ_2.

It was assumed that the radiator is sufficiently narrow, and therefore, the transverse component of the surface current density can be neglected $\vec{\eta} = (0, \eta_y)$. A source of electromotive force is connected to the gap of the radiator of width $2b$, ensuring continuity of the surface current density function in the gap and on the surface of the radiator. It was also assumed that the tangential component of the strength of the external field has only one longitudinal component $\vec{E}_\tau^{\text{ext}} = \{0, E_y^{\text{ext}}, 0\}$. On the surface of a flat radiator, the following boundary conditions are assumed:

$$\begin{aligned} &\eta_y(x, -l) = \eta_y(x, +l) = 0 \\ &\vec{E}_\tau(x, y) = 0 \text{ when } x \in [-a, a],\ y \in [-l, -b] \cup [b, l] \\ &\vec{E}_\tau(x, y) = -\vec{E}_\tau^{\text{ext}} \text{ when } x \in [-a, a],\ \ y \in [-b, b] \end{aligned} \tag{1}$$

To solve the problems of electrodynamic analysis, the surface impedances method was used. The essence of the surface impedances method is to determine the surface impedances matrix that connects the Fourier transform $\vec{T}_\tau = \{T_x, T_y\}$ with the tangential component of the electric field intensity $\vec{E}_\tau$ with the Fourier image $\vec{F} = \{F_x, F_y\}$ of the surface current density on the radiator $\vec{\eta}$:

$$\begin{bmatrix} T_y \\ T_x \end{bmatrix} = \begin{bmatrix} Z_{11} & Z_{12} \\ Z_{21} & Z_{22} \end{bmatrix} \begin{bmatrix} F_y \\ F_x \end{bmatrix} \tag{2}$$

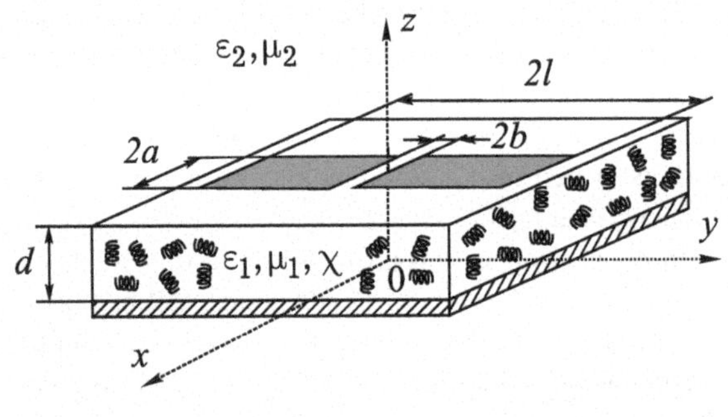

Fig. 1 Structure of MSA

where Z_{ij} $(i, j = 1, 2)$ are elements of the matrix of surface impedances $[Z]$, which are functions of variables β and h Fourier spaces $Z_{ij} = Z_{ij}(\beta, h)$.

In this case, it becomes necessary to decompose the strengths of the electric $\vec{E}$ and magnetic $\vec{H}$ fields, as well as the surface density of the current $\vec{\eta}$ into the Fourier integral:

$$\vec{E}(x, y, z) = \int\limits_{-\infty}^{\infty}\int\limits_{-\infty}^{\infty} \vec{T}(z, \beta, h)e^{-i\beta x - ihy} d\beta dh$$

$$\vec{H}(x, y, z) = \int\limits_{-\infty}^{\infty}\int\limits_{-\infty}^{\infty} \vec{M}(z, \beta, h)e^{-i\beta x - ihy} d\beta dh$$

$$\vec{\eta}(x, y) = \int\limits_{-\infty}^{\infty}\int\limits_{-\infty}^{\infty} \vec{F}(\beta, h)e^{-i\beta x - ihy} d\beta dh \quad (3)$$

where

$$\vec{T}(z, \beta, h) = \frac{1}{4\pi^2}\int\limits_{-\infty}^{\infty}\int\limits_{-\infty}^{\infty} \vec{E}(x', y', z)e^{i\beta x' + ihy'} dx'dy'$$

$$\vec{M}(z, \beta, h) = \frac{1}{4\pi^2}\int\limits_{-\infty}^{\infty}\int\limits_{-\infty}^{\infty} \vec{H}(x', y', z)e^{i\beta x' + ihy'} dx'dy'$$

$$\vec{F}(\beta, h) = \frac{1}{4\pi^2}\int\limits_{-a}^{a}\int\limits_{-l}^{l} \vec{\eta}(x', y')e^{i\beta x' + ihy'} dx'dy'$$

$$\vec{E}(x', y', z) = \vec{x}_0 E_x(x', y', z) + \vec{y}_0 E_y(x', y', z) + \vec{z}_0 E_z(x', y', z)$$

$$\vec{H}(x', y', z) = \vec{x}_0 H_x(x', y', z) + \vec{y}_0 H_y(x', y', z) + \vec{z}_0 H_z(x', y', z)$$

$$\vec{\eta}(x', y') = \left[\vec{n}_0, \vec{H}^{(2)} - \vec{H}^{(1)}\right] = \vec{x}_0\eta_x(x', y') + \vec{y}_0\eta_y(x', y')$$

where $\vec{x}_0, \vec{y}_0, \vec{z}_0$ are unit vectors (unit vectors) on the coordinate axes, $\vec{n}_0$ is unit vector of the normal to the surface of the radiator, directed from the chiral layer to the dielectric half-space; $\vec{H}^{(1)}, \vec{H}^{(2)}$ are vectors of magnetic field strengths in a chiral metamaterial and a dielectric half-space, respectively. $\{T_x, T_y, T_z\}, \{M_x, M_y, M_z\}, \{F_x, F_y\}$ are component vectors of the Fourier transforms of the electric field strength $\vec{E}$, magnetic field strength $\vec{H}$ and the surface current density $\vec{\eta}$, respectively. $\{E_x, E_y, E_z\}$, $\{H_x, H_y, H_z\}$, $\{\eta_x, \eta_y\}$ are the components of the electric field strength $\vec{E}$, magnetic field strength $\vec{H}$ and surface current density $\vec{\eta}$, respectively. In expression (3) it is taken into account that the surface density of the electric current $\vec{\eta}$ is different from zero only at the surface of the vibrator: $x \in [-a, a]$, $y \in [-l, l]$.

Elements of the matrix of surface impedances Z_{ij} $(i, j = 1, 2)$ are determined through the elements of the matrix of surface admittances Y_{ij} $(i, j = 1, 2)$, which, in turn, are determined through the elements of the admission matrix of the chiral layer (chiral substrate) $Y_{ij}^{(1)}$ $(i, j = 1, 2)$ and the dielectric half-space $Y_{ij}^{(2)}$ $(i, j = 1, 2)$. The expression for determining the elements of the surface admittance matrix has the following form:

$$Y_{ij} = Y_{ij}^{(2)} - Y_{ij}^{(1)} \tag{4}$$

where $Y_{ij}^{(1)}$ and $Y_{ij}^{(2)}$ $(i, j = 1, 2)$ are elements of the matrices of input admittances of the chiral and dielectric layers, respectively.

It should be noted that the matrix of surface admittances is the inverse matrix of surface impedances, so that the elements of the surface impedance matrix will be determined as follows:

$$Z_{11} = Y_{22}/\Delta, \quad Z_{12} = -Y_{12}/\Delta, \quad Z_{21} = -Y_{21}/\Delta, \quad Z_{22} = Y_{11}/\Delta \tag{5}$$

where $\Delta = Y_{11}Y_{22} - Y_{12}Y_{21}$.

Below is an expression for determining the elements of the matrix of input admittances:

$$\begin{bmatrix} M_x^{(1,2)} \\ M_y^{(1,2)} \end{bmatrix} = \begin{bmatrix} Y_{11}^{(1,2)} & Y_{12}^{(1,2)} \\ Y_{21}^{(1,2)} & Y_{22}^{(1,2)} \end{bmatrix} \begin{bmatrix} T_y^{(1,2)} \\ T_x^{(1,2)} \end{bmatrix} \tag{6}$$

where $T_x^{(1)}$, $T_y^{(1)}$, $M_x^{(1)}$, $M_y^{(1)}$ are Fourier transforms of the tangential components of the electric and magnetic field strength, respectively, in the chiral metamaterial. $T_x^{(2)}$, $T_y^{(2)}$, $M_x^{(2)}$, $M_y^{(2)}$ are Fourier transforms of the tangential components of the electric and magnetic field strength, respectively, in the dielectric half-space.

1.1 Determination of Elements of the Matrix of Input Impedances

In most cases, the basis for the investigation of chiral media is the phenomenological theory, which involves the use of the following material equations [4, 13, 14].

$$\begin{aligned} \vec{D} &= \varepsilon_0 \varepsilon_1 \vec{E} \mp i\chi\sqrt{\varepsilon_0\mu_0}\vec{H} \\ \vec{B} &= \mu_0\mu_1\vec{H} \pm i\chi\sqrt{\varepsilon_0\mu_0}\vec{E} \end{aligned} \tag{7}$$

where ε_0, μ_0 are electric and magnetic constants.

In these expressions, there is a chirality parameter χ characterizing the degree of interrelation between the processes of electric and magnetic polarizations in the medium. Here and below, the upper signs in the expression (7) correspond to the chiral medium on the basis of the "right-hand" elements (for example, right-handed spirals), and the lower signs to the chiral medium on the basis of "left-handed" elements (respectively, left-handed spirals).

Taking into account the above material equations, Maxwell's differential equations for the chiral layer will look as follows:

$$\begin{aligned} rot\vec{E} &= -i\omega\mu_0\mu_1\vec{H} \pm k\chi\vec{E} \\ rot\vec{H} &= i\omega\varepsilon_0\varepsilon_1\vec{E} \pm k\chi\vec{H} \end{aligned} \tag{8}$$

где ω is cyclic frequency; $k = \omega\sqrt{\varepsilon_0\mu_0}$ is wave number for a plane electromagnetic wave in free space.

We rewrite the system of Maxwell's equations (8) in the following form:

$$\begin{aligned} &\frac{\partial E_z^{(1)}}{\partial y} - \frac{\partial E_y^{(1)}}{\partial z} = -i\omega\mu_0\mu_1 H_x^{(1)} \pm k\chi E_x^{(1)} \quad \frac{\partial H_z^{(1)}}{\partial y} - \frac{\partial H_y^{(1)}}{\partial z} = i\omega\varepsilon_0\varepsilon_1 E_x^{(1)} \pm k\chi H_x^{(1)} \\ &\frac{\partial E_x^{(1)}}{\partial z} - \frac{\partial E_z^{(1)}}{\partial x} = -i\omega\mu_0\mu_1 H_y^{(1)} \pm k\chi E_y^{(1)} \quad \frac{\partial H_x^{(1)}}{\partial z} - \frac{\partial H_z^{(1)}}{\partial x} = i\omega\varepsilon_0\varepsilon_1 E_y^{(1)} \pm k\chi H_y^{(1)} \\ &\frac{\partial E_y^{(1)}}{\partial x} - \frac{\partial E_x^{(1)}}{\partial y} = -i\omega\mu_0\mu_1 H_z^{(1)} \pm k\chi E_z^{(1)} \quad \frac{\partial H_y^{(1)}}{\partial x} - \frac{\partial H_x^{(1)}}{\partial y} = i\omega\varepsilon_0\varepsilon_1 E_z^{(1)} \pm k\chi H_z^{(1)} \end{aligned} \tag{9}$$

From the system of (9), by means of algebraic transformations, we express $E_x^{(1)}$, $H_x^{(1)}$ by $E_y^{(1)}$, $H_y^{(1)}$:

$$\begin{aligned} &\frac{\partial^2 H_y^{(1)}}{\partial x \partial y} + i\omega\varepsilon_0\varepsilon_1\left(-\frac{\partial E_y^{(1)}}{\partial z} \mp 2k\chi E_x^{(1)}\right) \mp k\chi\frac{\partial H_y^{(1)}}{\partial z} = \frac{\partial^2 H_x^{(1)}}{\partial y^2} + k^2\left(\varepsilon_1\mu_1 + \chi^2\right)H_x^{(1)} \\ &\frac{\partial^2 E_y^{(1)}}{\partial x \partial y} + i\omega\mu_0\mu_1\left(\frac{\partial H_y^{(1)}}{\partial z} \pm 2k\chi H_x^{(1)}\right) \mp k\chi\frac{\partial E_y^{(1)}}{\partial z} = \frac{\partial^2 E_x^{(1)}}{\partial y^2} + k^2\left(\varepsilon_1\mu_1 + \chi^2\right)E_x^{(1)} \end{aligned} \tag{10}$$

Using the expressions (3), we expand the components $E_x^{(1)}$, $H_x^{(1)}$, $E_y^{(1)}$, $H_y^{(1)}$ into the Fourier integral and substitute them in (10). As a result, we obtain the coupling expressions for the tangential components of the Fourier transforms of the electric and magnetic field strengths ($T_x^{(1)}$, $M_x^{(1)}$, с $T_y^{(1)}$, $M_y^{(1)}$):

$$\begin{aligned} M_x^{(1)} &= a(h)\left[-\beta h M_y^{(1)} - i\omega\varepsilon_0\varepsilon_1 s(h)\frac{\partial T_y^{(1)}}{\partial z} \mp q(h)\frac{\partial M_y^{(1)}}{\partial z} \pm \zeta(h) i\omega\varepsilon_0\varepsilon_1\beta h T_y^{(1)}\right] \\ T_x^{(1)} &= a(h)\left[-\beta h T_y^{(1)} + i\omega\mu_0\mu_1 s(h)\frac{\partial M_y^{(1)}}{\partial z} \mp q(h)\frac{\partial T_y^{(1)}}{\partial z} \mp \zeta(h) i\omega\mu_0\mu_1\beta h M_y^{(1)}\right] \end{aligned} \tag{11}$$

where

$$a(h) = \frac{1}{\left(k^2\varepsilon_1\mu_1 + (k\chi)^2 - h^2\right)}\left(1 - \frac{4(k\chi)^2k^2\varepsilon_1\mu_1}{\left(k^2\varepsilon_1\mu_1 + (k\chi)^2 - h^2\right)^2}\right)^{-1},$$

$$\zeta(h) = \frac{2k\chi}{\left(k^2\varepsilon_1\mu_1 + (k\chi)^2 - h^2\right)},$$

$$s(h) = 1 - \frac{2(k\chi)^2}{\left(k^2\varepsilon_1\mu_1 + (k\chi)^2 - h^2\right)},\ q(h) = k\chi - \frac{2k^2\varepsilon_1\mu_1 k\chi}{\left(k^2\varepsilon_1\mu_1 + (k\chi)^2 - h^2\right)_1}.$$

$T_y^{(1)}$ and $M_y^{(1)}$ are determined from the following system of differential equations:

$$\begin{aligned} &\nabla^2\vec{E} + k^2\left(\varepsilon_1\mu_1 + \chi^2\right)\vec{E} \mp 2i\omega\mu_0\mu_1 k\chi\vec{H} = 0 \\ &\nabla^2\vec{H} + k^2\left(\varepsilon_1\mu_1 + \chi^2\right)\vec{H} \pm 2i\omega\varepsilon_0\varepsilon_1 k\chi\vec{E} = 0 \end{aligned} \tag{12}$$

In connection with the fact that it is required to determine the Fourier transforms of the components of the electric and magnetic fields $T_y^{(1)}$, $M_y^{(1)}$, we also expand expression (12) in the Fourier integral (3), and then take the derivatives with respect to x and y:

$$\begin{aligned} &\frac{\partial^2\vec{T}}{\partial z^2} + \left[k^2\left(\varepsilon_1\mu_1 + \chi^2\right) - \beta^2 - h^2\right]\vec{T} \mp 2i\omega\mu_0\mu_1 k\chi\vec{M} = 0 \\ &\frac{\partial^2\vec{M}}{\partial z^2} + \left[k^2\left(\varepsilon_1\mu_1 + \chi^2\right) - \beta^2 - h^2\right]\vec{M} \pm 2i\omega\varepsilon_0\varepsilon_1 k\chi\vec{T} = 0 \end{aligned} \tag{13}$$

Expressions for $T_y^{(1)}$ and $M_y^{(1)}$ are determined from the solution of a given system of differential equations with respect to the Fourier transforms (13). In this case, we must take into account the boundary condition: $T_y^{(1)}(z = 0) = 0$.

$$\begin{aligned} T_y^{(1)} &= \pm(C_R\sin(\gamma_R z) + C_L\sin(\gamma_L z)) \\ M_y^{(1)} &= i\sqrt{\frac{\varepsilon_0\varepsilon_1}{\mu_0\mu_1}}(C_R\sin(\gamma_R z) - C_L\sin(\gamma_L z)) \end{aligned} \tag{14}$$

where $\gamma_R = \sqrt{k^2(n+\chi)^2 - \beta^2 - h^2}$; $\gamma_L = \sqrt{k^2(n-\chi)^2 - \beta^2 - h^2}$; $n = \sqrt{\varepsilon_1\mu_1}$. C_R, C_L are some constants.

As a result, the matrix elements (6) were used to derive the expressions for the elements of the admission matrix of the chiral layer:

$$Y_{11}^{(1)}(\beta, h) = \frac{a(h)\left[i\omega\varepsilon_0\varepsilon_1 v^+(\beta, h) - Y_{12}^{(1)}(\beta, h)w^+(\beta, h)\right]}{\pm(\sin(\gamma_R z) + \sin(\gamma_L z))}$$

$$Y_{12}^{(1)}(\beta, h) = \frac{w^{-}(\beta, h)}{i\omega\mu_0\mu_1 v^{-}(\beta, h)} \tag{15}$$

$$Y_{21}^{(1)}(\beta, h) = \frac{a(h)Y_{22}^{(1)}(\beta, h)}{\mp(\sin(\gamma_R z) + \sin(\gamma_L z))} w^{+}(\beta, h)$$

$$Y_{22}^{(1)}(\beta, h) = \frac{\sin(\gamma_R z) - \sin(\gamma_L z)}{i\omega\mu_0\mu_1 v^{-}(\beta, h)a(h)}$$

where

$$v^{+}(\beta, h) = -s(h)(\pm(\gamma_R \cos(\gamma_R z) + \gamma_L \cos(\gamma_L z))) + \zeta(h)\beta h(\sin(\gamma_R z) + \sin(\gamma_L z))$$
$$v^{-}(\beta, h) = -s(h)(\gamma_R \cos(\gamma_R z) - \gamma_L \cos(\gamma_L z)) \pm \zeta(h)\beta h(\sin(\gamma_R z) - \sin(\gamma_L z))$$
$$w^{+}(\beta, h) = -\beta h(\pm(\sin(\gamma_R z) + \sin(\gamma_L z))) \mp q(h)(\pm(\gamma_R \cos(\gamma_R z) + \gamma_L \cos(\gamma_L z)))$$
$$w^{-}(\beta, h) = \beta h(\sin(\gamma_R z) - \sin(\gamma_L z)) \pm q(h)(\gamma_R \cos(\gamma_R z) - \gamma_L \cos(\gamma_L z))$$

Thus, these expressions allow us to determine the elements of the matrix of input admittances of the chiral layer on the basis of "right-" and "left-sided" conducting inclusions of the mirror-asymmetric shape.

The expressions for the elements of the matrix of the input admittances ($Y_{11}^{(2)}$, $Y_{12}^{(2)}$, $Y_{21}^{(2)}$, $Y_{22}^{(2)}$) of the dielectric half-space are determined in an analogous way and are given below:

$$Y_{11}^{(2)} = -\frac{k^2\varepsilon_2\mu_2 - \beta^2}{\omega\mu_0\mu_2 r_2}; \quad Y_{12}^{(2)} = -\frac{h\beta}{\omega\mu_0\mu_2 r_2};$$
$$Y_{21}^{(2)} = \frac{h\beta}{\omega\mu_0\mu_2 r_2}; \quad Y_{22}^{(2)} = \frac{k^2\varepsilon_2\mu_2 - h^2}{\omega\mu_0\mu_2 r_2} \tag{16}$$

where $r_2 = \sqrt{k^2\varepsilon_2\mu_2 - \beta^2 - h^2}$.

1.2 The Surface Impedance Matrix Elements

Substituting expressions (15), (16) into formula (4), we obtain expressions for determining the elements of the matrix of surface admittances ($Y_{11}, Y_{12}, Y_{21}, Y_{22}$), and then substitute them in (5). Finally, we obtain expressions for determining the elements of the matrix of surface impedances ($Z_{11}, Z_{12}, Z_{21}, Z_{22}$) of the dielectric-chiral interface on the basis of "left-" and "right-handed" conducting elements of mirror-asymmetric shape, metallized on one side.

2 Mathematical Model of a Microstrip Antenna with a Chiral Substrate. The Method of Singular Integral Representations of the Field

On the surface of the radiator and the chiral medium, the following boundary conditions must be satisfied:

$$\begin{aligned} E_x^{(1)} &= E_x^{(2)},\ E_y^{(1)} = E_y^{(2)} \\ \eta_x &= H_y^{(2)} - H_y^{(1)},\ \eta_y = H_x^{(2)} - H_x^{(1)} \end{aligned} \tag{17}$$

where $E_x^{(1)}$ and $E_x^{(2)}$, $E_y^{(1)}$ and $E_y^{(2)}$ are respectively, the *x*- and *y*-components of the electric field strength vector at the interface in the chiral medium and in the upper half-space, $H_x^{(1)}$ and $H_x^{(2)}$, $H_y^{(1)}$ and $H_y^{(2)}$ are *x*- and *y*-components of the magnetic field strength vector at the interface in the chiral medium and in the upper half-space, η_x, η_y are *x*- and *y*-components of the current density vector on the surface of the radiator (chiral medium).

Since the MSA radiator is narrow, the transverse component η_x of the current density is much smaller than the longitudinal component η_y, so only the elements are determined, Z_{11} and Z_{21}, as follows from (2), these elements bind the Fourier transforms of the longitudinal and transverse components of the electric field strength with the Fourier image of the longitudinal surface current density component:

$$\begin{aligned} T_y &= Z_{11} F_y \\ T_x &= Z_{21} F_y \end{aligned} \tag{18}$$

Since the expressions (18) are written with respect to the Fourier transforms, we apply the inverse Fourier transform to them, and also take into account the fact that the current density is zero everywhere except for the emitter. As a result, we obtain the following expression for determining the field strength on the surface of the radiator:

$$E_y^{cm}(x, y, d) = \int_{-a}^{a} \int_{-l}^{l} \eta_y(x', y') Z^{\Sigma}(x', y'; x, y) dx' dy' \tag{19}$$

where

$$Z^{\Sigma}(x', y', x, y) = \frac{1}{4\pi^2} \int_{-\infty}^{\infty} \int_{-\infty}^{\infty} Z_{11}(\beta, h) e^{-i\beta(x-x')} e^{-ih(y-y')} d\beta dh$$

Since the radiator is rather narrow, the transverse variation of the longitudinal component of the surface current density can be represented in the form of a quasi-static approximation: $\eta_y(x') = 1/\sqrt{1-(x'/a)^2}$, therefore, the surface current density distribution function takes the following form:

$$\eta_y(x', y') = \frac{f(y')}{\sqrt{1-(x'/a)^2}} \tag{20}$$

where $f(y')$—an unknown function describing the longitudinal distribution of the current density.

The improper integral in (19) is divergent; therefore, to eliminate the divergence, we subtract and add the term with the asymptotic factor $Z_{11}^{\infty}(h)$, and then we substitute the expression (20) into (19), taking into account the boundary condition $f(-l) = f(l) = 0$ and integrating the resulting expression by parts in y', and also carrying out a number of algebraic operations, we obtain an integral equation with respect to the unknown function $f'(y') = df(y')/dy'$:

$$E_y(x, y, d) = \int_{-a}^{a}\int_{-l}^{l} \frac{f'(y')}{\sqrt{1-(x'/a)^2}} \Delta Z_{11}^{\Sigma}(x', y'; x, y)dx'dy' + \\ + \int_{-a}^{a}\int_{-l}^{l} \frac{f'(y')}{\sqrt{1-(x'/a)^2}} Z_{11}^{\infty}(x', y'; x, y)dx'dy' \tag{21}$$

where $\Delta Z_{11}^{\Sigma}(x', y', x, y) = -\frac{1}{4\pi^2}\int_{-\infty}^{\infty}\int_{-\infty}^{\infty}\frac{1}{ih}\Delta Z_{11}(\beta, h)e^{-i\beta(x-x')}e^{-ih(y-y')}d\beta dh.$

$$\Delta Z_{11}(\beta, h) = Z_{11}(\beta, h) - Z_{11}^{\infty}(h)$$

$$Z_{11}^{\infty}(h) - Z_{11}(\beta, h)\ when\ |h| \to \infty$$

Asymptotic representation of the element of the surface impedance matrix $Z_{11}(\beta, h)$ when $|h| \to \infty$ has the following form:

$$Z_{11}(\beta, h) \underset{|h|\to\infty}{\longrightarrow} Z_{11}^{\infty}(h) = \frac{i\omega\mu_0(\mu_1+\mu_2)}{\left(k^2(\varepsilon_1+\varepsilon_2)(\mu_1+\mu_2)-(k\chi)^2\right)}|h| \tag{22}$$

In the above expressions, there are some table integrals, namely:

$$\int_{-a}^{a}\frac{e^{i\beta x'}}{\sqrt{1-(x'/a)^2}}dx' = \pi a J_0(\beta a)$$

$$\int_{-\infty}^{\infty} J_0(\beta a) e^{-i\beta x} d\beta = 2\int_{0}^{\infty} J_0(\beta a)\cos(\beta x) d\beta = 2\zeta(x)$$

$$\int_{-\infty}^{\infty} \operatorname{sgn}(h) e^{-ih(y-y')} dh = \frac{2i}{y'-y} \tag{23}$$

where $J_0(\beta a)$ is Bessel function of the first kind of zero order;

$$\zeta(x) = \begin{cases} \frac{1}{\sqrt{a^2-x^2}}, & |x| < a \\ 0, & |x| > a \end{cases}; \quad \operatorname{sgn}(h) = \frac{|h|}{h} = \begin{cases} -1, & h < 0 \\ 1, & h \geq 0 \end{cases}$$

Substituting the asymptotic representation (22) into the expression (21), taking into account the values of the table integrals (23), we obtain a singular integral representation of the tangential component of the electric field $E_y(x, y, z)$:

$$\begin{aligned} E_y(x, y, d) = \\ = -\frac{1}{4\pi}\int_{-l}^{l} af'(y')\int_{-\infty}^{\infty}\int_{-\infty}^{\infty} J_0(\beta a)\frac{\Delta Z_{11}(\beta, h)}{ih} e^{-i\beta x} e^{-ih(y-y')} d\beta dh dy' \\ -\frac{C_{\varepsilon,\mu,\chi}}{\pi}\zeta(x)\int_{-l}^{l}\frac{af'(y')}{y'-y} dy' \end{aligned} \tag{24}$$

where $C_{\varepsilon,\mu,\chi} = \dfrac{i\omega\mu_0(\mu_1+\mu_2)}{\left(k^2(\varepsilon_1+\varepsilon_2)(\mu_1+\mu_2)-(k\chi)^2\right)}$.

In a similar way, we obtained a singular integral representation of the field for the transverse component $E_x(x, y, d)$:

$$\begin{aligned} E_x(x, y, d) = -\tfrac{1}{4\pi}\int_{-l}^{l} af'(y')\int_{-\infty}^{\infty}\int_{-\infty}^{\infty} J_0(\beta a)\tfrac{\Delta Z_{21}}{ih} e^{-i\beta x} e^{-ih(y-y')} d\beta dh.dy' - \\ -\tfrac{C_{\varepsilon,\mu,\chi}}{2\pi i}\int_{-l}^{l} af'(y')\left(2\ln(y'-y)(\mu_1+\mu_2)\zeta'(x) + \left(2\theta(y-y')-1\right)\pi i\mu_2 k\chi\zeta(x)\right)dy' \end{aligned} \tag{25}$$

where $\Delta Z_{21} = Z_{21}(\beta, h) - Z_{21}^{\infty}(\beta, h)$; $Z_{21}^{\infty}(h) - Z_{21}(\beta, h)$ *when* $|h| \to \infty$.

$$\zeta'(x) = \begin{cases} i\frac{x}{(a^2-x^2)^{3/2}}, & |x| < a \\ 0, & |x| > a \end{cases} \quad \theta(y-y') = \begin{cases} 0, & y < y' \\ 0.5, & y = y' \\ 1, & y > y' \end{cases}$$

Substituting in (24) the boundary conditions (2), which, generally speaking, are valid at any point of the radiator, we obtain a singular integral equation with a Cauchy singularity with respect to an unknown function $f'(y')$ characterizing the longitudinal distribution of the current density along the radiator.

$$\frac{1}{\pi}\int_{-l}^{l}\frac{af'(y')}{y'-y}dy' = F(y) \tag{26}$$

where $F(y) = \sigma v(y) - \frac{1}{C_{\varepsilon,\mu,\chi}}\frac{a}{4\pi}\int_{-l}^{l} af'(y')K(y,y')dy'$.

$$K(y,y') = \frac{1}{C_{\varepsilon,\mu,\chi}}\frac{a}{4\pi}\int_{-\infty}^{\infty}\int_{-\infty}^{\infty} J_0(\beta a)\frac{\Delta Z_{11}(\beta,h)}{ih}e^{-ih(y-y')}d\beta dh$$

$$\sigma = i\frac{1}{2}\frac{2\pi a}{\lambda}\frac{1}{Z_c}\frac{l}{a}\frac{1}{C_{\varepsilon,\mu,\chi}}V$$

$v(y)$ is voltage profile in the vibrator gap;
$V = 2bE_0$ is a stress value in the gap;
Z_c is wave impedance of the medium.

In the future, numerical calculations assume the use of dimensionless quantities: a/λ, l/λ, d/λ, b/l, so we introduce new variables: $\alpha = \beta a$, $\xi = ha$, $t = y/l$, $t' = y'/l$.

This equation was solved by several methods. First, the widely known method of moments (MoM) was used. In this case, the basis functions are Chebyshev polynomials of the first kind:

$$af'(t') = \sum_{n=1}^{\infty}\frac{A_n T_n(t')}{\sqrt{1-t'^2}} \tag{27}$$

The choice of such a basis function was due to the availability of an analytical calculation of the following integrals:

$$\int_{-1}^{1}\frac{T_n(t')}{\sqrt{1-t'^2}(t'-t)}dt' = \pi U_{n-1}(t) \qquad \int_{-1}^{1}\frac{T_n(t')T_p(t')}{\sqrt{1-t'^2}}dt' = \begin{cases}\pi, & p=n=0\\ \pi/2, & p=n\\ 0, & p\neq n\end{cases} \tag{28}$$

For the same reason, the Chebyshev polynomials of the second kind were chosen as weight functions:

$$\int\limits_{-1}^{1} U_{m-1}(t)U_{k-1}(t)\sqrt{1-t^2}dt = \begin{cases} \pi/2, & m = k \\ 0, & m \neq k \end{cases} \quad (29)$$

The exponentials in the core of the SIE on the Bessel functions and the Chebyshev polynomials of the first and second kind:

$$\begin{aligned} e^{-i\frac{l}{a}\xi t} &= \frac{2i}{(\xi l/a)}\sum_{k=1}^{\infty} i^{-k} k J_k(\xi l/a)U_{k-1}(t) \\ e^{i\frac{l}{a}\xi t'} &= 2\sum_{p=1}^{\infty}\frac{i^p}{(1+\delta_{0,p})}J_p(\xi l/a)T_p(t') \end{aligned} \quad (30)$$

where $\delta_{0,p}$ is Kronecker symbol, $\delta_{0,p} = \begin{cases} 1, & p = 0 \\ 0, & p \neq 0 \end{cases}$.

In view of the foregoing, the problem has been reduced to the following system of linear algebraic equations (SLAE), with respect to unknown coefficients A_m:

$$\frac{2}{\pi}\sigma \int\limits_{-b/l}^{b/l} \sqrt{1-t^2}U_{m-1}(t)v(t)dt = A_m + \sum_{n=1}^{\infty} A_n \frac{mi^{n-m+1}}{(1+\delta_{0,n})}\varsigma_{m,n} \quad (31)$$

The function of the longitudinal distribution of the current density at the emitter is determined as follows:

$$af(t) = \int_{-1}^{t} af'(t')dt' = \sum_{n=1}^{\infty}\int_{-1}^{t}\frac{A_n T_n(t')}{\sqrt{1-t'^2}}dt' = -\sqrt{1-t^2}\sum_{n=1}^{\infty}\frac{A_n}{n}U_{n-1}(t) \quad (32)$$

The current function is defined as:

$$I_y(t) = \int\limits_{-a}^{a} \eta_y(x,t)dx = \int\limits_{-a}^{a}\frac{f(t)}{\sqrt{1-(x/a)^2}}dx = a\pi f(t) \quad (33)$$

In Fig. 2 shows the results of the calculation of the real (Fig. 2a) and imaginary (Fig. 2b) parts of the current function for different numbers of terms of the approximating series (32).

It can be seen from the graphs that the number of terms in the series has a rather strong effect on the current distribution function.

This equation was also solved with the help of the partial operator rotation method (POAM), whose essence consists in reducing the singular equation (26) to the integral Fredholm equation of the second kind by means of the inversion formula for an integral of Cauchy type not bounded on the interval [−1; 1]:

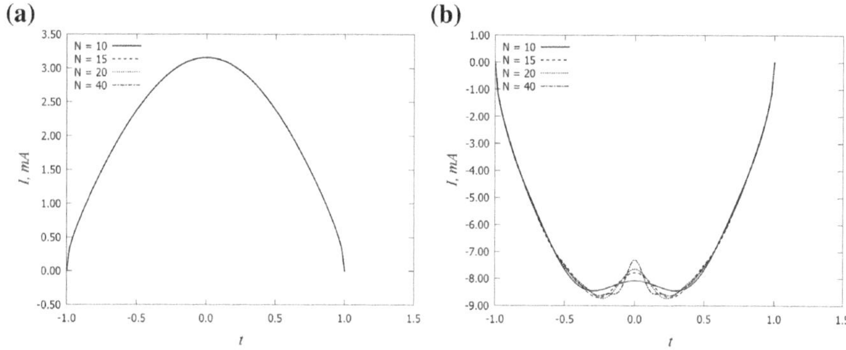

Fig. 2 Current distribution by the MSA radiator (MoM)

$$af'(t) = \frac{1}{\pi\sqrt{1-t^2}}\left[a_0\pi - \int\limits_{-1}^{1}\frac{\sqrt{1-t'^2}}{t'-t}F(t')dt'\right] \tag{34}$$

where $a_0 = 0$ is a null constant determined from condition $f(-l) = f(l) = 0$.

Denoting $Int(t) = \int_{(l_0-b)/l}^{(l_0+b)/l}\frac{\sqrt{1-t'^2}}{t'-t}v(t')dt'$, where $l_0 \in [-l; l]$ is point of excitation, and taking (34) into account, we obtain the Fredholm integral equation of the second kind with respect to the unknown function $af'(t)$:

$$af'(t) = -\frac{\sigma}{\pi}\frac{Int(t)}{\sqrt{1-t^2}} - \frac{1}{\pi\sqrt{1-t^2}}\int\limits_{-1}^{1}\int\limits_{-1}^{1}\frac{\sqrt{1-t'^2}}{t'-t}af'(t'')K(t',t'')dt''dt' \tag{35}$$

The second term on the right-hand side of the integral equation (35) is expressed through a function $\varphi(t)$, as a result of which expression (35) can be rewritten as follows:

$$af'(t) = -\frac{\sigma}{\pi}\frac{Int(t) + \varphi(t)}{\sqrt{1-t^2}} \tag{36}$$

Taking into account expression (36), we obtain the Fredholm integral equation of the second kind with respect to the unknown function $\varphi(t)$:

$$\varphi(t) = -\frac{1}{\pi}\int\limits_{-1}^{1}\int\limits_{-1}^{1}\frac{\sqrt{1-t'^2}}{t'-t}\frac{Int(t'') + \varphi(t'')}{\sqrt{1-t''^2}}K(t',t'')dt''dt' \tag{37}$$

Similarly, as in MoM, we expand the exponential functions in a series in the Chebyshev polynomials and the Bessel functions (30), and also take into account the integrals (28), (29) and, as a result, we obtain the following expression:

$$\varphi(t) = \frac{1}{\pi}4i\sum_{\substack{k=1\\n=0}}^{\infty}\frac{ki^{n-k}}{1+\delta_{0,n}}\varsigma_{n,k}\int\limits_{-1}^{1}\frac{Int(t'')+\varphi(t'')}{\sqrt{1-t''^2}}T_n(t'')dt''\times T_k(t) \tag{38}$$

where $\varsigma_{n,k} = \frac{1}{2}\int_{-\infty}^{\infty}\frac{J_n(\xi l/a)J_k(\xi l/a)}{\xi}\Delta g(\xi)d\xi$.

An unknown function $\varphi(t)$ can be represented in the form of an expansion in a series of Chebyshev polynomials of the first kind:

$$\varphi(t) = \sum_{k=1}^{\infty}A_kT_k(t) \tag{39}$$

As a result, we obtain a system of linear algebraic equations for determining the coefficients A_k:

$$A_k = \frac{4i}{\pi}\sum_{n=1}^{N}ki^{n-k}\varsigma_{n,k}\left(Int_n + \frac{\pi}{2}A_n\right),\quad k=\overline{1\ldots N} \tag{40}$$

where $Int_n = \int_{-1}^{1}\frac{Int(t'')T_n(t'')}{\sqrt{1-t''^2}}dt''$.

Using the integrals (28), we finally obtain an expression for the function $af(t)$:

$$af(t) = -\frac{\sigma}{\pi}\left[\int\limits_{-1}^{t}\frac{Int(t')}{\sqrt{1-t'^2}}dt' - \sqrt{1-t^2}\sum_{k=1}^{N}\frac{A_kU_{k-1}(t)}{k}\right] \tag{41}$$

It should be noted that the excitation is given by a function $v(t)$. Generally speaking, this function is chosen arbitrarily, but a number of requirements are raised to it, connected with questions of convergence of the integral in the expression for $Int(t)$. In particular, the function $v(t)$ must be continuous in the gap region and equal to zero on its edges, which, in turn, corresponds to physical reality. There are no other restrictions to the selection of the excitation function. With this in mind, a function $v(t)$ of the form was chosen as such a function $v'(t)$:

$$v(t) = v'(t) = \begin{cases} 0, & |t-l_0| > \frac{b}{l} \\ \sqrt{1-((t-l_0)l/b)^2} & |t-l_0| < \frac{b}{l} \end{cases} \tag{42}$$

In the course of numerical calculations, it was revealed that the main parameters that significantly affect the convergence of the algorithm are $\xi_{\max}$—the limiting value of the integration variable of the improper integral with respect to h in (24), at which the required accuracy of the computations is achieved and N is the number of terms of the series, consisting of Chebyshev polynomials, multiplied by the corresponding coefficients (expression (41)). Of course, the accuracy of the calculations will depend on the choice of the values of these parameters, and, in addition, the convergence

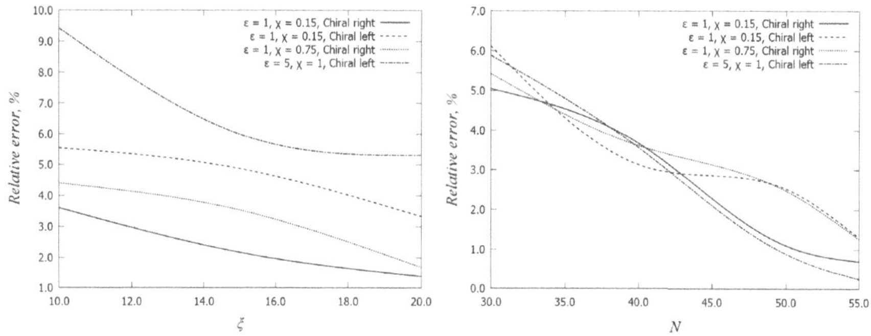

Fig. 3 Dependencies of relative errors from ξ_{max}(left) and from N (right)

of the whole method as a whole, so it is quite important to determine their optimal value.

To study the numerical algorithm for convergence, it is proposed to use the values of the current distribution function determined at the points corresponding to the center of the radiator ($t = 0$) and calculated for different normalized lengths of the radiator l/λ (from 0.25λ to 2λ, where λ is the wavelength) for different values of the parameter chirality and relative permittivity, as well as for various types of chiral substrate. Under the types of substrates are meant chiral substrates made on the basis of "left-handed" and "right-handed" inclusions.

As a result of this study, the relative error of computations was estimated and the optimal values of the parameters ξ_{max} and N were determined for different emitter lengths, as well as for various values of the chirality parameter. By the true value we will understand the values of the current distribution function, determined at the maximum values ξ_{max} and N. The graphs of the dependences of the errors on, as ξ_{max} well as on N, are shown in Fig. 3.

The method of the POAM relative to the parameter ξ_{max} has a sufficiently fast convergence. Proceeding from considerations of the resource-consuming nature of the method and the accuracy of the calculation, in numerical calculations, depending on the macroscopic parameters of the chiral medium, of which the chirality parameter is most important, and also the type of substrate, the parameter ξ_{max} value should be chosen to be no more than 20.

On the other hand, the number of members of the approximating row N exerts a much stronger influence.

Achieving the convergence of the final solution depends on such factors as the dimensions of the radiator l/λ, the values of the macroscopic parameters of the chiral medium (ε_1 and χ), and also the type of the chiral substrate.

The most significant of these is the normalized length of the radiator l/λ. So, for example, for a small radiator, you can limit the minimum values given ($l/\lambda \leq 0.25$), which in turn allows you to significantly reduce the resource intensity of the method in question and, as a result, increase the calculation time. Accordingly, for radiators of a larger size, a larger value of the parameter N should also be selected.

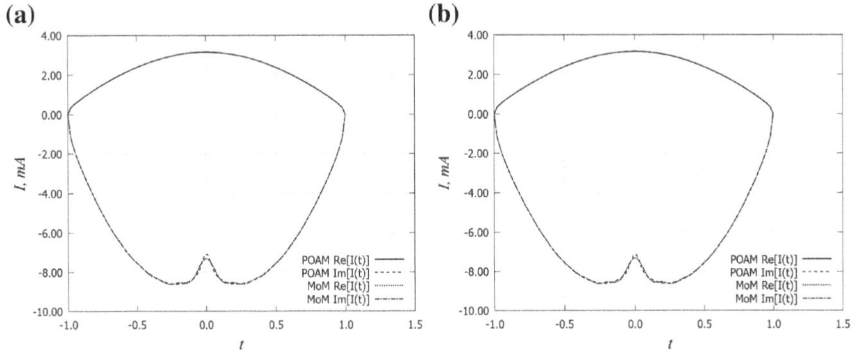

Fig. 4 Current distribution by the MSA radiator (MoM and POAM)

The values of the macroscopic parameters of the chiral medium also affect the convergence of this method, but less. As the study showed, with an increase in the chirality parameter by a factor of 5 to ensure convergence for sufficiently large dimensions of the radiator ($1 > l/\lambda > 0.5$), it is necessary to increase the value of the parameter N by an average of 10. It should be noted that for small radiator lengths ($l/\lambda \leq 0.5$), the value of the parameter N can be kept the same. The type of chiral substrate also affects the convergence of the method with respect to the parameter N. As in the previous case, this effect is most noticeable for large dimensions of the radiator ($l/\lambda > 0.5$).

In Fig. 4 shows the current distributions along the MSA radiator, calculated with the help of the POAM at $N = 10$ (Fig. 4a) and at $N = 40$ (Fig. 4b). Current distributions calculated by MoM are also presented $N = 40$.

As can be seen, these graphs coincide, however, when using the POAM, the number of necessary terms of the approximating series for accurate determination of the current function is only 10, while for MoM it is 40, which in turn will undoubtedly affect the computation time. However, the use of MoM is a simpler solution from a mathematical point of view, which is important in solving complex problems.

In Fig. 5 shows the dependence of the input impedance on the length of the emitter arm normalized to the wavelength, calculated by different methods (MoM and POAM) for $\varepsilon_1 = 1$ and for $\chi = 0.3$ a chiral substrate based on right-handed elements. In this case, the number of members of the series was selected as constant and equal to 30. Figure 5a shows the real (left) and imaginary (right) parts of the input impedance of the MSA with the chiral substrate, calculated using MoM with the number of terms of the series equal to 30, and in Fig. 5b—with the number of members of the series, equal to 55.

In this case, the input resistance of the microstrip antenna is determined as follows:

$$Z_{in} = \frac{V}{I(t = 0)} \tag{43}$$

where $I(t = 0)$ is a value of the current distribution function, in the vibrator gap.

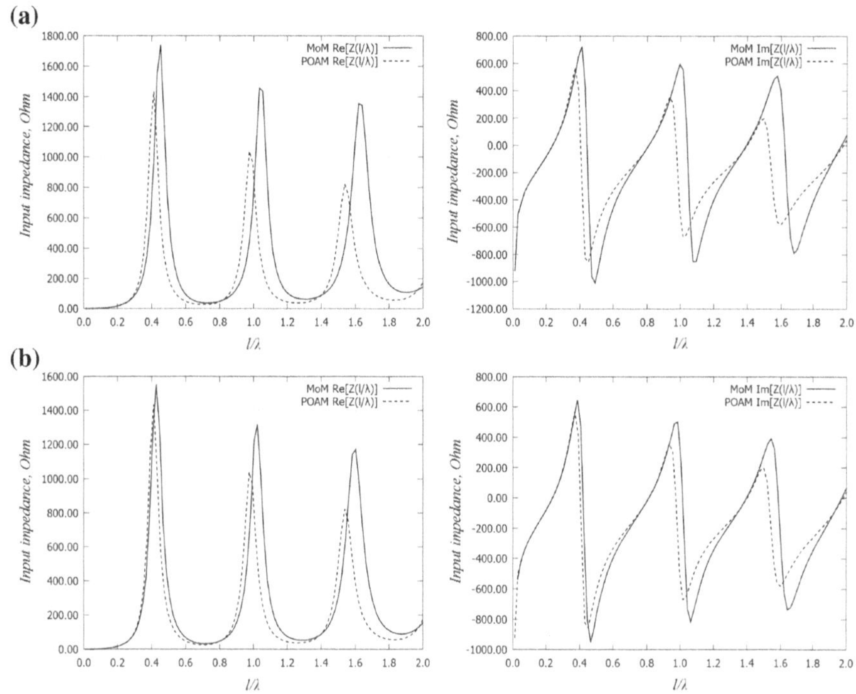

Fig. 5 The dependence of the input impedance on l/λ (MoM and POAM)

When calculating the characteristics of a microstrip antenna with a chiral substrate, the value of the voltage in the gap of the vibrator was set equal to 1 V ($V = 2bE_0 = 1$).

Obviously, with a further increase in the number of terms in the approximating series, these functions will coincide.

Thus, the proposed approach to the analysis of MSA with substrates from chiral metamaterial allows correctly calculating the current distribution. Knowledge of the current function, in turn, allows you to calculate the field at any point in space, including the near zone, and in addition, to determine other characteristics of the antenna. The proposed approach can be generalized to the case of different types of radiators (spiral, ring, fractal, etc.), as well as to the case of multilayer structures.

3 Characteristics of Microstrip Antennas with Chiral Substrates

3.1 Impedance Characteristics of MSA with a Chiral Substrate

In Figs. 6, 7, 8 and 9, the graphs of the dependences of the real (left) and imaginary (right) parts of the input impedance of the MSA with a chiral substrate based on the "left-" and "right-hand" elements on the arm length normalized to the wavelength l/λ with the following general parameters: $\varepsilon_2 = 1$, $\mu_1 = \mu_2 = 1$, $a/\lambda = 0.025$, $b/\lambda = 0.01$, $d/\lambda = 0.1$.

As can be seen from the graphs given, in the case when the chiral sub-spoon is made on the basis of left-handed spirals, the number of resonances in the interval from 0 to 2λ increases. There is also a shortening effect (displacement of resonances in frequency). With an increase in the dielectric constant, the quality factor of the

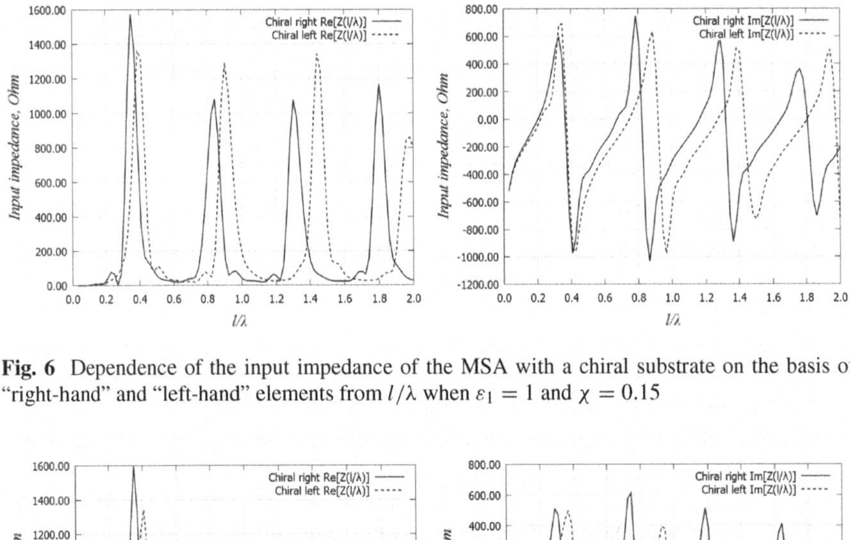

Fig. 6 Dependence of the input impedance of the MSA with a chiral substrate on the basis of "right-hand" and "left-hand" elements from l/λ when $\varepsilon_1 = 1$ and $\chi = 0.15$

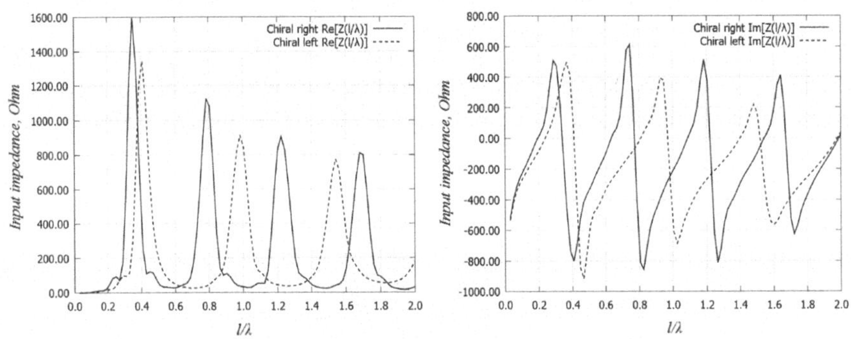

Fig. 7 Dependence of the input impedance of the MSA with a chiral substrate on the basis of "right-hand" and "left-hand" elements from l/λ when $\varepsilon_1 = 1$ and $\chi = 0.3$

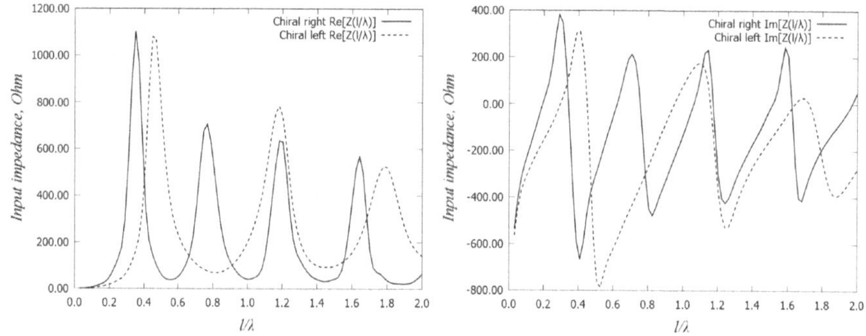

Fig. 8 Dependence of the input impedance of the MSA with a chiral substrate on the basis of "right-hand" and "left-hand" elements from l/λ when $\varepsilon_1 = 1$ and $\chi = 0.5$

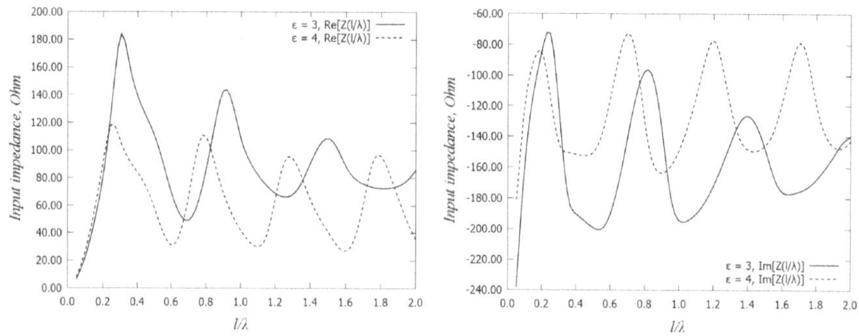

Fig. 9 Dependence of the input impedance of the MSA with a chiral substrate on the basis of "left-hand" elements from l/λ when $\chi = 1$ and $\varepsilon_1 = 3$, $\varepsilon_1 = 4$

system is significantly reduced. It should also be noted that at $\varepsilon_1 > 3$, the reactance is purely capacitive in nature. The resulted results of calculations of the input impedance are quite accurately (at least qualitatively) consistent with the calculations of other authors.

At some frequencies, a situation is possible where the effective permittivity of the chiral substrate is zero. It is of interest to estimate the impedance characteristics of MSA at such or close dielectric permittivity in the limiting case (ie, the macroscopic parameters are constant throughout the frequency band). A structure was chosen for which the effective permittivity is $\varepsilon_1 = 0.01$, and the chirality parameter is $\chi = 0.5$. The chiral elements of this structure are "left-handed".

In Fig. 10 shows the dependence of the real (left) and imaginary (right) parts of the input impedance on the normalized wavelength with the following parameters: $\varepsilon_2 = 1$, $\mu_1 = \mu_2 = 1$, $a/\lambda = 0.025$, $b/\lambda = 0.01$, $d/\lambda = 0.1$.

Analyzing these graphs, we can say that the microstrip antenna with practically zero effective dielectric constant is high-good, and its reactive input resistance is predominantly capacitive.

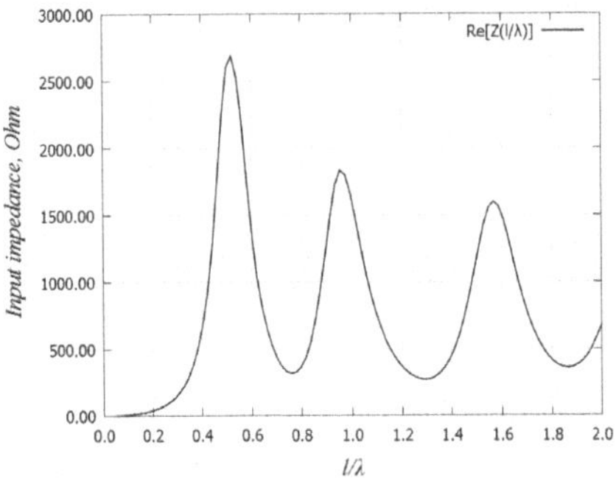

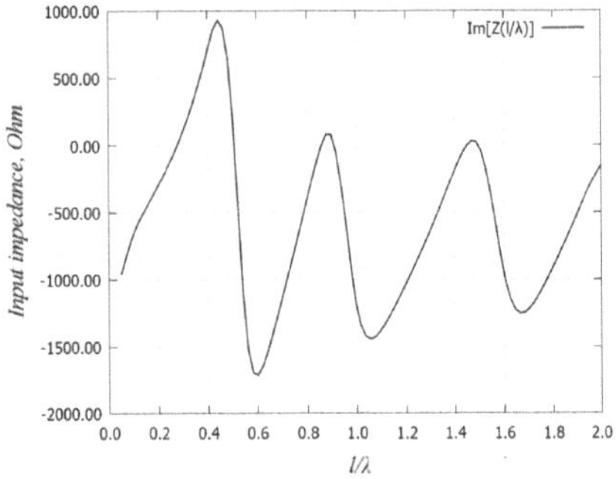

Fig. 10 Dependence of the input impedance of the MSA with a chiral substrate on the basis of "left-hand" elements from l/λ when $\varepsilon_1 = 0.01$ and $\chi = 0.5$

Thus, in this subsection we showed the dependences of the input impedance of the MSA with the chiral substrate on the length of the arm normalized to the wavelength. Compared with the MSA with a dielectric substrate, especially at high values of the relative permittivity and the chirality parameter, resonance frequency shift is observed, as well as a significant decrease in the quality factor. It is established that in the case when the chiral substrate is made on the basis of "right-hand" elements, the number of resonances increases. It is shown that, as the dielectric constant increases, there is a decrease in the Q of the system, and at $\varepsilon_1 \geq 3$ and $\varepsilon_1 = 0.01$, the reactance is mainly capacitive in nature.

3.2 Calculation of the Electric Field Produced by a Microstrip Antenna with a Chiral Substrate

Now let us consider the problem of determining the electric field produced by MSA with a chiral substrate. The electric field strength generated by the MSA has two components $\vec{E} = \{E_x, E_y\}$. In this regard, the expression for determining the field is written as follows:

$$E_x(x, y, z) = \int_{-a}^{a}\int_{-l}^{l} \eta_y(x', y')Z^x(x', y', x, y)dx'dy'$$

$$E_y(x, y, z) = \int_{-a}^{a}\int_{-l}^{l} \eta_y(x', y')Z^y(x', y', x, y)dx'dy' \qquad (44)$$

where

$$Z^x(x', y', x, y) = \frac{1}{4\pi^2} \int_{-\infty}^{\infty} \int_{-\infty}^{\infty} Z_{21}(\beta, h) e^{-ir_2(z-d)} e^{-i\beta(x-x')} e^{-ih(y-y')} d\beta dh$$

$$Z^y(x', y', x, y) = \frac{1}{4\pi^2} \int_{-\infty}^{\infty} \int_{-\infty}^{\infty} Z_{11}(\beta, h) e^{-ir_2(z-d)} e^{-i\beta(x-x')} e^{-ih(y-y')} d\beta dh$$

Below we present the data of normalized expressions, taking into account the integrals (23):

$$E_y(p, t, w) = \frac{iZ_c}{4\pi} \left(\frac{2\pi a}{\lambda}\right) \int_{-l}^{l} af(t') \times \int_{-\infty}^{\infty} \int_{-\infty}^{\infty} J_0(\alpha) Z'_{11}(\alpha, \xi) e^{-ir_2\left(w\frac{l}{a} - \frac{d}{a}\right)} e^{-i\frac{l}{a}\alpha p} e^{-i\frac{l}{a}\xi(t-t')} d\alpha d\xi dt'$$

$$E_x(p, t, w) = \frac{iZ_c}{4\pi} \left(\frac{2\pi a}{\lambda}\right) \int_{-l}^{l} af(t') \times \int_{-\infty}^{\infty} \int_{-\infty}^{\infty} J_0(\alpha) Z'_{21}(\alpha, \xi) e^{-ir_2\left(w\frac{l}{a} - \frac{d}{a}\right)} e^{-i\frac{l}{a}\alpha p} e^{-i\frac{l}{a}\xi(t-t')} d\alpha d\xi dt' \quad (45)$$

The function $af'(t')$ is also considered known, the method of finding which is described in Sect. 2.

In Fig. 11 shows plots of the dependences of the electric field modulus x and y-components (Fig. 11a) and phase (Fig. 11b) on the coordinate z, with the following parameters: $\varepsilon_2 = 1, \mu_1 = \mu_2 = 1, a/\lambda = 0.025, l/\lambda = 0.5, b/\lambda = 0.01, d/\lambda = 0.1,$

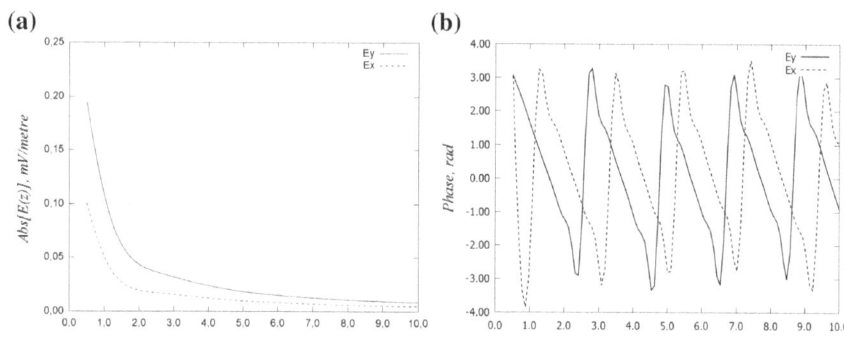

Fig. 11 Dependences of the modulus and phase of the x and y components of the electric field strengths of the electric field on the coordinate when $\varepsilon_1 = 1$ and $\chi = 0.75$

$\varepsilon_1 = 1$ and $\chi = 0.75$. The chiral support of the MSA is based on the "right-hand" elements. And in Fig. 12 shows similar graphs for a chiral substrate made on the basis of "left-hand" elements.

In Fig. 13 are the graphs of the dependences of the electric field modulus x and y-components (Fig. 13a) and phase (Fig. 13b) on the z coordinate, with the following parameters: $\varepsilon_2 = 1$, $\mu_1 = \mu_2 = 1$, $a/\lambda = 0.025$, $l/\lambda = 0.25$, $b/\lambda = 0.01$, $d/\lambda = 0.1$, $\varepsilon_1 = 3$ and $\chi = 1$. In this case, the chiral substrate of the MSA is made on the basis of "left-hand" elements.

As can be seen from the graphs shown, E_y the component of the electric field produced by the microstrip antenna with a chiral substrate is commensurate with E_x, and their phases are shifted by some amount. This testifies that the emitted electromagnetic waves have elliptical polarization.

Below, in Fig. 14, the normalized directionality diagrams (DD) of a given microstrip antenna with a chiral substrate are given with the following parameters: $\varepsilon_2 = 1$, $\mu_1 = \mu_2 = 1$, $a/\lambda = 0.025$, $l/\lambda = 0.5$, $b/\lambda = 0.01$, $d/\lambda = 0.1$, $\varepsilon_1 = 1$

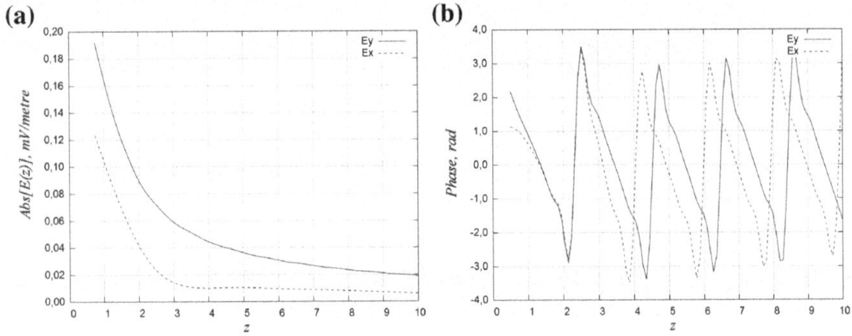

Fig. 12 Dependences of the modulus and phase of the x and y components of the electric field strengths of the electric field on the coordinate when $\varepsilon_1 = 1$ and $\chi = 0.75$

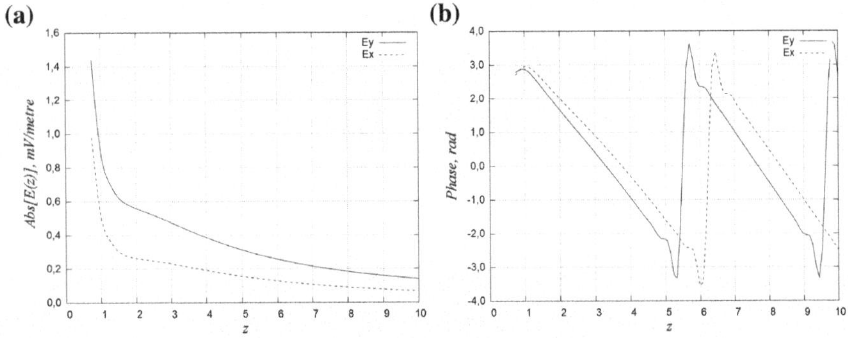

Fig. 13 Dependences of the modulus and phase of the x and y components of the electric field strengths of the electric field on the coordinate when $\varepsilon_1 = 3$ and $\chi = 1$

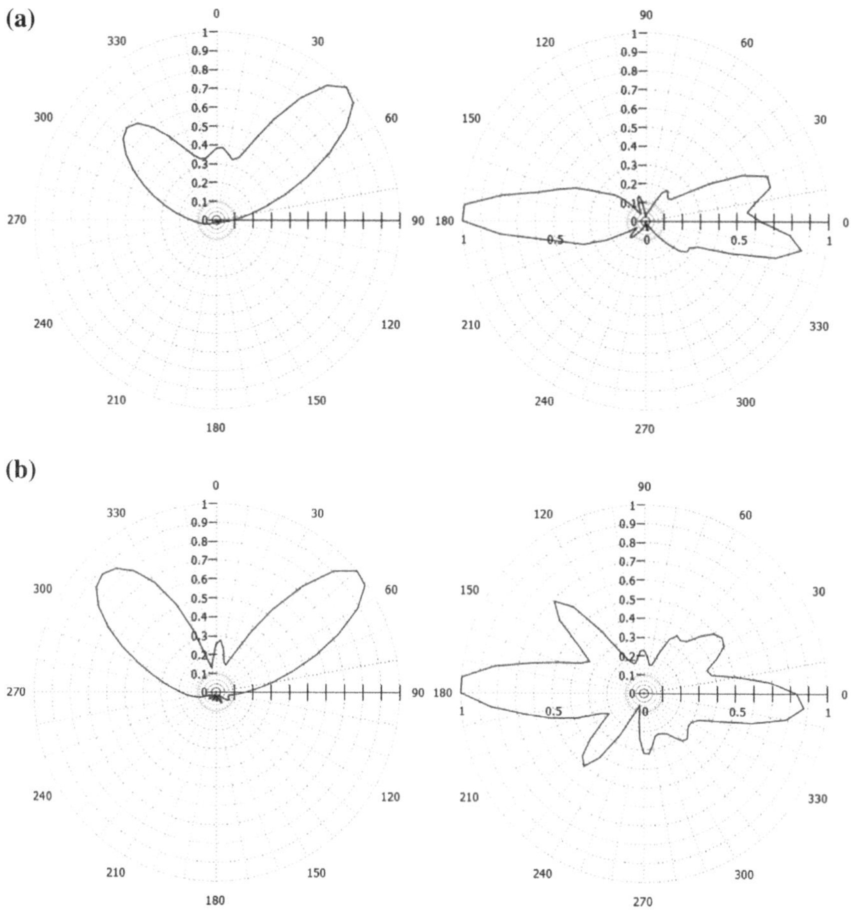

Fig. 14 Normalized DD in the meridional (left) and azimuthal (right) planes of MSA with a chiral substrate when $\varepsilon_1 = 1$ and $\chi = 0.75$

and $\chi = 0.75$. In this case, the chiral substrate of the MSA is made on the basis of the "right" (a) and "left-side" (b) elements. When constructing the DD data in the above expressions, a transition to a spherical coordinate system was carried out. In addition, since it was revealed that the emitted field has both components, therefore the expressions for E_θ and E_φ will be written as follows

$$\begin{aligned} E_\theta &= E_x \cos\theta \cos\varphi + E_y \cos\theta \sin\varphi \\ E_\varphi &= -E_x \sin\varphi + E_y \cos\varphi \end{aligned} \tag{46}$$

4 Diffraction of a Plane Electromagnetic Wave on a Microstrip Antenna with a Chiral Substrate

This subsection is devoted to the study of the diffraction of a plane wave on an MSA with a chiral substrate, which is a conducting plate located between the dielectric half-space and the chiral layer metallized from the lower side [15].

A plane monochromatic electromagnetic wave (EMW) of linear polarization falls at this MSA at an angle of θ_0. We will assume that the electromagnetic field (EMF), formed as a result of the distribution of currents over the plate surface in a certain way, which in turn arises under the action of a plane EMW, satisfies the Maxwell equations, the boundary conditions (47), and also the radiation condition at infinity. It is also assumed that the width of the plate must be so narrow ($2a << l, \lambda$) that the transverse component of the current density can be neglected $\vec{\eta} = \left(\eta_x = 0, \eta_y\right)$. The appearance of this structure is shown in Fig. 15

$$\begin{aligned} &\eta_y(x, -l) = \eta_y(x, +l) = 0 \\ &E_y(x, y) = 0 \; when \; x \in [-a, a], \;\; y \in [-l, l] \end{aligned} \tag{47}$$

The total strength of the electric field is defined as the superposition of the electric field strengths created by the incident and reflected waves:

$$\vec{E} = \vec{E}^{inc} + \vec{E}^{ref}. \tag{48}$$

In view of the fact that the incident EMW has linear polarization ($\vec{E}^{inc}$ always lies in a plane parallel to ZOY ($\varphi_0 = \pi/2$)), the projection of its vector on the XOY plane has only the component E_y^{inc}, the expression for which has the following form:

$$E_y^{inc}(y, z) = E_0 e^{ikz\cos\theta_0 + iky\sin\theta_0} \cos\theta_0 \tag{49}$$

where E_0 is a value of the amplitude of the field strength of the EMW; $k = \omega\sqrt{\mu_0\varepsilon_0} = 2\pi/\lambda$ is wave number for free space; $i = \sqrt{-1}$—imaginary unit.

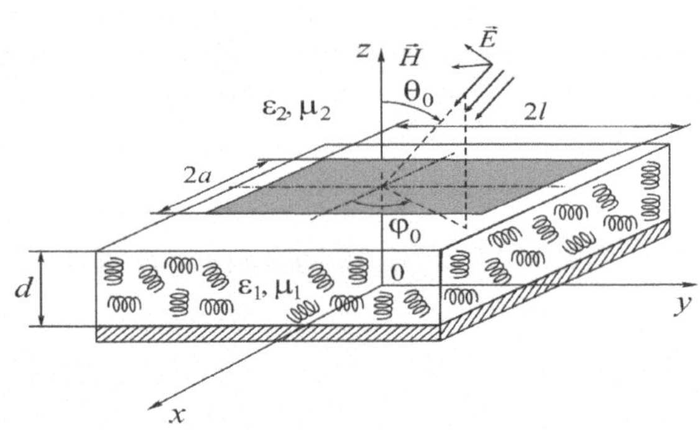

Fig. 15 Geometry of the diffraction problem

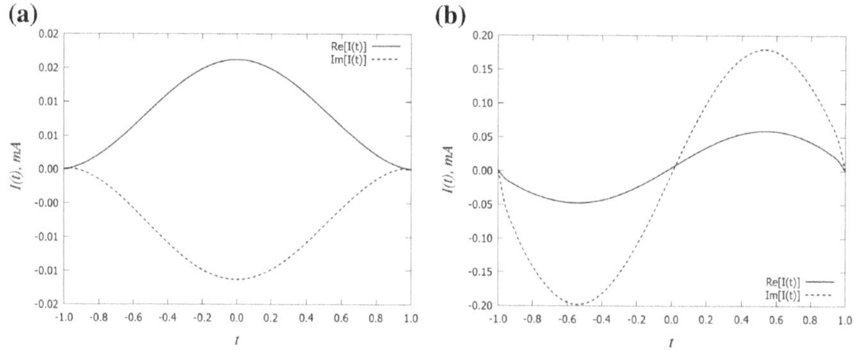

Fig. 16 Current distribution by chiral structure when $\theta_0 = 0$ (**a**) and $\theta_0 = \pi/4$ (б)

The method of determining the singular integral representation of the field (SIR) of a similar chiral structure is described in detail above. When the boundary conditions (47) and the expressions for the strength of a plane electromagnetic wave (49) are substituted in the quasi-static approximation of the transverse distribution of the surface current density $\eta_y(x', y') = f(y')/\sqrt{1-(x'/a)^2}$, we obtain a singular integral equation with a Cauchy singularity with respect to the unknown function $f(y')$ describing the longitudinal current distribution on the plate.

$$-E_y^{inc}(y,z) = -\frac{a}{4\pi}\int\limits_{-l}^{l} f'(y')\int\limits_{-\infty}^{\infty}\int\limits_{-\infty}^{\infty} J_0(\beta a)\frac{\Delta Z_{11}(\beta,h)}{ih}e^{-ih(y-y')}d\beta dh dy' - \frac{C_{\varepsilon,\mu,\chi}}{\pi}\int\limits_{-l}^{l}\frac{f'(y')}{y'-y}dy' \qquad (50)$$

In Fig. 16 are graphs of current distribution along the chiral structure on the basis of left-handed elements with parameters: $\varepsilon_1 = \varepsilon_2 = 1$, $\mu_1 = \mu_2 = 1$, $l/\lambda = 0.5$, $a/\lambda = 0.025$, $b/\lambda = 0.01$, $d/\lambda = 0.1$, $\chi = 0.15$ and various θ_0.

Thus, in this chapter we consider a method of electrodynamics analysis of MSA with substrates from a chiral metamaterial. Expressions are given for the elements of the admittance matrix of the chiral layer, based on the “left” and “right-hand” elements.

A mathematical model of MSA with a chiral substrate was developed and investigated. A singular integral representation of the electric field for a given antenna is obtained. These representations make it possible to calculate the radiation field at any point of space, incl. near zone.

The internal electrodynamics problem of diffraction of a plane electromagnetic wave on a metal strip located in a layer-like chiral structure is solved.

The characteristics of MSA with chiral substrates are investigated. It is shown that such MDAs emit waves with elliptical polarization, have asymmetric radiation patterns and have a lower Q-factor than MSA with dielectric substrates. In addition, the use of chiral substrates can improve the electrodynamics characteristics of such antennas and reduce their size.

References

1. J.R. James, P.S. Hall, *Handbook of Microstrip Antenna* (Peter Peregrinus Ltd., London, United Kingdom, 1989)
2. D.M. Pozar, D.H. Schaubert, *Microstrip Antennas, the Analysis and Design of Microstrip Antennas and Arrays* (IEEE, New York, 1995)
3. N. Engheta, *Metamaterials: Physics and Engineering Explorations*, N. Engheta, R.W. Ziolkowski (eds.) (Wiley-IEEE Press, 2006), 414 pp.
4. I.V. Lindell, A.H. Sihvola, S.A. Tretyakov, A.J. Viitanen, *Electromagneticwaves in chiral and bi-isotropic media* (Artech House, London, 1994), p. 291
5. D.M. Pozar, Microstrip antennas and arrays on chiral substrates. IEEE Trans. Antennas Propag. **40**(10), 1260–1263 (1992)
6. A. Erentok, R.W. Ziolkowski, Metamaterial-inspired efficient electrically small antennas. IEEE Trans. Antennas Propag. **56**(3), 691–707 (2008)
7. C. Zebiri, S. Daoudi, F. Benabdelaziz, M. Lashab, D. Sayad, N.T. Ali, R.A. Abd-Alhameed, Gyro-chirality effect of bianisotropic substrate on the operational of rectangular microstrip patch antenna. Int. J. Appl. Electromag. Mech. **51**, 249–260 (2016)
8. L.-W. Li, T.-X. Zhao, M.-S. Leong, T.-S. Yeo, An spatial-domain method of moments analysis of a cylindrical-rectangular chirostrip. Prog. Electromag. Res. PIER **35**, 165–182 (2002)
9. C. Caloz, T. Itoh, *Electromagnetic Metamaterials: Transmission Line Theory and Microwave Applications* (Wiley, New York, NY, USA, 2005)
10. S. Adekola, A. Mowete, A. Ayorinde, Compact theory of the broadband elliptical helical antenna. Eur. J. Sci. Res. **31**(3), 446–490 (2009). ISSN 1450-216X
11. C. Zebiri, S. Daoudi, F. Benabdelaziz, et al., Gyro-chirality effect of bianisotropic substrate on the operational of rectangular microstrip patch antenna. Int. J. Appl. Electromag. Mech. **51**, 249–260 (2016)
12. A.N. Tikhonov, V.Y. Arsenin, *Solutions of Ill-Posed Problems* (Wiley, Washington, DC, 1977). [in Russian]
13. A. Lakhtakia, V.K. Varadan, V.V. Varadan, Time-harmonic electromagnetic fields in chiral media. Lecture Notes in Physics. (Springer-Verlag, Berlin, Heidelberg and Boston, 1989), 121 pp.
14. M.A. Buzova, D.S. Klyuev, M.A. Minkin et al., Solving the electrodynamic problem for a microstrip radiating structure with a chiral substrate. Tech. Phys. Lett. **6**, 491–493 (2018)
15. M.A. Buzova, A.N. Dementyev, D.S. Klyuev et al., Current distribution in a gyrotropic microstrip structure when excited by a plane electromagnetic wave. Dokl. Phys. **6**, 231–234 (2018)

Microstrip and Fractal Antennas Based on Chiral Metamaterials in MIMO Systems

Dmitriy S. Klyuev, Anatoly M. Neshcheret, Oleg V. Osipov, Alexander A. Potapov and Julia V. Sokolova

Abstract This chapter describes the possibilities of increasing the throughput in MIMO systems using metamaterial substrates in antenna systems. It is shown that the using of such substrates makes it possible to reduce the mutual influence between the emitters. Estimates of wins in bandwidth are compared with traditional solutions. In addition, the results of studies of microstrip antennas (MSA) with a fractal emitter located on a chiral substrate are presented. It is shown that a similar approach is an opportunity to improve the characteristics of MSA.

Abbreviations

MIMO Multiple Input Multiple Output
MSA Microstrip antenna

D. S. Klyuev · A. M. Neshcheret · O. V. Osipov · J. V. Sokolova
L. Tolstoy str., 23, 443010 Samara, Russia
e-mail: klyuevd@yandex.ru

A. M. Neshcheret
e-mail: neshchereta@gmail.com

O. V. Osipov
e-mail: nanometa@yandex.ru

J. V. Sokolova
e-mail: sokolova-yu-v@yandex.ru

A. A. Potapov (✉)
V. A. Kotelnikov Institute of Radio Engineering and Electronics, Russian Academy of Sciences, Moscow, Russia
e-mail: potapov@cplire.ru

C. H. Skiadas and Y. Dimotikalis (eds.), *12th Chaotic Modeling and Simulation International Conference*, Springer Proceedings in Complexity,
https://doi.org/10.1007/978-3-030-39515-5_23

1 Possibilities of Increasing Throughput, MIMO Systems, By the Using of Antenna Based on Metamaterials

One of the important tasks in mobile radio networks is to increase the quality of service by applying an integrated approach to the modernization of the entire system as a whole. This is especially true at the present time, since there are already requirements for signaling not only telephony, but also multimedia data in real time. At the same time, with improvements in the software part of networks (the use of new signal-code structures, various types of modulation, etc.), there is a need for improving the hardware. One of the examples of this approach is MIMO technology [1, 2], the essence of which is the organization of several poorly correlated subchannels for information exchange between the source and the receiver, thereby allowing to significantly increase the spectral efficiency of the entire system. This technology is of particular interest in the case of the multipath nature of signal propagation observed in conditions of high and super-high density of construction, as well as in conditions of complex urban terrain.

In terms of electrodynamics, the use of new types of antenna systems, including those based on metamaterials, can further improve the quality of service. Metamaterial is an artificial composite material with spatial dispersion. This metamaterial possesses electrophysical properties that are uncharacteristic of natural materials, and are caused by a periodic structure of macroscopic elements of a certain shape. The peculiarity of metamaterials lies in the fact that their effective permeabilities can simultaneously take negative values [3]. Among the metamaterials, chiral metamaterials, representing a dielectric container, in which the conductive inclusions of the mirror-asymmetric shape are uniformly distributed [4, 5] should be distinguished apart.

The using of metamaterials in antenna technology makes it possible to reduce their weight and linear dimensions, improve the directional properties, and also compensate for the electrically small antennas reactivity. In addition, based on metamaterials, it is possible high-resistance surfaces creating, which serve to reduce the influence between the antenna arrays radiators [6–8]. Chiral metamaterials based on spiral elements are considered in this chapter.

In the framework of the approach, which involves the antennas new type using based on metamaterials to improve the efficiency of radio engineering systems, it is of interest to determine the winnings in comparison with traditional solutions.

A sufficiently tangible impact on the capacity of MIMO systems is exerted by the mutual influence of the antenna array elements [9, 10]. In general, to assess the MIMO system capacity, taking into account the mutual influence of the antenna system radiators, the variation of the Shannon formula allows:

$$C = \log_2 \det\left(\boldsymbol{I} + \frac{1}{D_N} \boldsymbol{Z}_{RX}^{-1} \boldsymbol{H} \boldsymbol{Z}_{TX} \boldsymbol{R} \boldsymbol{Z}_{TX}^{H} \boldsymbol{H}^{H} \left(\boldsymbol{Z}_{RX}^{-1}\right)^{H} \right), \tag{1}$$

where

D_N is a total noise power in each receiving antenna;
$\boldsymbol{Z}_{RX}, \boldsymbol{Z}_{TX}$ are the matrix impedances of the receiving and transmitting antenna systems, respectively;
$\boldsymbol{H}$ is a channel matrix;
$\boldsymbol{H}^H$ is the Hermitian conjugation operation of the channel matrix;
$\boldsymbol{R}$ is a correlation matrix of the input signal.

It is seen from the above formula that the mutual influence of the radiators is described by the impedance matrices of the pedestrian and receiving antenna systems. The general form of the antenna system impedance matrix having 3 radiators is recorded as follows:

$$\boldsymbol{Z} = \begin{pmatrix} z_{11} & z_{12} & z_{13} \\ z_{21} & z_{22} & z_{23} \\ z_{31} & z_{32} & z_{33} \end{pmatrix}, \tag{2}$$

In (2), along the diagonal of the matrix are the intrinsic impedances of each of the antenna system inputs. The off-diagonal elements are calculated as the ratio of the voltage induced on the passive input of the antenna system radiator to the current at the active input of the antenna system radiator:

$$z_{ij} = \frac{\dot{U}_i}{\dot{I}_j}, \tag{3}$$

where

$\dot{U}_i$ is a complex voltage on a passive radiator,
$\dot{I}_j$ is a complex current on the active radiator.

To estimate the degree of mutual influence between the emitters, we will conditionally assume that there is no correlation of the signals in the partial MIMO channels, there is also no attenuation and distortion of the signals in the medium, and the same antenna systems are installed in the source and receiver. On this basis, the expression for calculating the throughput can be rewritten as follows:

$$C = \log_2 \det\left(\boldsymbol{I} + \frac{q}{N_{TX}} \boldsymbol{Z}^{-1} \boldsymbol{H} \boldsymbol{Z} \boldsymbol{Z}^H \boldsymbol{H}^H \left(\boldsymbol{Z}^{-1}\right)^H \right), \tag{4}$$

where

q is a signal strength;
N_{TX} is a number of partial channels.

To ensure the equality of the energy characteristics of the MIMO system, it is necessary to normalize the impedance matrix as follows:

$$Z^{norm} = \frac{Z}{\sqrt{\mathrm{tr}(ZZ^H)}} \quad (5)$$

where $\mathrm{tr}(X)$ is matrix X trace.

As an example of the implementation of the MIMO 2 × 2 system, several pairs of microstrip antenna arrays (one receiving, the other, respectively, transmitting) were considered, the substrates of one of the pairs were made of a dielectric with a relative permittivity of 1 ($\varepsilon = 1$), and for other pairs—from various types of chiral metamaterials. On the substrates of the antenna arrays, parallel to each other at a distance equal to the working wavelength λ_0, there are two symmetrical half-wave radiators.

The following types of chiral metamaterials were considered:

- Biisotropic structures based on chaotically oriented right- and left-handed elements (Fig. 1);
- Bianisotropic structures based on right- and left-handed elements, oriented in the same way. Several configurations of such structures were considered, in particular, the structure in which the helices are oriented vertically (configuration 1) (Fig. 2a), and also for the structure in which the spirals are oriented horizontally (configuration 2) (Fig. 2b).

Right-and left-hand elements are understood to be right-handed and, respectively, left-twisted helices.

As a homogeneous dielectric container, in which the described conductive inclusions are arranged in a certain way, a material with a small dielectric constant, for example, expanded polystyrene, is usually used.

To determine the dispersion of the macroscopic parameters of the chiral structure, the Maxwell-Garnett model is used [11]. Expressions for finding the effective permittivity and the chirality parameter are given below:

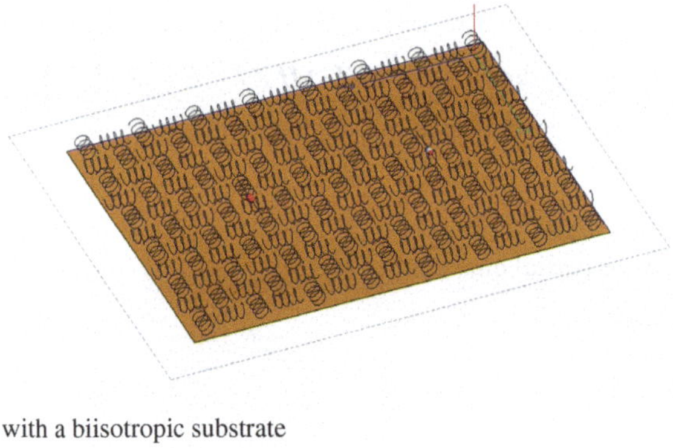

Fig. 1 Antenna with a biisotropic substrate

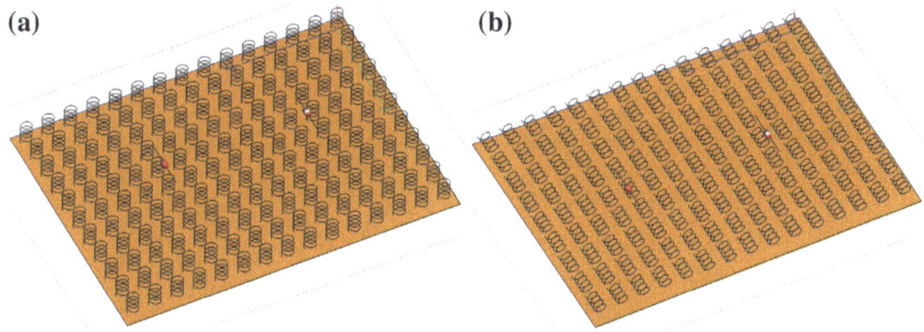

Fig. 2 MSA with bianisotropic chiral substrates configurations 1 (**a**) и 2 (**b**)

$$\varepsilon(f) = \varepsilon_c + \frac{\beta_0^2}{4\pi^2\left(f_0^2 - f^2\right)} \tag{6}$$

$$\chi(f) = A_0 \frac{\beta_0^2 f}{2\pi c\left(f_0^2 - f^2\right)} \tag{7}$$

where ε_c is the dielectric constant of the container in which the conducting inclusions of the mirror-asymmetric shape are placed; β_0 is a parameter having a frequency dimension and associated with internal processes in the chiral structure; A_0 is a distance between adjacent conducting inclusions; $c = 1/\sqrt{\varepsilon_0\mu_0}$ is the light speed in vacuum; f_0 is resonant frequency of the chiral structure.

The resonant frequency of the macroscopic parameters of the chiral metamaterial is generally determined by the resonance properties of the conducting inclusions underlying this metamaterial, in accordance with Thomson's formula:

$$f_0 = \frac{1}{2\pi\sqrt{LC}} \tag{8}$$

where L is chiral elements inductance; C is elements capacity.

Consider, for example, a chiral structure based on thin wires. The form of a single fine-wire helix is shown in Fig. 3. This spiral is made of a thin wire, a radius r_H. The helix consists of turns N_H, located at a distance h_H from each other, and its internal radius is R_H. The length of the helix in the unfolded state (the length of the thin wire) is l_H. The height of the helix is H_H, and the angle of wrapping is α_H.

The formula for calculating the inductance of the helix coincides with the formula for determining the inductance of the solenoid and has the following form:

$$L_H = \mu_0\mu_c \frac{S_H N_H^2}{H_H} = \mu_0\mu_c \frac{\pi R_H^2 N_H^2}{H_H} \tag{9}$$

where μ_c is relative magnetic permeability; $S_H = \pi R_H^2$ is an area of the helix coil.

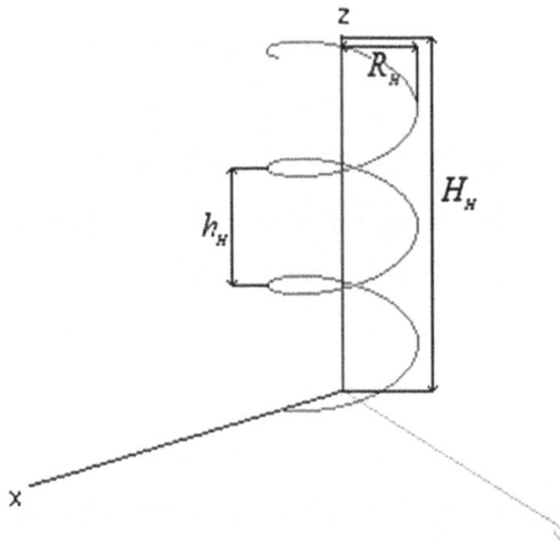

Fig. 3 Helix

The helix capacity is determined by the capacity of the wire C_L, from which it is made, as well as inter-turn capacity. If the capacity of a group of helices located close to each other is determined, there is also an inter-element capacitance C_{IE}.

The expression for determining the total capacity of a group of spirals is as follows:

$$C_H = C_L + C_{IC} + C_{IE} \tag{10}$$

The formula for determining the capacity of the helical wire coincides with the formula for determining the capacitance of a straight conductor:

$$C_L = \varepsilon_0 \varepsilon_c \frac{l_H}{18 \ln\left(\frac{2l_H}{r_H}\right) - 1} 10^{-9}, \tag{11}$$

where $l_H = 2\pi(R_H + 2r_H)N_H + H_H$ is the length of the elements in the unfolded state.

Formula for determining the interturn capacity:

$$C_{IC} = \varepsilon_0 \varepsilon_c \frac{S_{IC}(N_H - 1)}{h_H}, \tag{12}$$

where $S_{IC} = \pi\left[(R_H + 2r_H)^2 - R_H^2\right]$ is the area of the ring formed by a single turn of wire.

To determine the distance between adjacent turns h_H we use the following expression:

$$h_H = \frac{H_H}{N_H + 1}, \tag{13}$$

The inter-element capacity is determined in accordance with the following expression:

$$C_{IE} = \varepsilon_0 \varepsilon_c k_H \frac{S_{IE}}{A_0}, \tag{14}$$

where $S_{IE} = \frac{N_H(2R_H+2r_H)}{\cos(\alpha_H)} 2r_H$ is area of space occupied by spiral elements; k_H is spatial coefficient; $\alpha_H = \frac{\pi}{2(N_H+1)}$ is angle of wire wrapping.

The spatial coefficient k_H, depends on the location of the spirals. For the case of such a chiral structure, this coefficient is equal to $k_H = 0.25$.

Substituting the expressions for determining the capacity of the wire, to determine the interturn capacity and the expression for determining the interleum capacitance in (10), we finally obtain expressions for determining the total capacity of the chiral structure based on the helices.

In Fig. 4 shows the bandwidth versus frequency for right- and left-side bian-isotropic substrates for the vertical orientation of chiral elements (configuration 1) (Fig. 4a) and for horizontal orientation (configuration 2). The graphs also show the dependence of the capacity for a dielectric sub-spoon.

In Fig. 5a shows the bandwidth versus frequency for right-handed and left-handed elements of bi-isotropic substrates, and Fig. 5b—the dependence of the capacity for biisotropic and bianisotropic (configuration 1) substrates of various types, as well as for a dielectric substrate.

It can be seen from the graphs that the throughput in the case of using chiral substrates on the basis of left-handed elements is generally higher than on the basis of right-handed ones. The lowest throughput is achieved with the use of a dielectric substrate, and the largest is achieved when using biisotropic chiral substrates based on left-handed spirals. In the case of bianisotropic structures on the basis of left-handed elements, there is an additional resonance at higher frequencies. When using bianisotropic chiral substrates with a horizontal orientation of the elements, the throughput practically does not change.

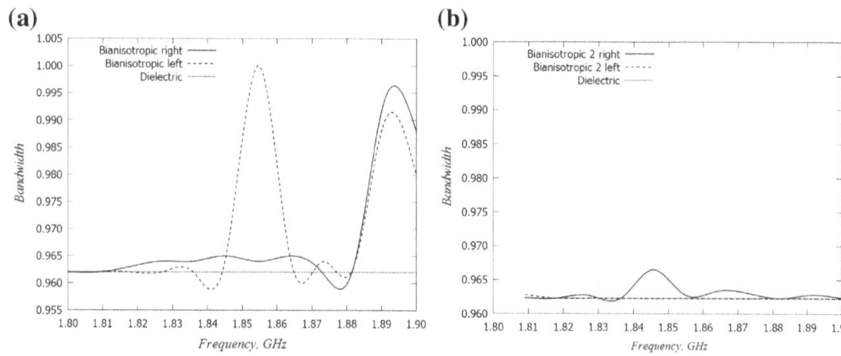

Fig. 4 Bandwidth versus frequency for bianisotropic substrates

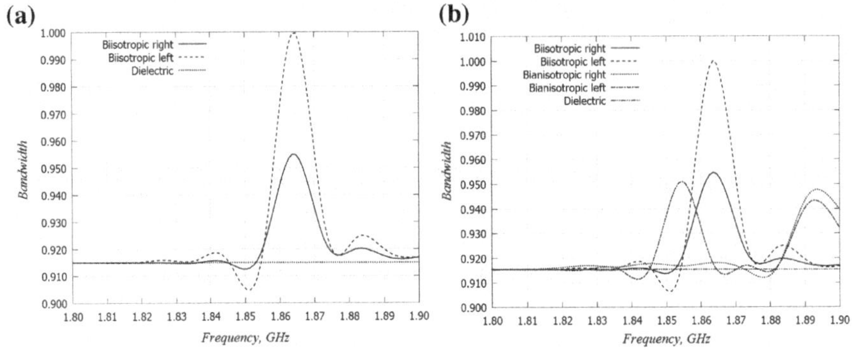

Fig. 5 Bandwidth versus frequency

In this chapter, the MIMO 3 × 3 system was also considered, since it is optimal. Calculation of the characteristics of antenna systems was carried out in the frequency range 1.8–1.9 GHz.

In Fig. 6 shows the dependence of the reflection coefficients on the frequency for each antenna systems radiator with a substrate of a biisotropic chiral metamaterial (Fig. 6a), based on the right-handed spirals, and also from the dielectric (Fig. 6b). In the case where the substrate is filled from a chiral metamaterial, an abrupt change in the reflection coefficient.

The throughput was estimated for the MIMO system when using these antenna systems. In Fig. 7 shows the bandwidth versus frequency.

It can be seen from the presented graph that, in general, the throughput in the case of an antenna system with a substrate of chiral metamaterial is much higher, so the use of such antennas in radio communication networks using MIMO technology is very promising. The spasmodic nature is also explained by the resonance properties of the chiral metamaterial.

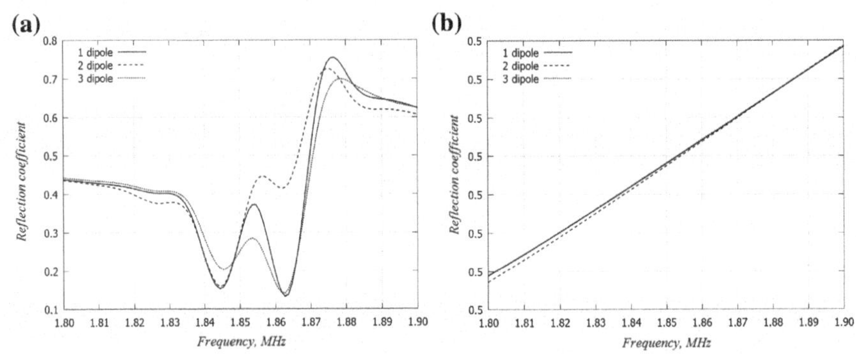

Fig. 6 Dependence of the reflection coefficients on the frequency

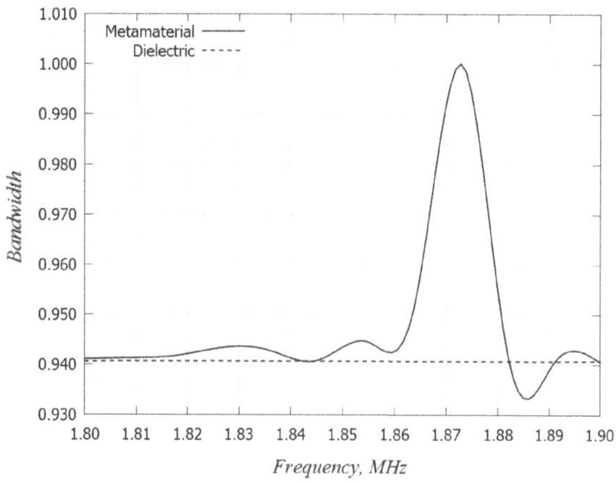

Fig. 7 Bandwidth versus frequency for MIMO 3 × 3

Thus, as a result of the conducted studies, the estimation of the throughput of MIMO systems was carried out using traditional antenna systems (antennas with dielectric substrates), and also with the use of antenna systems with substrates from chiral metamaterials. Based on the calculated estimates, it was concluded that the use of such antenna systems is promising in MIMO systems of radio communication networks.

2 Characteristics of a Microstrip Antenna with a Fractal Radiator Located on a Chiral Substrate

Currently, there is an increasingly active development of non-classical approaches to the creation of antenna technology associated with the use in their design of artificial composite structures—metamaterials, in order to improve the electrical and mass-dimensional characteristics. As shown by the results of the research, the use of substrates from chiral metamaterial in microstrip antennas (MSA) allows to increase the amplification factor, to significantly reduce the quality factor, to reduce the dimensions of the radiator. When studying the data of MSA with chiral substrates, a thin rectangular half-wave radiator was used. It should be noted that even with such a radiator, the MSA data can emit electromagnetic waves of elliptical polarization.

Similar MSA with substrates from chiral metamaterial are not without drawbacks, in particular, to the main it is necessary to include a small working frequency band, where all the above-described effects manifest themselves. One approach to extending the working frequency band is to use multi-layer chiral substrates, each layer of which is "tuned" to a certain frequency. It is obvious that for a sufficiently large separation of the resonant frequencies, several resonances will occur and, under certain conditions, an increase in the working band is possible. However, in this case, large

Fig. 8 Geometry of the structure

thermal losses are unavoidable, which in turn can lead to extremely low radiation efficiency.

To further improve the characteristics of the antennas, another approach is proposed that involves the use of fractal emitters located on chiral substrates in MSA. In Fig. 8 shows the geometry of the design of MSA, where the radiator is a multi-band Sierpinski dipole located on a substrate of chiral metamaterial based on right-spiral spirals.

The calculation was carried out in the Feko 7.0 software package in the frequency range from 0.3 to 4 GHz. It was found that when using a chiral substrate, such an antenna has better directional properties (has one narrow lobe) at a frequency of 3.5 GHz. At other frequencies, the directional properties of MSA with dielectric and chiral substrates are almost identical.

In Fig. 9 shows the directional patterns of a microstrip antenna with a fractal emitter located on a dielectric (solid line) and chiral (dashed line) substrates, at frequencies of 0.55 GHz (a), 2.45 GHz (b) and 3.5 GHz (c), in Vertical (left) and azimuth (right) planes.

It should be noted that, as in the case of an MPA with a rectangular radiator located on a chiral substrate, when using a fractal radiator, there is the possibility of emitting electromagnetic waves of elliptical polarization.

Thus, the proposed approach associated with the use of fractal emitters in MPA with chiral substrates, as a whole, allows improving their electrical characteristics and is very promising.

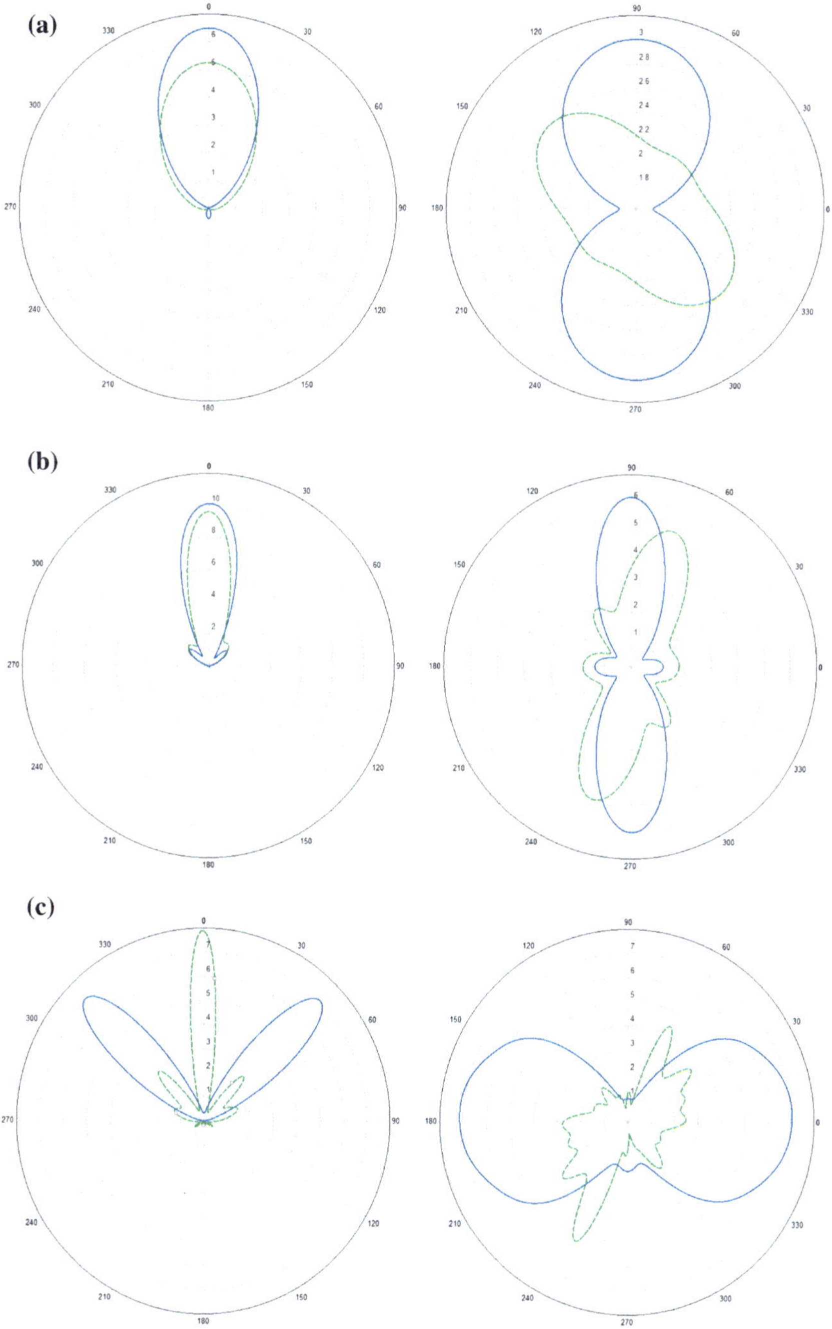

Fig. 9 Antenna directivity diagrams

References

1. A.B. Gershman, N.D. Sidiropoulos, *Space-Time Processing for MIMO Communication* (Wiley, 2006)
2. V. Kuhn, *Wireless Communications Over MIMO Channels: Applications to CDMA and Multiple Antenna Systems* (Wiley, 2006)
3. N. Engheta, R.W. Ziolkowski, A positive future for double-negative metamaterials. IEEE Trans. Microw. Theory Tech. **53**(4), 1535–1556 (2005)
4. A. Lakhtakia, V.K. Varadan, V.V. Varadan, *Time-Harmonic Electromagnetic Fields in Chiral Media*. Lecture Notes in Physics, vol. 335 (Springer-Verlag, Berlin, 1989)
5. I.V. Lindell, A.H. Sihvola, S.A. Tretyakov et al., *Electromagneticwaves in Chiral and Bi-isotropic Media* (Artech House, Boston, 1994)
6. M. Hirvonen, J.C.-E. Sten, Power and Q of a horizontal dipole over a metamaterial coated conducting surface. IEEE Trans. Antennas Propag. **56**(3), 684–690 (2008)
7. D.M. Pozar, Microstrip antennas and arrays on chiral substrates. IEEE Trans. Antennas Propag. **40**(10), 1260–1263 (1992)
8. C. Zebiri, S. Daoudi, F. Benabdelaziz, M. Lashab, D. Sayad, N.T. Ali, R.A. Abd-Alhameed, Gyro-chirality effect of bianisotropic substrate on the operational of rectangular microstrip patch antenna. Int. J. Appl. Electromag. Mech. **51**, 249–260 (2016)
9. I.E. Telatar, Capacity of multi-antenna Gaussian channels. Eur. Trans. Telecom. **10**, 585 (1999)
10. Yu.N. Parshin, A.V. Komissarov, Throughput of the telecommunication system in the conditions of changing spatial structure of the radio path with artificial multipath, in *Digital Signal Processing* (vol. 1, 2012), pp. 50–55 [in Russian]
11. I.V. Semchenko, S.A. Tretyakov, A.N. Serdyukov, Research on chiral and bianisotropic media in byelorussia and russia in the last ten years. Prog. Electromag. Res. **12**, 335–370 (1996)

The manufacturer's authorised representative in the EU is Springer Nature Customer Service Centre GmbH, Europaplatz 3, 69115 Heidelberg, Germany. If you have any concerns regarding our products, please contact ProductSafety@springernature.com

Printed and bound by CPI Group (UK) Ltd, Croydon, CR0 4YY

25/09/2026

02222901-0007